Biomass as a Nonfossil Fuel Source

Biomass as a Nonfossil Fuel Source

Donald L. Klass, EDITOR

Institute of Gas Technology

Based on a symposium sponsored

by the Division of

Petroleum Chemistry at the

ACS/CSJ Chemical Congress

(177th ACS National Meeting),

Honolulu, Hawaii, April 2, 1979.

ACS SYMPOSIUM SERIES 144

AMERICAN CHEMICAL SOCIETY
WASHINGTON, D.C.   1981

Library of Congress CIP Data

Biomass as a nonfossil fuel source.
  (ACS symposium series; 144 ISSN 0097–6156)
  Includes bibliographies and index.

  1. Biomass energy—Congresses.

  I. Klass, Donald L. II. American Chemical Society.
Division of Petroleum Chemistry. III. ACS/CSJ
Chemical Congress, Honolulu, 1979. IV. Series: Amer-
ican Chemical Society. ACS symposium series; 144.

TP360.B586   662'.8   80-26044
ISBN 0–8412–0599–X   ACSMC8   144 1-564 1981

Copyright © 1981

American Chemical Society

All Rights Reserved. The appearance of the code at the bottom of the first page of each
article in this volume indicates the copyright owner's consent that reprographic copies of
the article may be made for personal or internal use or for the personal or internal use of
specific clients. This consent is given on the condition, however, that the copier pay the
stated per copy fee through the Copyright Clearance Center, Inc. for copying beyond that
permitted by Sections 107 or 108 of the U.S. Copyright Law. This consent does not extend
to copying or transmission by any means—graphic or electronic—for any other purpose,
such as for general distribution, for advertising or promotional purposes, for creating new
collective works, for resale, or for information storage and retrieval systems.

The citation of trade names and/or names of manufacturers in this publication is not to be
construed as an endorsement or as approval by ACS of the commercial products or services
referenced herein; nor should the mere reference herein to any drawing, specification,
chemical process, or other data be regarded as a license or as a conveyance of any right or
permission, to the holder, reader, or any other person or corporation, to manufacture, reproduce, use, or sell any patented invention or copyrighted work that may in any way be
related thereto.

PRINTED IN THE UNITED STATES OF AMERICA

TP 360
B 586
CHEM

ACS Symposium Series

M. Joan Comstock, *Series Editor*

Advisory Board

David L. Allara	James P. Lodge
Kenneth B. Bischoff	Marvin Margoshes
Donald D. Dollberg	Leon Petrakis
Robert E. Feeney	Theodore Provder
Jack Halpern	F. Sherwood Rowland
Brian M. Harney	Dennis Schuetzle
W. Jeffrey Howe	Davis L. Temple, Jr.
James D. Idol, Jr.	Gunter Zweig

C4653

FOREWORD

The ACS SYMPOSIUM SERIES was founded in 1974 to provide a medium for publishing symposia quickly in book form. The format of the Series parallels that of the continuing ADVANCES IN CHEMISTRY SERIES except that in order to save time the papers are not typeset but are reproduced as they are submitted by the authors in camera-ready form. Papers are reviewed under the supervision of the Editors with the assistance of the Series Advisory Board and are selected to maintain the integrity of the symposia; however, verbatim reproductions of previously published papers are not accepted. Both reviews and reports of research are acceptable since symposia may embrace both types of presentation.

CONTENTS

PREFACE

Excluding most of the contribution made by biomass, which is defined as organic waste such as agricultural residues and urban refuse, and land- and water-based plant material such as trees, grasses, and algae, the United States consumed about 78.2 quads (1 quad = 10^{15} Btu) of primary energy in 1979. The contribution of each energy component was 37.1 quads for petroleum, 19.8 quads for natural gas, 15.2 quads for coal, 3.2 quads for hydroelectric power, 2.8 quads for nuclear electric power, and 0.1 quad for electric power production from wood and waste and geothermal sources. Few realize that the biomass contribution, in all its forms, for the production of heat, steam, electric power, and synfuels for 1979 was about 1.9 quads, or a contribution of about 2.3% to the total primary energy consumption. Thus, biomass energy consumption is equivalent to about one million barrels of oil per day, so it is obviously a commercial reality now. Indeed, as the costs of fossil energy increase and the available supplies shrink, especially petroleum and natural gas, we will begin to return to a renewable source of fixed carbon in the form of biomass to assure a continuous supply of organic liquid and gaseous fuels and chemicals.

The concept of using biomass as a primary energy source is not new. Wood was a major source of primary energy and chemicals in the United States only a relatively few years ago. As late as 1880, over 50% of the U.S. energy demand was supplied by wood. After 1880, fossil fuels began to dominate as a primary energy supply and have continued to be our largest source of energy to the present time.

In the 1970s, a major effort was launched in the United States to develop modern technology for the utilization of biomass energy. The symposium on biomass as a nonfossil fuel source, presented in Honolulu, Hawaii in April 1979 by the Division of Petroleum Chemistry at the American Chemical Society/Chemical Society of Japan Joint Chemical Congress, was devoted to this subject. Twelve basic and applied research papers were presented at this symposium on biomass energy. This book contains updated versions of ten of these papers and fifteen additional papers to balance the treatment of the subject. These are grouped into the categories of biomass production, liquid fuels, gaseous fuels, economics and energetics, and systems analysis. It will become apparent to the reader who is being introduced to the subject for the first time that there are many routes for the utilization of biomass energy and that many activities are underway to develop commercial processes and systems. Substitute natural gas in the form of methane from landfills, liquid alcohol fuels to replace gasoline, and direct biomass combustion for steam and electric power production are typical technologies now in use and under development. For the reader who already has been involved in biomass energy, many of the papers have extensive bibliographies that serve as a reference source.

It should be emphasized that, though this book is edited and all the papers reviewed by independent referees, I have not attempted to convert an author's views with which I disagree to my own way of thinking. However, these instances are in the minority. Universal agreement on a given biomass subject does not exist necessarily among those who have been in the field, mainly because some of the work has not yet progressed to the point where the ultimate answers are in hand.

Finally, I would like to briefly state my personal opinions on the present and future prospects of biomass energy. It is not a panacea for all of our energy problems, but it will find a logical place in the commercial energy market. Further, suitable biomass energy supplies, because of their generally dispersed nature, will be used initially in small-scale, localized applications. Large-scale central utility systems and synfuel plants supplied with biomass raw materials will be the exception rather than the rule in the 1980s and are not expected to reach commercial status to any significant extent until after 1990. Nevertheless, biomass will continue to contribute more to our energy and chemical needs as time passes.

Because of the multitude of organic residues and plant species, and the many processing combinations that yield solid, liquid, and gaseous fuels, the selection of the best technology and raw materials for specific applications seems very difficult. Many factors must be examined in depth to choose and develop systems that are technically feasible, energetically and economically practical, and environmentally acceptable. These factors are particularly important for large-scale biomass energy farms where continuity and efficiency of operation and synfuel production are paramount. The problem is not so intractable that it defies solution. But there are several major barriers to be overcome or at least reduced in size to facilitate commercial use of biomass energy technology on a scale that will satisfy a large portion of our energy demand. These barriers, none of which is insurmountable in my judgment, include such factors as excessive cost of biomass-derived synfuels, low or negative net energy production efficiencies for some systems, the problem of acquiring sufficient and suitable land for biomass production, conflicts with foodstuffs production, obtaining advance approvals and permits from state and federal agencies, and dependence on forgiven taxes and subsidies for economic success.

At the present time, the commercialization of biomass energy is proceeding at the proverbial snail's pace. The excessive cost of synfuels from biomass in integrated growth, harvesting, and conversion systems, and from integrated waste collection and conversion systems, is the prime reason for the low commercialization rate. Although synfuel production capacity (plant size) and financing conditions impact directly on synfuel costs, the estimated and actual manufacturing costs of most biomass-derived synfuels are not presently competitive with fossil fuels. Examples are SNG from manure and natural gas, and ethanol from sugarcane for gasohol and gasoline. As the price of crude oil continues to increase, I expect the cost of fuels and chemicals from biomass will become competitive with conventional petroleum derivatives.

At this time, the major factor influencing synfuel costs from biomass is biomass cost itself; conversion and other associated costs are often a smaller part of the total cost. Plant biomass production costs are affected most by independent inputs such as the costs of planting, fertilization, irrigation, and harvesting. An incremental increase in biomass yield often cannot be justified based on the additional cost of achieving this yield improvement. For organic wastes that are debited against conversion process cost, the delivered cost of the waste, which includes the costs of collection and transport, is sometimes too high to justify synfuel manufacture. Credits must be taken for the by-products and if they cannot be sold at certain minimum prices, the operation is not profitable. Finally, alternative biomass uses such as those for materials of construction, foodstuffs, animal feeds, and soil conditioning that offer a higher profit margin than synfuel must be considered. The potential owners and operators of a biomass energy system cannot be expected to undertake a business venture to commercialize biomass energy if the profits are too small in comparison with other alternatives. Tax incentives and other forms of subsidy already have been suggested to reduce synfuel costs and thereby stimulate the investment of private capital. Whether or not this approach can be effective remains to be established. In any case, biomass costs should be reduced to help make commercial synfuel manufacture economically attractive on its own merits.

I would like to express my appreciation to the Division of Petroleum Chemistry for sponsoring this somewhat "alien" symposium. (After all, biomass will displace a significant portion of petroleum if my projections are accurate.) I especially want to thank all of the speakers who somehow managed to be in Hawaii at the appointed time despite the airline travel problems prevalent during the symposium, and also all of the contributors of other articles that I requested to try to provide a more balanced treatment of biomass energy. The authors' individual efforts were indispensable in assembling a book of this type.

Institute of Gas Technology   DONALD L. KLASS
Chicago, Illinois

June 1980

INTRODUCTION

Industrial Development of Biomass Energy Sources

GEORGE P. SCHAEFER

Booz–Allen & Hamilton, Incorporated, 4330 East–West Highway, Bethesda, MD 20014

A wide diversity of companies has entered into the development of biomass resources to solve non-energy and energy-related problems. These companies can be grouped as follows:

- Companies currently utilizing or producing biomass or biomass-derived materials and products (e.g., paper, lumber, food, and distilled spirits) are attempting to recover and use greater amounts of the resources and by-products available to them to reduce costs, develop new products, and produce energy.

- Companies which have large amounts of wastes (e.g., animal manures) are developing new ways of reducing and disposing of the wastes, reducing operating costs, and producing energy.

- Manufacturers and entrepreneurs are conducting research and development, production, and marketing of equipment to convert biomass feedstocks into energy. The goal of these activities, primarily, is to develop new products and processes which can be marketed to potential biomass users.

- Utilities which have large demands for fuels on a continuing basis are supporting the development of new, renewable, supply sources to help satisfy this demand.

0097-6156/81/0144-0003$05.00/0
© 1981 American Chemical Society

The composition of these companies and their motivations are important to government policy makers and companies considering entry into the industry, for the motivations provide a framework with which to evaluate alternative options. It is important to know, for example, that the development of energy from biomass is of secondary importance to many companies when developing a new marketing plan or tax incentive program.

In recognition of this factor, the Office of Policy and Analysis within the U.S. Department of Energy (DOE) asked Booz, Allen & Hamilton Inc. to assess the nature of industrial activities in the utilization of biomass for energy. This assessment, performed in the summer and fall of 1979, focused upon identifying the structure of the industry, the types of companies active in the industry, what they are doing, and what the motivations are for these activities.

Initially, an extensive literature research was performed to determine the companies actively pursuing biomass energy development, the issues critical to the expansion of the industry, and market perspectives which exist. Based upon the data collected, Booz, Allen interviewed 100 executives of companies nationwide to determine the scope of private sector involvement in biomass energy development. These companies, shown in Figure 1, represent a cross section of companies active in the development of energy from biomass and are representative of companies in the field.

The interviews were performed by two-person teams utilizing a standardized interview form developed by Booz, Allen and reviewed by the client. They were conducted on-site and were considered confidential. The interviews focused upon the current and planned activities of the companies in the development of energy from biomass, the motivations for their activities, the financial commitments which the companies were making, and their market outlook.

The data collected indicate that private sector involvement in biomass energy development is extensive despite industry's perception that federally sponsored work has had little impact. Another key finding was that government regulatory policies generally had a greater effect upon industry than DOE and these policies often contradicted DOE's position.

INDUSTRY STRUCTURE

There is no single biomass for the energy industry. Rather, many companies are active in utilizing a variety of biomass resources. In most cases, the principal line of business of these companies is not biomass development but agricultural production, wood products manufacturing, distilling, and similar

```
┌─────────────────────────────────────────────────────────────────────────────┐
│ FOREST PRODUCTS COMPANIES                                                     │
│                                                                               │
│ . Boise-Cascade          . Georgia-Pacific         . Scott Paper              │
│ . Champion Paper         . International Paper      . Union Camp               │
│ . Crown-Zellerbach       . Interstate Paper         . Weyerhaeuser            │
│                                                                               │
│ AGRICULTURAL PRODUCTS COMPANIES                                               │
│                                                                               │
│ . Anheuser-Busch         . Diamond-Sunsweet        . Land O'Lakes             │
│ . Archer-Daniels-Midland . First Colony Farms       . National Distillers     │
│ . Brown & Williamson     . Grain Processing         . New Life Farm           │
│ . Cajun Sugar Cooperative   Corporation            . Pioneer Hi-Bred          │
│ . Castle and Cook        . Jack Daniels               International           │
│ . C. P. Brewer           . Kaplan Industries        . Publicker Distillers    │
│ . Dekalb AgResearch      . Kelco Corporation        . Smith Bowman Distillers │
│                                                     . Sunny Time Foods         │
│                                                                               │
│ EQUIPMENT MANUFACTURERS                                                        │
│                                                                               │
│ . American Can           . General Electric        . A. O. Smith              │
│ . American Fry-Feeder    . Halcyon                 . A. E. Stanley            │
│ . Bio-Gas of Colorado    . Hamilton-Standard       . Thermonetics             │
│ . Bio-Solar R&D          . Johnson Energy Systems  . Vermont Wood Energy      │
│ . Chromoloy Corporation  . Oneida Heater              Corporation             │
│ . Combustion Power       . PyroSol                 . Wheelabrator Clean-fuels │
│ . Evans Products         . Rexnard-Envirex         . Yukon Industries         │
│ . Forest Fuels                                                                │
│                                                                               │
│ ELECTRIC AND GAS UTILITIES                                                     │
│                                                                               │
│ . Bonneville Power       . Pacific Gas & Electric  . San Diego Water and      │
│    Administration        . Seattle Power & Light      Utilities Depart-       │
│ . Burlington Electric    . Southern California         ment                    │
│ . Eugene Water and Electric  Edison                . Southern California       │
│    Board                 . University of Oregon        Gas                     │
│ . Lamar Utility Board    . Natural Gas Pipeline    . United Gas Pipeline       │
│                                                                               │
│ PETROLEUM COMPANIES AND DISTRIBUTORS                                           │
│                                                                               │
│ . American Oil Company   . Gulf Oil                . Mobil Oil                 │
│ . Bohler Brothers        . MarCom Industries       . Occidental Petroleum      │
│ . Fannon Oil             . Mid-West Solvents                                   │
│                                                                               │
│ RESEARCH & ENGINEERING                                                         │
│                                                                               │
│ . Arthur D. Little       . Institute of Gas        . WED Enterprises          │
│ . CPR Forest Products       Technology             . Wright-Malta             │
│ . Energy Resources       . Intertechnology         . Bechtel                  │
│    Company               . Marelco, Inc.           . Chemapac                 │
│ . Garrett Energy R&D     . Oasis 2000              . Ultrasystems             │
│ . Gas Research Institute . SRI International                                   │
│ . IE Associates          . Touche-Ross                                        │
│                                                                               │
│ TRADE ORGANIZATIONS                                                            │
│                                                                               │
│ . Alternative Alcohol    . Northwest Pine          . Wood Energy              │
│    Fuels Institute          Association               Institute               │
│ . Distilled Spirits      . Western Wood Products                              │
│    Council of the U.S.      Association                                        │
│ . National Gasohol       . Wood Energy Corpora-                               │
│    Commission               tion                                              │
└─────────────────────────────────────────────────────────────────────────────┘
```

Figure 1.   Companies interviewed for the study

endeavors. In this assessment, these companies were examined according to the types of biomass energy products which they produce. Based upon the data and information collected, the industry was divided into four parts:

● Alcohol fuels
● Thermal energy from wood
● Thermal energy from agricultural wastes
● Gaseous fuels.

Each segment has a different set of feedstocks, conversion technologies, and products associated with it, as shown in Figure 2.

Alcohol Fuels

The alcohol fuels segment is receiving significant attention from the public and private sectors at the present time. The use of biomass resources — primarily herbaceous crops, such as corn, wheat, grain sorghum, and wood mill residues — is viewed as a means for reducing our dependence on imported oil and for using excess crops. The key characteristics of the companies active in this segment are that they have:

● Access to ethanol feedstocks
● Access to existing gasoline marketing systems
● Experience in designing and building fermentation units
● An interest in reducing dependence upon others for fuel.

By utilizing standard fermentation and distillation processes, companies in this sector are producing anhydrous ethanol. It can be blended with unleaded gasoline to form gasohol, which is currently marketed throughout the nation, although most sales are in the Midwest.

Thermal Energy From Wood

More biomass-derived energy is produced from wood than any other source. The use of wood for thermal energy production is motivated primarily by a desire to reduce waste disposal problems and oil and gas usage. Companies active in this sector generally have:

● Experience in handling biomass materials and/or solid fuels
● Access to wood and wood wastes
● Experience in building and utilizing direct combustion units
● External support for, or entrepreneurial interest in, developing particular equipment.

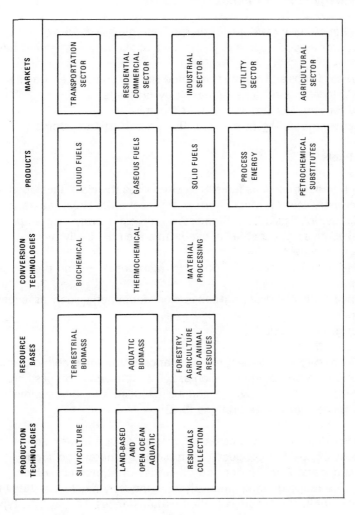

Figure 2.   Biomass overview (nonfossil, primary organic materials)

The direct combustion of wood to produce thermal energy, which can be used as process steam or heat, is the most frequent application in this sector. Some research and development is being performed in other areas, especially on the gasification of wood.

Thermal Energy From Agricultural Residues

The use of agricultural residues for thermal energy is very similar to the thermal-energy-from-wood segment. Companies active in this segment have:

- A need to dispose of agricultural process by-products and access to a centrally located stock of residues or by-products
- A need for low-cost process heat or steam
- Experience in building or utilizing direct combustion systems.

Gaseous Fuels

The development of gaseous fuels from biomass is the least developed of the biomass industry sectors. Most efforts in this sector are experimental and the commercial use of gases from biomass is still years away in the opinion of industry executives. Companies active in this area generally have:

- Access to a resource base — primarily manures — which present a disposal problem
- Experience with constructing conversion equipment, such as anaerobic digesters
- Need for gaseous fuels.

The most widespread approach to gasification is the anaerobic digestion of manures and land-based aquatic biomass. This conversion process produces either a medium-Btu gas, which can be used on-site, or in some cases, upgraded to a substitute natural gas (SNG).

The biomass industry is composed of groups of companies active in the development of specific products or uses for biomass. The focus of their activities is generally not to develop processes or equipment which can be used in a wide range of applications. Rather, industry has focused its efforts on particular applications suited to their circumstances. The specific types of companies and their activities are described in the following section.

CORPORATE ACTIVITIES AND OUTLOOK

Corporate activities in biomass include a wide range of efforts related to meeting a number of internal needs and problems which, in many cases, are not energy-related. In addition, many companies are active in the development of new products and markets from biomass. Within the four industry segments described above, various types of activities are being pursued in the development of biomass.

Alcohol Fuels

Although the alcohol fuels segment is in its infancy, many companies are active in its development. At the present time, there is no significant vertical integration in the industry. That is, firms are generally active in only one of the three component areas described in Figure 2.

Companies presently engaged in the large-scale producton of ethanol generally have access to a continuing resource base which can be converted to ethanol. These companies are normally agriprocessors, food processors, and distilleries. At the present time, one large agriprocessor, Archer-Daniels-Midland, is producing large quantities of ethanol from by-products generated during the production of corn sweetener. Many of the other companies active in this segment are also exploring alternative means to produce ethanol from their feedstocks, wastes, or by-products by the addition of new distillation units, as indicated in Figure 3. These large-scale producers consider the minimum efficient size of an ethanol plant to be 10 million gallons a year, based upon current costs for feedstocks, by-products and ethanol. In virtually every case, these activities have been initiated within the past two to three years and represent new operations for the companies in recognition of the market opportunities which are developing in alcohol fuels.

Small- and medium-sized farms are interested in the development of small-scale ethanol production facilities (approximately 250,000 gallons per year) as a means of developing additional uses for their crops, and as a means to develop independence from oil suppliers. While, to date, there has been little construction of on-farm units, these farms represent a large potential market if current trends continue.

A third group of companies in the alcohol fuels segment of the biomass industry are architect/engineering firms, fuel distributors, and oil companies. These companies are attempting to build upon their expertise in construction, fuel handling and retailing to develop new markets for their products. In most cases, gasohol distributors are independent wholesalers and retailers,

10

BIOMASS AS A NONFOSSIL FUEL SOURCE

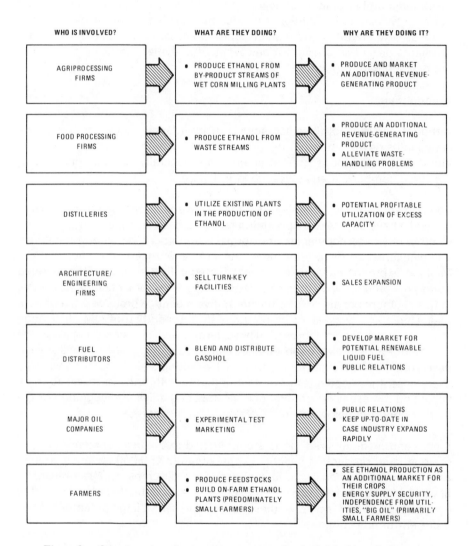

Figure 3.   Overview of motivations for entry into the alcohol fuel/gasohol segment

although large oil companies, including Amoco, Exxon, and Gulf, are beginning to market gasohol.

Industry executives anticipate that the alcohol fuels business will grow more rapidly in the next few years than other segments of the industry. Systems currently exist, or are expanding, in many regions for collecting the resource, converting it to ethanol, and transporting and marketing it. In addition, current government policies encourage the use of gasohol. Two major barriers to alcohol fuels are uncertainty regarding government policies, including a problem recently addressed in part by President Carter, who proposed that tax exemptions be extended to the year 2000, and the lack of a pipeline system to distribute large volumes of gasohol.

Thermal Energy From Wood

The thermal energy from wood segment of the biomass industry is composed of a broad mix of companies that are dominated by the forest products and pulp and paper industries. These companies are active in all the various component areas of the industry from ownership of the resource base to the use of wood for products and energy. Generally, the main thrust of these companies is to produce products from wood rather than thermal energy. Only small wood distributors and hardware-oriented compaines are active in the development of energy from wood, as shown in Figure 4.

Forest resource companies are actively pursuing the development of wood resources for a number of applications — to increase the growth rate of wood, to improve the efficiency of harvesting and collecting wood, to develop more efficient combustion systems. The production of energy from waste wood is growing, and close to 50 percent of the total energy requirements of these companies is produced from biomass in the form of heat, steam, and cogenerated electricity.

Wood fuel distribution systems are beginning to emerge in most regions of the country, although many potential users indicate that they still have difficulty acquiring wood due to the lack of adequate distribution systems. With the exception of some large distributors in parts of the Northeast, Southeast, and Northwest, the existing distribution system of fuel wood for the industrial and residential markets is composed of small operators. Most industrial users of wood for energy are transporting wood themselves, or contracting for its transportation.

Large wood energy users include manufacturing firms, utilities, and other facilities, such as universities, which are beginning to utilize wood as a cheaper, more reliable source of energy. These users are burning wood both

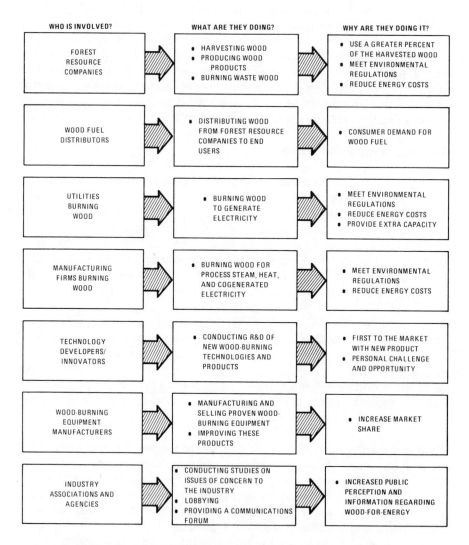

Figure 4.   Motivations for entry into the thermal energy from wood sector

alone or with coal to lower emissions from their fuel-burning installations. Another group of companies is active in the development of wood combustion systems to meet the growing demand for utilizing wood energy. Companies specializing in the development, manufacture and marketing of wood-burning equipment for residential, commercial, and industrial uses are emerging in those regions where wood is most heavily used.

Business executives expect growth of the wood-for-energy business to occur slowly. While the means exist to collect and convert wood into useful products, the lack of wood fuel distribution systems is perceived to be a major barrier to growth. A further concern of executives is the lack of assurances regarding the long-term availability of supplies.

Thermal Energy From Agricultural Residues

The segment concerned with thermal energy from agricultural residues is small and localized at the present time. Only a few types of agricultural residues, such as nut shells and corn cobs, are being utilized for the production of energy. Corporate activities are focused upon the use of agricultural residues which are low in moisture content, where the residues are a by-product of another operation, and where they are accumulated in a central location.

Agricultural processing companies are entering into this segment to find a means to dispose of their by-products and wastes, as well as to develop a cheap, independent source of energy, as shown in Figure 5. To meet this growing demand, equipment manufacturers are developing boilers and other combustion equipment which can be utilized easily by agricultural processing companies. Manufacturers with experience in hardware development and fabrication are attempting to spin off their experience in other markets to this new market.

In addition to the generation of steam and heat, agricultural processors are developing cogeneration systems to produce electricity which can be used on-site. In most cases, these systems generate more electricity than can be consumed by the processor. Local electric utilities in these cases may serve as markets for this power and some companies are purchasing this electricity from the processor. In California, two major utilities are also entering into agreement to purchase steam from the processors to generate electricity.

Business executives see the barriers to widespread use of agricultural residues for thermal energy as similar to those in the wood segment. The lack of collection and distribution systems for the feedstock is a key barrier, as is the cyclical nature of its availability.

IMPACT OF FEDERAL GOVERNMENT PROGRAMS

The Federal Government has developed numerous programs designed to accelerate the conversion of biomass to energy. In some cases cited by industry, these programs conflict with one another. Industry executives indicated that Federal incentives programs and regulatory policies are having a major impact on the growth of biomass activities, while DOE programs are having little impact.

A major factor that affects the economic viability of the alcohol fuels industry is the continuation of gasohol tax exemptions provided by Federal and some state governments. At the present time, gasohol is exempt from the 4-cent per gallon ($16.80 per barrel of alcohol) excise tax on gasoline and state exemptions range from 3 to 9.5 cents per gallon ($12.60 to 39.90 per barrel of alcohol). Extension of the Federal tax exemption to the year 2000 has been proposed; this is viewed by executives as a major step towards accelerating the development of the industry.

The Bureau of Alcohol, Tobacco and Firearms (ATF) is responsible for monitoring and regulating all processes which result in the production of ethanol or other alcoholic substances. Currently, regulations which apply to the production of alcohol for human consumption also apply to ethanol produced for fuel and industrial uses. The regulations are cumbersome and retard the development of the industry. Development of streamlined procedures would help accelerate the development of the industry according to industry executives.

The U.S. Forest Service, through its control over public forests and its policies, has a significant impact on the amount, location, and price of wood that is available for all uses, including energy. Of greatest importance are the regulations regarding cutting, harvesting, and management of the forest, especially the amount that can be cut. Clearer guidelines regarding the use of public lands could help improve the availability of wood for energy.

The DOE has three major offices which are active in biomass technology development and commercialization. The Biomass Energy Systems Branch and Wood Commercialization Program within the Office of the Assistant Secretary for Conservation and Solar Applications are the major funders of biomass projects in DOE. In addition, DOE provides funds to the Solar Energy Research Institute for biomass research and development activities. Presently, DOE funding is focused on the production, conversion, and/or utilization of biomass, rather than on programs which focus upon a particular type of biomass or industry segment. Industry executives felt that the impact of

Gaseous Fuels

According to the interviews performed as part of the assessment of the biomass industry, the gaseous-fuels-from-biomass industry is small. Corporate activities are focused on the anaerobic digestion of manures to produce gaseous fuel. Minimal activities are directed towards the development of other feedstocks and processes.

Despite the current industry attention to these systems and their potential to meet site-specific needs, industry experience indicates that the economics of the anaerobic digestion of manures are not competitive at the present time. In fact, many of the companies interviewed indicated that the refeed materials (by-products which can be fed to animals) produced by anaerobic digestion have higher economic value than the gas which is produced.

Resource owners, such as cattle feedlot operators, farmers, agricultural processors and wastewater treatment plant operators, produce large quantities of wastes that could serve as feedstocks. They are exploring ways to convert these wastes to energy as shown in Figure 6. A number of anaerobic digesters have been built at facilities such as feedlots, but the results are mixed. Gaseous fuels are produced, but the amount is small and high in cost.

Equipment companies, industrial research organizations, and utilities are providing support for research and development in this segment. Manufacturers are attempting to develop equipment which can convert the feedstock economically into gases and are currently building demonstration facilities. Utilities and industrial research organizations foresee the development of biogasification as a potential renewable source of energy and are actively supporting efforts in this area. Individual utilities have built anaerobic digestion facilities to begin testing systems, and the gas industry is supporting R&D on new feedstock production and conversion. Electric utilities foresee the possibility of converting the product gas into electricity and are supporting demonstration facilities to test the efficiency of the process.

The outlook for the gaseous fuels segment is that limited development of the resource will continue until company executives are convinced that these fuels can be produced economically. Some uncertainty exists regarding the conversion technology, which also hinders its development. In addition, if the product gases are not upgraded to a high-Btu gas, new distribution systems must be developed.

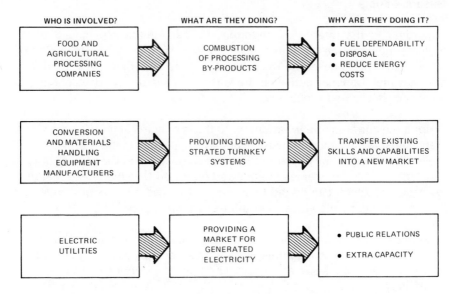

Figure 5.   Motivations for entry into the agricultural residues industry segment

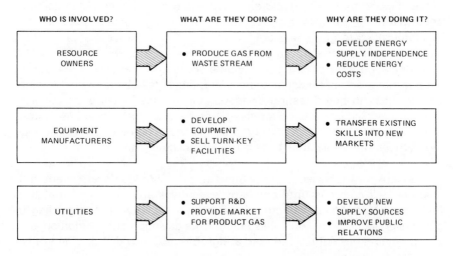

Figure 6.   Motivations for entry into the gaseous fuels industry segment

these programs could be enhanced by placing more emphasis on the development of near- and mid-term technologies and applications.

Other Federal agencies, such as the Federal Energy Regulatory Commission (FERC) and the U.S. Department of Agriculture, affect the growth of specific sectors. In some cases, Federal programs have contradictory impacts upon the industry. For example:

- DOE is encouraging the development of cognerated electric power. However, FERC has not implemented regulations which would permit utilities to buy this power for their consumers.

- DOE's encouragement to utilize wood as an energy source is not consistent with Forest Service policies on managing wood availability.

These findings suggest that Federal policies could be more effective if they could be coordinated to enhance development of biomass energy. A reorganization of DOE's program along product lines would help to improve communications with industry and would increase DOE's effectiveness.

ACKNOWLEDGEMENT

The following people participated in Booz, Allen's Biomass Industry Assessment: James F. Lowry, Scott D. Moeller, Kenneth G. Salveson, Michael R. Sedmak, Satish Suryawanshi.

RECEIVED MAY 12, 1980.

BIOMASS PRODUCTION

Forest Biomass for Energy

A Perspective

R. L. SAJDAK, Y. Z. LAI, G. D. MROZ, and M. F. JURGENSEN

Department of Forestry, Michigan Technological University, Houghton, MI 49931

A primary challenge of the near future is the development of alternative energy sources to make this nation less dependent on imported oil. Increasing the use of wood for energy production has been suggested as one method of meeting this goal (1,2). There are a number of advantages for developing a wood-related energy base in this country. Most importantly, wood is a renewable resource and its production normally has a relatively low environmental impact. Wood can be burned and thus converted directly into energy. Used in this way, it is a relatively clean fuel and the residual ash is useful as a fertilizer. The technology also exists for converting wood into other energy forms such as oil, gas, alcohol, charcoal, and electricity.

The forest resource of the United States can make a significant contribution toward meeting national energy needs. Forests occupy about one third of our land area and the wood inventory of this resource is enormous. Of the total annual biomass produced in these forests, only about thirty percent is presently used (3). Better utilization of our annual forest production could make significant quantities of wood material available for energy purposes. However, annual forest growth is well below what is currently possible. More intensive management systems could double the productivity of our forest lands within fifty years (4).

Wood now supplies about two percent of total U.S. energy needs, primarily through the use of manufacturing wastes and mill residues for boiler fuel.

0097-6156/81/0144-0021$07.00/0
© 1981 American Chemical Society

Various studies suggest wood could supply up to 10 percent of the Nation's current energy needs within the next decade. Depending upon the strategies used, eventually it may be possible to supply 20 percent of our total energy budget (5). However, the use of wood for energy production must be kept in proper perspective. Wood is not the only product of our forests. These lands play a vital role in providing various social and cultural benefits such as wilderness, outdoor recreation, wildlife, fish, and clean water. Therefore, no single resource or forest use can be examined in isolation from the others. Energy uses will have to be balanced against the growing demand on our forests for lumber, fiber products, and recreational opportunities.

This paper will analyze the feasibility and implications of increased utilization of our forests as a source of energy. Consideration will also be given to the production of biomass from intensively cultured plantations as well as the quality of the biomass produced by different management techniques.

THE U. S. FOREST RESOURCE

About 740 million acres or 33 percent of the land area of the United States is classified as forest land. In order to be classified as forest, at least 10 percent of the land must be stocked with trees of any size. Also in this category are lands that formerly had tree cover but have not been developed for other purposes, as well as lands whose primary use is not timber production. Nearly two-thirds of this area, or 488 million acres, is classified as commercial forest land. Commercial forests are defined as forested land capable of producing at least 20 cubic feet of industrial wood per acre per year, and is not reserved for uses which are incompatible with timber production (6). Thus, National Parks, wilderness areas, and other special use areas are not included in this category.

To help deal with the diversity and complexity of vegetational and environmental differences in various parts of the country, the forest resource is discussed by four major geographic regions: Northern, Southern Rocky Mountains — Great Plains, and the Pacific Coast. The Northern Region includes Maryland, West Virginia, Kentucky, Missouri, and all states north and west to the Great Plains. The Southern Region encompasses Virginia, Tennessee, Arkansas, Oklahoma, and states to the south. The Rocky Mountains — Great Plains Region includes Western South Dakota and all states west of the Great Plains except those bordering on the Pacific Ocean. The Pacific Coast Region has the four states along the Pacific Ocean, including Hawaii.

The Northern Forest Region

This region, which contains over one-half of the Nation's population (53 percent), is the second most densely forested area with 35 percent of the total land in commercial forests (Table I). It is also the only region to register an increase in commercial forest acreage during the period 1952-1977. Private, non-industrial land ownership in the states of New York, Pennsylvania, and West Virginia contributed to most of this increase. This ownership makes up 71 percent of the forest holdings in the Northern Region.

Table I. DISTRIBUTION OF COMMERCIAL FOREST LAND IN THE UNITED STATES BY REGION AND BY OWNERSHIP CLASS (6)

Region	Area[a]	Distribution[b] (%)	Public (%)	Ownership[c] Industry (%)	Private (%)
North	170,769	35	19	10	71
South	188,433	39	9	19	72
Rocky Mountains — Great Plains	57,765	12	75	4	21
Pacific Coast	70,758	14	63	17	20
TOTAL	487,725				

a Thousand acres.

b Percentages of the total U.S. commercial forest land.

c Percentages of the region.

The Northern Region's forests vary considerably as do their uses. Seventy-five percent of the timber volume is in hardwoods, which are utilized for furniture, veneer, pulp, pallets, and railroad ties (Table II). Softwood volume is the smallest of any region (25 percent). The stocking levels are the lowest in the country and the average annual growth per acre is quite low (Table III). It should be noted that these volume figures are based on commercial-sized timber and do not include wood present in small, non-merchantable trees and in the tops and limbs of merchantable trees.

Table II. TIMBER PRODUCTION ON UNITED STATES COMMERCIAL FOREST LAND IN 1977 (6)

Region	Total Volume (million cu. ft.)	Average Volume Per Acre (cubic feet)	Timber Type Softwood (%)	Hardwood (%)
North	200,337	1173	25	75
South	230,037	1221	43	57
Rocky Mountains— Great Plains	112,405	1946	94	6
Pacific Coast	258,024	3646	92	8
TOTAL	800,803	1997	62	38

Table III. NET ANNUAL GROWTH AND HARVEST ON COMMERCIAL TIMBERLANDS IN THE UNITED STATES (6)

Region	Growth (1000 cu. ft.)	Harvest (1000 cu. ft.)	Average Growth Per Acre (cubic feet)
North	5,927,587	2,739,535	34.7
South	10,826,042	6,571,223	57.4
Rocky Mountains — Great Plains	1,689,553	845,786	29.2
Pacific Coast	3,431,151	4,278,868	48.5
TOTAL	21,874,333	14,425,230	44.8

The Southern Region

This region is the most densely forested in the Nation with 39 percent of the land area in commercial forests. It is also the second most densely populated. Commercial forest acreage declined during the 25 year period, 1952 to 1977, primarily due to conversion of forest lands to agricultural uses. Forest ownership, as in the North, is primarily in private non-industrial holdings (72 percent). Industry ownership is the largest of any region (19 percent) and public ownership is the smallest (9 percent).

Southern forests are quite equally divided between hardwood and softwood tree species. Forty-three percent of the timber volume is in softwoods. Average annual growth per acre of forest land is the highest of any region. The South is projected to supply over one-half of the Nation's softwood requirements by the year 2030, nearly doubling its 1976 output.

The Rocky Mountains-Great Plains Region

This region contains the smallest percentage (12 percent) of this country's commercial forests and is the least populated. Most of the forest land is in public ownership (75 percent). During the period 1952 to 1977, this region lost over 10 percent of its forest land, most of which was entered into the wilderness and National Park system.

These forests are predominantly softwoods which comprise 94 percent of the timber volume. About one-third of this volume is in the pinyon pine/juniper timber type which is relatively unimportant for wood production. Annual growth averages 29 cubic feet per acre per year, the lowest of all regions. These forests are very important for watershed, recreation, and livestock grazing.

The Pacific Coast Region

There is a tremendous amount of diversity of the forests in this region. The climate ranges from arctic to tropical and some of the most productive forests in the world are found along the coast from Northern California to Washington. Only 14 percent of the total land area is in commercial forests and like the Rocky Mountain Region, most is in public ownership. This region also recorded a decline in commercial forest land due to transfers into wilderness areas and parks.

Most of the timber volume is in softwoods (92 percent) and the average annual growth is second highest in the Nation. Average growing stock per acre is the highest in any region, and annual timber cut is more than net ingrowth. Overcutting is due to the accelerated removal of old growth, overmature stands. This region now supplies over one-half of our softwood timber needs. The projections are for this region's softwood output to decline after 1990.

FOREST PRODUCTIVITY ASSESSMENT

In evaluating the forest resource to determine how much wood is available for energy production, a multitude of factors needs to be considered. Most of the commercial forest land in the Northern and Southern Regions is in private, non-industrial ownership. It is difficult to assess the contribution these lands will make toward supplying our energy needs. Objectives of forest ownership vary considerably and timber production is often second to non-tangible goals. Private forest lands are often managed on an opportunistic basis with little regard to a regulated and sustained timber yield.

Therefore, it may be difficult to obtain a dependable supply of wood from a particular geographic area. Also, most private forest holdings are small and the most efficient total tree harvest-systems cannot operate economically in such situations. When conducting conventional roundwood harvests, the removal of logging residue from small, widely scattered operations poses a difficult problem.

The private forest resource can be very productive. A recent study indicated that wood production on these lands was 61 percent of capacity under current management practice. In contrast, the National Forests grew wood at 49 percent of capacity (7). Improved management of small forests for increased biomass production may be the most difficult problem of all. Usually lacking is the willingness and capacity of the small land-owner to make investments for a return which will be 20-30 years away (8). Substantial increases in the supply of wood from these ownerships can be achieved only through governmental assistance, such as cost-sharing and technical assistance programs (9).

Conceivably, much of the wood from private forests may be used for home heating purposes, particularly in the more populated forested areas. As the cost of heating oil increases, heating with wood becomes more attractive and more private forests will be dedicated for fuelwood production. In Maine, for example, over 50 percent of the homes are currently being heated with wood. The impact of this type of management on future supplies of high value sawlogs in a particular region could be significant.

Increasing Timber Output

With better management, U.S. timber supplies could be dramatically increased in the future. Average annual growth on commercial forest land in 1976 was 45 cubic feet per acre. If the forests were fully stocked, average growth would average 75 cubic feet per acre per year. This increase in annual growth would roughly equal the total volume harvested from all forests in 1976 (6).

The possibilities for intensifying management exist for all ownerships and in all regions of the country except for the Rocky Mountains-Great Plains. Preliminary results of the U.S. Forest Service and Forest Industries Council studies indicate there are economic opportunities for intensified management on 160 million acres of commercial timberland or about 34 percent of the Nation's total. The opportunities are most concentrated in the 113 million acres of the Southern Region. About three-fourths of the treatment strategies involve regeneration of non-stocked areas, harvesting mature forests, regenerating higher yielding young stands, and converting existing stands to more productive species. The estimated total cost of treating the Southern acreage is $8.8 billion. This investment would increase annual growth in this region by more than 8.5 billion cubic feet. The National Forests, particularly those in the West, also have the capability of supporting larger harvests. Additional investments would be needed for road construction, stand improvement, reforestation, and salvage (9).

Increased forest yields can be obtained by using known and proven intensive culture techniques. Use of genetically improved planting stock can increase yields by 10 to 20 percent. Fertilization and thinning programs, better tree spacing, and increased fire, insect, and disease protection can all improve yields significantly. The limits to increasing wood yields by intensive cultural techniques are not known. A reasonable estimate is that the growth could double on half of the commercial forest land in 50 years (3).

Availability of Wood for Energy Production

We can only speculate on the true size of the total timber resource of the United States. To date, all of the inventories and surveys on a national scale have been based on volume measurements of the merchantable parts of trees. Tables I, II, and III reflect this. Merchantable volume is a vague term, particularly since merchantability limits are rapidly changing. The concept of whole-tree utilization has reinforced this confusion. With the development of whole-tree harvesting methods, previously non-merchantable parts of the tree are chipped and used for pulp and paper, composite products, and fuel. These new concepts of utilization make the whole tree the basic unit of measurement. Since accurate volume determination is difficult on irregular shaped objects, weight of biomass is the new standard of measure for all tree components.

There have been numerous estimates made on the total biomass and biomass potential of our forests (3,10). The inventory procedure used is based on estimates and averages, and is fully described by Wahlgren and Ellis (11). Forest surveys based on biomass measurement techniques are needed to accurately determine the quantities and location of our wood resource. Many studies on measuring the weight of individual trees, and to a lesser extent forest stands, have been made. This work has been summarized by Keays (12), and Hitchcock and McDonnell (13). As work in forest biomass measurement is refined, regional weight tables can be developed and accurate biomass inventories compiled.

Work has already begun in this direction. Pioneering work initiated by Young and others in Maine has resulted in the completion of a forest biomass inventory on nearly two million acres in that state (14). The next forest survey of Maine, to be started in 1980, by the U.S. Forest Service will be in terms of both merchantable volume and total wood biomass. Information such as this, as well as measurement of annual dry matter production, will determine the availability of wood supplies on a regional basis for industrial and energy purposes. Biomass inventories must also undergo economic assessments since in many situations, the cost of collecting, processing, and transporting biomass materials would exceed any reasonable value anticipated for fuel or other products.

Forest Residue

Forest residue is defined as the biomass left in the woods after harvest and includes tree tops, limbs, cull material, and all present and future non-merchantable trees. This differs from mill residue such as bark, edgings, and sawdust, which is often fully utilized as boiler fuel. The amount of residue remaining after harvest will vary according to merchantability standards, method of harvest, and forest stand composition and quality. In a typical hardwood sawlog harvest as much as 50 percent of the potential usable biomass may be left as residue. If the sawlog stand is overmature or of poor quality, additional residue may be left. In contrast, if the stand is whole-tree chipped for pulp material, very little residue will remain.

Throughout the United States there are a number of ongoing studies whose objectives are to determine, on a regional basis, the real potential and economic availability of woody biomass for energy. The Department of Energy in cooperation with the U.S. Forest Service is in the process of developing a National Wood Energy Data System. This system will identify amounts and locations of wood fuels in excess of commercial needs. The information will be delineated by state and in some cases down to the county level. In addition, the location of current and potential large wood burning systems within each unit will be identified (14).

The Maryland Department of Natural Resources recently completed an assessment of the availability, cost, and reliability of wood fuels on the Delmarva Peninsula (14). The study concluded that cull trees and timber harvest residue could provide over 1.9 million tons of wood fuel annually at a cost to users of $12.50 per green ton. Other studies in Minnesota, New York, Oregon, and Washington are involved in similar utilization and biomass assessments to determine the availability of wood fuels for energy.

Recently, the Northern Wisconsin and Upper Michigan region was intensively studied by the U.S. Forest Service to determine the amount of residue available in the region (15). Using available forest survey information and computer simulation, the harvest amounts and delivered cost of biomass were determined. A "Managed Harvest" procedure was used to determine the amount of wood product and residue that should be removed each year for a 10-year period. Only two products were considered: sawlogs and wood chips. The assumption was made that excess chips not needed for paper products would be available for energy purposes or more succinctly, "pulp the best and burn the rest." The overall objective of the "Managed Harvest" procedure was to move the forests to a fully-regulated and more productive condition.

Several harvesting strategies were examined and significant differences were found to exist in the deliverable cost and amount of recovered biomass among the harvesting systems used. One harvesting strategy involved the mechanized thinning of overstocked, small diameter forest stands. Over three million dry tons of biomass was projected to be deliverable from the entire study area in 1980 at a cost of $15.91 per ton. This volume would also be available in each succeeding year.

A second strategy involved the clearcutting of mature and overmature stands and stands too poorly stocked to be carried to a normal cutting age. Sawlogs were removed before the residual stand was chip-harvested. This strategy is used where even-aged forest management is practiced and also where stands would be converted to a fully-stocked situation. Ten million tons could be produced annually, exclusive of sawlogs, at an average cost of $12.35 per ton delivered.

The costs per delivered ton are averages for the entire tonnage available. Substantial volumes of forest biomass are available at lower costs, but as an attempt is made to recover increasing amounts of the biomass, the costs increase.

The study concluded that significant forest biomass quantities are available in this region. For both Northern Wisconsin and Upper Michigan, 30.5 million dry tons annually would be deliverable at a 1980 projected cost of $16.07 per ton. These biomass quantities would be available each year for the next 10-years. At the end of the 10-year period, a new assessment will have to be made.

The costs for woody residue materials will change as more efficient harvesting equipment is developed. Koch and Nicholson (16) described a mobile chipper designed to effectively harvest biomass on relatively flat terrain. Their studies show that if a minimum of 25 tons (green weight) of biomass is available per acre, about 85 percent of such biomass could be recovered and delivered to the forest roadside for about $11.82 per green ton. This machine was designed to recover forest residue that is ordinarily bulldozed and burned during the preparation of a site for planting.

INTENSIVE PLANTATION CULTURE

Traditional plantation culture in this country has approximated what occurs in nature. In this situation trees are planted at fairly wide spacing (6 feet or more) often with little or no site preparation. Occasionally, the trees are released from competing vegetation and the plantation may be thinned once or twice. The rotation length of these plantations may be 30 years or more, depending on the growth rate and end product desired. Coniferous trees, particularly the pines, have been planted far more often than hardwoods.

Recently, interest has developed in the intensive culture of plantations on a short rotation. However, this concept is now new. The Europeans have been managing plantations in this manner for decades and in the 1960's, McAlpine and his coworkers further developed the concept in the United States with sycamore (*Platanus occidentalis* L.) (17). Since that time, studies have been initiated with other tree species throughout the United States and Canada.

In concept, the intensive culture of plantations on a short rotation is essentially an agronomic system. Trees are planted on prepared sites at close spacing (4 feet or less), cultivated, fertilized, and irrigated during the rotation. Agricultural-type forage equipment would then harvest the crop at ages of less than 10 years. The entire aboveground portion of the tree is utilized. Regeneration of the plantation would be by coppice growth, thereby limiting the plantations mainly to hardwoods which stump or root sprout after cutting.

A number of advantages are evident in the intensive culture of plantations on a short rotation when contrasted to conventional plantation management (18,19). These are:

1. Higher yields per unit of land area, therefore less land would be needed to produce a given amount of biomass.

2. Early amortization of plantation establishment costs.

3. Increased efficiency of most cultural and harvesting operations because of complete mechanization.

4. Reduced plantation regeneration costs after the first rotation.

5. Genetically improved trees can be utilized quickly.

6. The biomass produced will be of more uniform quality.

There are some significant disadvantages, however:

1. Initial plantation establishment and management costs per acre are very high, thereby increasing the financial risk involved.

2. Site limitations are important for this type of forest practice. The land must be relatively flat and soil texture, structure, drainage and stoniness will be important considerations.

3. The relatively uniform genetic makeup of the trees increase epidemic disease and insect hazards.

Hardwoods have been preferred for intensive plantation management because of their sprouting capability and the fast growth of these sprouts for the first 10-20 years, as compared to conifers.There are exceptions, however, where conifers may be more desirable. Williford et al. (20) reported loblolly pine (*Pinus taeda* L.) to be superior in biomass production on many sites in the south. Studies by the U.S. Forest Service at Rhinelander, Wisconsin, indicate conifers may have advantages under certain site conditions (21). For example, jack pine (*Pinus banksiana* Lamb.) is well adapted to the North, has few serious insect and disease problems, and is less demanding of nutrients and moisture than many hardwoods.

Numerous species trials are underway in various parts of the country to identify the best species for localized biomass production (14). Several of the more intensively studied candidate species are discussed in more detail as follows:

American Sycamore

This was the first species advocated for short rotation intensive culture. Sycamore plantations are established using seedlings or from cuttings. If cuttings are used, a clonal plantation is established which results in a high degree of tree uniformity. Sycamore wood is moderately dense with a specific gravity of about 0.46 and a sapwood moisture content of 130 percent on an oven dry basis (40).

This species is most productive on rich alluvial land in the south. Plantations on upland sites have yielded poor survival and unacceptable levels of growth (22). Howlett and Gamache (18) reviewed numerous studies on the biomass productivity of sycamore in the south and found growth rates as high as 9 dry tons per acre per year. The average yield was about 4.5 dry tons per acre per year and in some studies yields were as low as 2 tons per acre per year. Usually, higher initial stand densities produced higher yields for rotation ages of up to four years. However, direct comparisons among sites are difficult because of the different cultural treatments applied in the various studies.

Poplars and Cottonwood

Various species and hybrids of the genus *Populus* are some of the more promising candidates for intensive biomass production. This group has long been cultivated in Europe and more recently in the Eastern United States and Canada. Poplar hybrids are easily developed and the resulting progeny are propagated vegetatively using stem cuttings. Consequently, there are literally hundreds of numbered or named clones established throughout the Eastern United States. The wood is moderately light as indicated by specific gravity values of 0.32 to 0.37 and the moisture content of the sapwood is about 146 percent (40).

Certain hybrid poplar clones show excellent response to intensive culture techniques. Dawson (21) reported a mean annual biomass yield of nearly 7 tons per acre on a 12 inch spacing during the first rotation. Anderson and Zsuffa (23) reported coppiced stands yielding 8.5 tons per acre per year on a two year rotation.

Eucalypts

Species of *Eucalyptus* appear to have promise as candidates for biomass production in intensively cultured plantations. However, a lack of cold hardiness would restrict their usage to the southeastern states, California and Hawaii. This evergreen hardwood genus contains over 500 species, most of which are native to Australia. The eucalypts have been planted throughout the world and display a wide adaptability to a variety of sites. Especially noteworthly is their capability to thrive on droughty and nutrient-deficient sites. Although primarily suited for frost-free areas, studies by industrial cooperators in the southeastern states indicated that considerable variation exists in resistance to freezing temperatures. (24). The genetic variation indicated was such that selection for freeze-tolerant eucalyptus species and races may be possible. Limited growth data is available but it has shown a mean annual biomass production of over 4 tons per acre per year. Greater productivity from eucalyptus plantations should be possible with more intensive culture and genetically improved stock.

Nitrogen-Fixing Species

Nitrogen is the most limiting macronutrient needed for tree growth on most forest sites. Concern over the high cost of nitrogen fertilizers for intensive plantation culture has prompted considerable interest in plants capable of using atmospheric nitrogen. Zavitkovski *et al.* (25) reviewed the studies which pertain to the use of nitrogen-fixing woody and herbaceous species for forestry purposes. Assessments were made on the possibility of using red alder (*Alnus rubra* Bong.) for biomass production, the use of nitrogen-fixing trees in mixtures with non-nitrogen fixing trees and the use of herbaceous legumes as nurse crops in intensively managed plantations. This review indicates significant management benefits are possible through the use of nitrogen-fixing trees. Red alder biomass yields are comparable to other fast growing species. The use of nitrogen-fixing trees and herbaceous material has given significant increases in biomass yields. Cost-benefit ratios have not been determined because of the variety of study conditions encountered. Much additional information is needed to fully assess the feasibility of using nitrogen-fixing species in short-rotation intensive plantation culture.

Economics of Intensive Plantation Culture

The yields from intensively cultured plantations are considerably greater than those from natural stands of similar tree species. It must be emphasized, however, that many of the reported plantation yields are from small study plots and indicate what is biologicaly possible. To project such yields over a larger area may be inappropriate. Several studies in Wisconsin, South Carolina and Georgia (26,27,28) are currently underway to determine the feasibility of intensive plantation culture on a large scale.

Rose and DeBell (29) assessed the economics of intensive plantation culture for wood fiber production. Spacing of trees and length of rotation appeared to be particularly cost sensitive in determining economic feasibility. Wide spacing (4 X 4 feet and 12 X 2 feet) and longer coppice rotations (4 year and 10 year, respectively) appeared feasible while two year coppice rotations did not.

Using a different perspective, Eimers (30) evaluated the economics of intensive plantation culture as an energy source from the standpoint of a company having extensive experience in growing and handling forest products. His conclusions were that fuel plantations are currently uneconomical. However, indications are that this could change if certain costs were reduced. The economic outlook for intensive management systems would improve if harvesting costs could be reduced and energy conversion costs lowered. The overall cost picture may also be more favorable if a dual purpose crop could be produced. For example, the biomass material produced from plantations would be sorted into high quality chips for fiber products and low quality chips converted into energy. A direct parallel already exists in many forest industries which currently use residues from the manufacture of forest products to fire boilers.

Eimers (30) also points out that planting at close spacing offers some management flexibility when thinning is considered. That is, thinning from intensively managed plantations can be used for energy, while the remaining trees can be grown for conventional forest products. The advantage to this management strategy is that decisions concerning wood use can be deferred to a future date when the economics of energy supply may be less turbulent than at present.

CHARACTERIZATION OF BIOMASS

The physical and chemical nature of biomass materials has a profound influence on their end-uses. For example, moisture content and specific gravity are important properties for the production of solid wood and paper products, as well as for conversion into energy. Other physical and mechanical properties of wood, such as fiber length, fibril angle and strength of individual fibers, are not important for energy applications. The current trends toward whole tree utilization and energy plantations through intensive management can result in substantial changes in the nature of wood resources available for industrial and commercial use.

The effects of various intensive management systems on wood properties have been reported in numerous publications and were recently summarized in three articles (31,32,33). It was clearly indicated that the major changes in wood properties are associated with shorter rotations, which result in a higher proportion of juvenile wood. This young wood, as compared to mature wood, contains a higher moisture content, lower specific gravity and a high proportion of reaction wood. The effects of changing these wood properties on the solid wood product and paper industry have been extensively reviewed by Bendtsen (31) and Einspaphr (32,33), respectively. In the following section, those wood properties which are important to the conversion of woody biomass into energy uses are discussed.

Distribution of Tree Components

Total above-ground tree biomass is generally divided into foliage, branch, and stem components. The branches and stems can be separated further into bark and wood portions. The importance of branches and foliage to total tree weight is dependent upon the age and/or size of the tree (34). It is estimated that the above-ground biomass currently produced on commercial forest land in the United States is approximately distributed as wood (80 percent), bark (12 percent), and foliage (8 percent) (11). Table IV shows some representative distributions of tree components by species and age.

The data clearly show that young trees have a substantially higher content of foliage and bark biomass than older, larger trees. These two components are not desirable for the production of fiber related products, such as paper, but are desirable for energy production due to their high energy content, particularly bark. The trend toward shorter stand rotation or the establishment of energy plantations would increase the availability of young tree biomass for energy use.

Moisture Content

Moisture content is a particularly important characteristic when using woody biomass as fuel, and is generally expressed as percent of the dry weight. Water contained in biomass adds to the cost of transportation, and lowers the efficiency of energy conversion by a direct combustion process, because of the energy required for evaporation of the water. It is estimated that about 15 percent of the total available heat in a wood or bark fuel is required for moisture evaporation, assuming a 100 percent moisture content (38). However, conversion by anaerobic digestion or fermentation methods are more efficient when using high-moisture biomass materials. The procedure for estimating effective heat values for both wood and bark fuels under varying moisture content and furnace enviroments has been summarized by Ince (39).

The moisture content of various forest biomass varies widely with species, geographic locations, genetic differences, tree components used, and tree age. Published data indicate that moisture content of mature wood may range from about 30 percent to more than 200 percent (40). Also, moisture content of the stem sapwood portion is usually higher than that of the associated heartwood. For young hardwood sprouts (6 to 15 years old), an average

Table IV. TREE COMPONENTS AS PERCENT OF ABOVE-GROUND BIOMASS

Species	Age, yr	Woody Biomass[a] Total %	Woody Biomass[a] Bark %[b]	Branches Total %	Branches Bark %	Stems Total %	Stems Bark %	Foliage %	Ref. No.
Acer saccharum Marsh. (sugar maple)	2	68	(19)	—	—	—	—	32	35
Acer rubrum L. (red maple)	2	64	(27)	—	—	—	—	36	35
	3	68	(22)	—	—	—	—	32	35
Robinia pseudoaccacia L. (black locust)	2	67	(29)	—	—	—	—	33	35
Pinus banksiana Lamb. (jack pine)	6	—	—	22	(38)	52	(20)	24	36
	40	—	—	16	—	80	(10)	4	37
Populus tremuloides Michx. (aspen)	40	—	—	10	—	88	(19)	2	37

[a] Sum of stems and branches

[b] Numbers in the parentheses are bark percentages of the tree component.

The data clearly show that young trees have a substantially higher content of foliage and bark biomass than older, larger trees. These two components are not desirable for the production of fiber-related products, such as paper, but are desirable for energy production due to their high energy content, particularly bark. The trend toward shorter stand rotation or the establishment of energy plantations would increase the availability of young-tree biomass for energy uses.

moisture content of 80 percent was reported for nine species collected in midsummer, ranging from 59 percent for green ash (*Fraxinus pennsylvanica* Marsh.) to 99 percent for yellow poplar (*Liriodendron tuplipifera* L.) (41).

It has been shown that wood moisture content has an inverse correlation with tree age (35,42). In general, wood from a young tree contains more water than wood from an older one of the same species (42). This is probably associated with the vigorous growth and the high proportion of sapwood in the younger tree.

There is no general consensus concerning a seasonal trend of moisture content in trees (43). Some researchers have reported larger seasonal variations while others have found very little. However, it appears that the young trees or sprouts, in contrast to mature trees, display a significant seasonal variation in moisture content, particularly for those growing in northern climates. Figure I shows seasonal variations in moisture content for 3-year old sugar maple (*Acer saccharum* Marsh) sprouts (35) and 50-year old yellow poplar (43). The maple, grown in the Upper Peninsula of Michigan, displays a marked seasonal variation in moisture content which reaches a maximum in June. This wood moisture pattern is essentially parallel to sprout growth activity. In contrast, the moisture content of mature yellow poplar grown in the Southern Appalachain Mountains of North Carolina does not vary significantly from season to season.

Specific Gravity

Specific gravity indicates the amount of solid material in a given volume, and is usually considered the best single index of intrinsic wood quality for fiber, wood products, and energy production. Specific gravity is inversely correlated with wood moisture content (44). Thus, it also varies with species, tree components, and the age of the tree. In general, juvenile wood has a relatively lower specific gravity than mature wood. In some species, the juvenile growht period may last for at least 10 years, during which there is a steady increase in specific gravity (45). Therefore, young trees will not provide the same yield of solid material or fiber per unit volume of wood as compared to older trees.

Chemical Composition

The chemical composition of biomass materials is generally discussed in terms of cell wall polysaccharides (cellulose and hemicelluloses), phenolics (lignin and polyphenols), extractives, and ash content. Wood normally contains small amounts of ash (1 percent) and various quantities of extractives

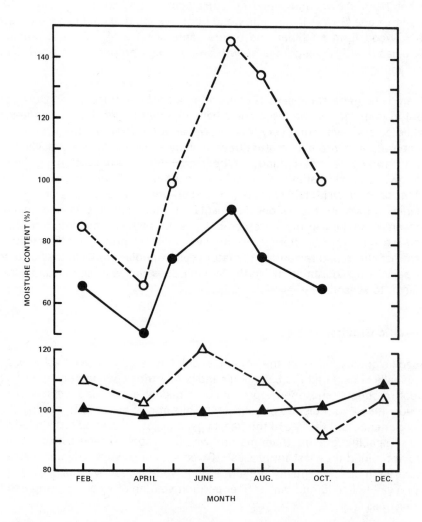

*Figure 1.   Season variation of moisture content of (○ bark, ● wood) 3-year-old
sugar maple and (△ bark, ▲ wood) 50-year-old yellow poplar (43)*

depending on tree species (46). Extractive-free hardwoods have a lignin content between 18 percent and 25 percent; it varies between 25 percent and 35 percent for the softwoods. The remaining materials in the wood are the polysaccharides (46).

There are definite changes in the chemical composition of reaction wood. Compression wood has a significant increase in lignin and a corresponding decrease in polysaccharides as compared to normal softwood. Tension wood has just the opposite relationship. Since juvenile wood tends to contain a high level of reaction wood, its chemical composition should differ from that of mature wood.

Other tree components, such as foliage and bark, contain a substantially higher content of extractives, and a slightly higher ash content than wood. Also, bark has a higher content of phenolics other than lignin, including phenolic acid and tannins as compared to wood.

Chemical composition, as discussed in the next section, is closely related to the caloric values of biomass, and also affects the efficiency of conversion, particularly when using a biological approach. For example, the presence of phenolics, particularly lignin, presents a major roadblock for enzymatic conversion of polysaccharides to alcohol. The conversion of juvenile biomass has been shown to have a higher moisture content and lower specific gravity than mature wood (47), and may respond more favorably to such a treatment process. The energy conversion of juvenile biomass materials by a thermal or biological methods needs to be explored.

Caloric Values*

The caloric value, or heat of combustion, of a natural fuel on weight basis is a function of the chemcial composition. It has been shown that a linear relationship exists between the heat of combustion and the carbon content of the substrate (48). Lignin has a higher heat of combustion than that of a polysaccharide (5884 vs. 3853 cal/g), because of its lower oxygen content. The extractives (terpenoid hydrocarbons or resin) with even lower oxygen contents have still higher heat contents (8124 and 9027 cal/g) (49). In contrast higher ash content will have a negative effect on the calorific value.

The caloric values of natural fuels reported in the literature have been summarized by various authors (38,29,41,48,50,51,52). The average values taken from these reviews are listed in Table V. An average heat content of 4781 and

---

* All heating values discussed in this section are high heating values.

Table V. AVERAGE CALORIC VALUES OF NATURAL FUELS (cal/g)

Author	Softwood			Hardwood			Ref. No.
	Wood	Bark	Needles	Wood	Bark	Needles	
Harder and Einspahr	—	5149 (15)[a]	—	—	4715 (9)	—	50
Harder and Einspahr	—	5241 (3)	—	—	4613 (15)	—	51
Corder	4886 (16)	5266 (9)	—	4596 (7)	—	—	38
Corder	—	5033 (24)	—	—	4688 (21)	—	52
Susott et al.	—	—	5145 (1)	—	—	4470 (4)	48
Ince	5133 (16)	5197 (13)	—	4971 (13)	4672 (9)	—	39
Koch	—	—	5016 (1)	—	—	—	49
Neenam and Steinbeck	—	—	—	4777 (9)	4631 (9)	5047 (9)	41
Average	5010	5177	5081	4781	4664	4759	

[a] Figures in parentheses are the number of species examined.

5010 cal/g was found for hardwood and softwood, respectively. It appears that little variation in caloric values exists among various hardwood species. Larger variations have been observed among softwood species because of the marked differences in extractive content. However, it can generally be concluded that the caloric value for a given volume of biomass material is primarily determined by its moisture content and specific gravity.

PERSPECTIVE

The near term and extended outlook of increasing use of wood for energy purposes is favorable. Commercial forests contain an abundance of biomass for fuel and intensively cultured plantations could add additional amounts. The question is how much and at what cost?

The use of wood for energy will compete with wood for material use. Wood as a construction material is much more important in the Nation's energy budget than as a fuel. The manufacture of lumber and plywood from wood is much less energy-intensive than the manufacture of metal and plastic products. Also, the energy savings due to the high insulating value of wood building components should be noted.

Forest survey statistics indicate we harvest and use upwards of 30 percent of current forest production for conventional forest products. The remaining 70 percent has been shown to be potentially available for various energy uses (3,10). However, before such increased wood use occurs, a number of factors need to be examined and hopefully resolved.

Conventional Wood Needs

It is apparent that increasing demands for traditional forest products will require some of the current biomass surplus. It is also apparent that some of the surplus biomass may be unavailable for economic reasons or unharvestable due to constraints imposed by other forest users or environmental factors. Consequently, what is then left of the surplus would be potentially available for energy use.

A recent study for the American Pulpwood Association evaluated the world wood supply and demand situation to the year 2000 (7). The United States was identified as one of the countries which will need more wood than it can supply. However, the study concluded that we have the potential to not only eliminate the expected deficit but also to become a net exporter of forest products. To accomplish this goal, we will need to increase utilization of current growth as well as increase productivity on all available lands.

Increasing Utilization

Wood currently unused because of poor markets and high utilization standards will help alleviate the projected U.S. wood deficit. Smaller logs must be harvested and the remaining portions of the trees chipped. High quality chips would likely be utilized for fiber products while low quality material can be burned.

Improved biomass inventories are also needed to identify the amounts, quality, locations, and deliverable costs of the excess forest biomass. Increasing the utilization of forest biomass is particularly cost-sensitive. As greater amounts of forest residue are removed from a given area, the costs will increase. In many cases, it may not be economically feasible to recover the residue generated by small landowner harvest operations and transport this to a large energy user. More likely, this residue and possibly higher value timber will be increasingly used for home heating in a direct response to rising conventional heating costs. Utilization will probably be most intense in the Northern and Southern Regions where most of our population exists and the forests are also most diverse.

Impact on Other Forest Uses

Increasing the utilization of our forests will have a definite impact on other forest uses such as wildlife, recreation, and water. Indeed, some segments of our society place timber production secondary as is evidenced by recent court decisions which limit management prerogatives on certain National and State Forests. The most efficient harvesting systems, i.e., clearcutting with whole-tree chipping, are especially in disfavor. However, it must be noted that the inherent nature of species like aspen *(Populus)* and jack pine requires that clearcutting be used for regeneration.

The removal of tops, limbs, and cull trees in a selective log harvest, while improving the visual impact, will destroy the cover needed for small animals and birds. A multiple use management approach is necessary when implementing increased forest utilization, but at some sacrifice in yield. Not every acre can be utilized to its fullest potential for biomass production.

Impact on Site Quality

More information is needed to determine the impact that increased utilization will have on site quality such as soil organic matter content, water holding capacity and fertility levels. (53) It appears that the more intensive the harvest, the greater the opportunity for soil deterioration. The increased removal of leaves and small branches represents a significant drain of nutrient capital on some sites and may cause some long-term reduction in site productivity.

Increasing Forest Productivity

As stated earlier, many forests are understocked. Some of the biomass surplus will need to be left on site to increase the amount of growing stock and therefore increase future yields. The increasing demand for wood products and fuelwood presents an opportunity for forest improvement practices as never before. Pre-commercial thinnings and improvement cuttings may now be profitable in supplying wood for energy uses while increasing the production of quality timber in the future.

It is expected that the forest products industries will lead the way toward increasing forest productivity. Industry lands are already the most productive as compared to private and public lands. At present, with about 14 percent of the commercial forest land area, the forest industries produce 33 percent of the wood in the United States. Less certain is the extent to which small private forests will increase productivity. The wood markets are assured but the investment required for forest improvement is unattractive for individuals, especially in the face of rising land costs, interest rates and taxes, and increasing regulations. Public lands are least productive because of location, terrain and land use history and because of the pressures from other users. Large increases in governmental funding will be needed and policy changes initiated before significant increases in productivity can be realized.

Biomass Energy Plantations

Intensively cultured short-rotation plantations have been advocated as providing additional sources of biomass for energy use. The land base needed for the implementation of this proposal on an extensive scale would seem to present a nearly insurmountable problem. For example, Evans (54) calculated that a 100 megawatt electric facility would require a forest biomass plantation nearly 200 square miles in size if oven dry yields of five tons per acre per year could be attained. It must be noted that the high yields obtained from small plot studies may be difficult to obtain on such a large scale. Biomass energy plantations would most likely be relegated to sub-marginal agricultural lands or areas where forests are not normally found. The inputs needed on these lands to achieve high yields of biomass may not be feasible or economical. As indicated earlier, the lower specific gravity and higher moisture content of short-rotation biomass may affect the energy conversion process.

Natural northern hardwood stands in Northern Michigan were found to produce nearly two dry tons of biomass per acre per year over a 50-year period. These yields were obtained with no management inputs whatsoever (35). Frederick and Coffman (55) reported mean annual dry weight production of nearly 2.5 tons per acre in a 25 year old unmanaged red pine (*Pinus resinosa* Ait.) plantation. Even greater yields are possible in the South and the Pacific Northwest. A more realistic approach might be to use the knowledge gained from the intensive culture studies and supply this to improved management of conventional forest plantations.

In conclusion, wood will play an increasingly important role in supplying a part of our Nation's energy needs. To what extent this will occur is unknown and will depend in part on the costs of more conventional fuels. Using wood for energy is the least profitable use of this resource and the differential should continue. The forest products industry will likely lead the way in increasing the use of wood for energy. This industry has had the expertise in growing, handling, and utilizing wood both for manufacturing products and for energy. The use of wood in generating electricity will increase but on a smaller scale as utilities take advantage of surplus wood in the heavily forested regions. It is apparent that the use of wood as a home heating fuel will continue to expand dramatically. The effects of these various wood energy uses on the environment are largely unknown.

ACKNOWLEDGEMENT

The authors wish to thank the U.S. Department of Energy for their partial support of this endeavor. Also appreciated is the encouragement received from the Environmental Sciences Division, Oak Ridge National Laboratory.

REFERENCES

1. Szego, G. C.; Kemp, C. C. *Chemtech* **1973,** *3,* 275.

2. Zerbe, J. *For. Farmer* **1977,** *37* (2), 12.

3. "Forest Biomass As an Energy Source", Task Force Study Report, Society of American Foresters, Washington, D. C., 1979, p 7.

4. Phelps, R. B. "Proceedings", Symposium on Intensive Culture of Northern Forest Types, USDA Forest Service, Gen. Tech. Rep. NE-29, 1977; p 17.

5. Ripley, T. H.; Doub, R. L. *Amer. For.* **1978,** *84,* (10), 16.

6. "Forest Statistics of the U.S., 1977", USDA Forest Service, 1978; p 133.

7. Clawson, M., "Proceedings", Annual Meeting, Society of American Foresters, 1978.

8. Miller, S. R. "Proceedings", Symposium on Principles of Maintaining Productivity on Prepared Sites, Mississippi State University, 1978, p 1.

9. N. A. "A Report to Congress on the Nation's Renewable Resources," USDA Forest Service: Washington, D.C., 1979; p 209.

10. Wahlgren, H. G. *Amer. For.* **1979,** *84* (10), 24.

11. Wahlgren, H. G.; Ellis, T. H. *Tappi* **1978,** *61* (11), 37.

12. Keays, J. L. "IUFR Forest Biomass Studies", University of Maine, Life Sci. and Agri. Expt. Station, 1971; p 93.

13. Hitchcock III, H. C.; McDonnell, J. P. "Proceedings", Forest Resource Inventions Workship, 1979; p 544.

14. Bente, Jr., P. F., Ed. "Bio-Energy Directory", The Bio-Energy Council: Washington, D.C., 1979; p 103.

15. N. A. "Forest Residues Energy Program", USDA Forest Service: St. Paul, Minn., 1978; p 297.

16. Koch, P.; Nicholson, T. "Proceedings", Second Annual Symposium on Fuels from Biomass, U.S. Department of Energy, 1978; p 227.

17. McAlpine, R. G.; Brown, C. L.; Herrick, A. M.; Ruark, H. E. *For. Farmer* **1966,** *26* (1), 6.

18. Howlett, K.; Gamache, A. "Silvicultural Biomass Farms", Mitre Tech. Reports Vol. II, 1977; p 136.

19. Heninger, R. L.; Murrey, M. D. Weyerhaeuser Co. Personal Communication.

20. Williford, M.; Kellison, R. C.; Frederick, D. J.; Gardner, W. E. "Conference Papers", Fifteenth Southern Forest Tree Improvement Conference, Mississippi State University, Starkville, June 19-21, 1979.

21. Dawson, D.; Zavitkovski, J.; Isebrands, J. G. "Proceedings", Second Annual Symposium on Fuels from Biomass, U.S. Department of Energy, 1978, p 151.

22. Kellison, R. C.; Gardner, W. E. "Proceedings", IUFRO Consultation on Fast-Growing Plantation Broad-Leaved Trees for Mediterranean and Temperate Zones, Lisbon, Portugal, 1979; in press.

23. Anderson, H. W.; Zsuffa, L. For. Res. Rep., Ministry of Nat. Res., Ontario No. 101., 1975; p 5.

24. Hunt, R. *Tappi* **1979,** *62,* (9), 79.

25. Zavitkowski, J.; Hansen, E. A.; McNeel, H. A. "Proceedings", Workshop on Symbiotic Nitrogen-fixation in the Management of Temperate Forests, 1979, p 389.

26. Dawson, D. H. "Proceedings", 3rd Annual Biomass Energy Systems Conference 1979; p 359.

27. Salo, D. J.; Henry, J. F.; Inman, R. E. "Design of a Pilot Silvicultural Biomass Farm at the Savannah River Plant," The Mitre Corp., McLean, Va., 1979; p 115.

28. Steinbeck, K. "Proceedings", 3rd Annual Biomass Energy Systems Conference, U.S. Department of Energy, 1979; p 47.

29. Rose, D. W.; DeBell, D. S. *J. of For.* **1978,** *11, 706.*

30. Eimers, K. L. *Chemtech* **1978,** *(4),* 212.

31. Bendtsen, B. A. *Forest Prod. J.* **1978,** *28,* 61.

32. Einspahr, D. W. *Tappi,* **1976,** *59* (10), 53.

33. Einspahr, D. W. *Tappi,* **1976,** *59* (11), 63.

34. Byram, T. D.; Van Bavel, C. H. M.; Van Buijtenen, J. D. *Tappi* **1978,** *61* (6) 65.

35. Lai, Y. Z.; Sajdak, R. L.; Mroz. G. D.; Jurgensen, M. F.; Schwandt, D. L. Unpublished Data.

36. Zavitkovski, J.; Dawson, D. H. "Tappi Conference Papers", Forest Biology — Wood Chemistry Conference, 1977; p 217.

37. Alban, D. H.; Perola, D. A. Schlaegel, B. E. *Can. J. For. Res.,* **1978,** *8,* 20.

38. Corder, S. E. "Wood and Bark as Fuel", Oregon State Univ., Forest Res. Lab., Corvallis, 1976, Research Bulletin 14, p 28.

39. Ince, P. J. "Estimating Effective Heating Value of Wood or Bark Fuels at Various Moisture Content", USDA Forest Service, Gen. Tech. Report, 1977; FPL 13, p 8.

40. "Wood Handbook; Wood as an Engineering Material," U.S. For. Prod. Lab., USDA Arg. Handbook 72, 1974; pp 3-6.

41. Neenan, M.; Steinbeck, K. *Forest Sci.* **1979,** *25* (3), 455.

42. Zobel, B.; Matthias, M.; Roberds, J. H.; Kellison, R. C. "Moisture Content of Southern Pine Trees", N. C. Sch. Forest Tech. Rep. 37, 1968; p 44.

43. Phillips, D. R.; Schroeder, J. G. *Wood Sci.* **1973,** *5* (4), 265.

44. Miller, S. R., Jr. "Proceedings", Fifth Southern Conference on Forest Tree Improvement, 1959; pp 97-106.

45. Zobel, B. J.; McElwee, R. L. *Tappi* **1958,** *41,* 167.

46. Rydholm, S. A. "Pulping Processes", Interscience: New York, 1965; pp 90-99.

47. Lai, Y. Z.; Sajdak, R. L.;Mroz, G. D.; Jurgensen, M. F.; Schwandt, D. L.; Steinhilb, H. M. *Tappi* **1979,** *62* (4), 84.

48. Susott, R. A.; DeGroot, W. F.; Shafizadeh, F. *J. Fire* and *Flammability*
     1975, *6,* 311.

49. Koch, P. "Utilization of the Southern Pines", USDA Agr. Handbook 420,
     1972; Vol. I, p 379.

50. Harder, M. L.; Einspahr, D. W. *Tappi* **1976,** *59* (12), 132.

51. Harder, M. L.; Einspahr, D. W. *Tappi* **1978,** *61* (12), 87.

52. Corder, S. E. "Properties and Uses of Bark as an Energy Source", Oregon
     State Univ. Forest Res. Lab, Corvallis, 1973; Bulletin 14, p 28.

53. "Proceedings", Impact of Intensive Harvesting on Forest Nutrient Cyc-
     ling, Syracuse, N.Y., 1979.

54. Evans, R. S. Information Report VP-X-129, Western Forest Prod. Lab.,
     Vancouver, B. C., 1974; p 15.

55. Frederick, D. J.; Coffman, M. S. *J. For* **1978,** *1,* 13.

RECEIVED JUNE 20, 1980.

Production of Nonwoody Land Plants as a Renewable Energy Source

ALEX G. ALEXANDER

University of Puerto Rico, P.O. Box H, Agricultural Experiment Station, Rio Piedras, PR 00928

Non-Woody Land Plants in Perspective

Literally thousands of terrestrial plant species can be regarded as potential energy sources. A majority of these are herbaceous seed plants which complete their growth and reproductive processes within a single growing season of a few months duration. They are widely distributed from arctic regions to the tropics (1-3). They are equally diverse with respect to their growth and anatomical characteristics, their cultural requirements, and their physiological and biochemical processes (2-9). Yet all have the capacity to convert sunlight to chemical energy and to store this energy in the form of biomass. An oven-dry ton of herbaceous biomass represents about 15×10^6 Btu's of stored energy. The direct firing of one such ton, in a stoker furnace with high-pressure boiler having a 70% conversion efficiency, would displace about two barrels of fuel oil.

In addition to their fibrous tissues, some species also produce sugar and starch in sufficient quantities to warrant extraction and conversion to ethanol. The latter can displace petroleum in the production of motor fuel or chemical feedstocks (10-19). Other species store additional energy in the form of natural hydrocarbons.

0097-6156/81/0144-0049$07.00/0
© 1981 American Chemical Society

While it is not correct to say that herbaceous land plants have been overlooked as a domestic energy resource, only a small number have been examined closely for this purpose. Among the latter are tropical grass species of *Zea, Sorghum, Saccharum,* and *Pennisetum* which were recognized for their high yields of fiber and fermentable solids long before the oil embargo of 1973. Throughout their history as cultivated crops, plants such as corn, sweet sorghum, sugarcane, and napier grass have evolved extensive technologies for their cultivation, harvest, post-harvest transport and storage, and for their processing and marketing. Yet, even for these plants major changes must be made in their management if they are to serve effectively as energy crops (5,6,23,24). Other tropical plants having very fine botanical or agronomic attributes and enjoying a year-round climate suited to biomass production have been generally ignored as energy resources. Pineapple, cassava, and a range of underutilized tropical species are appropriate examples (4,8,13).

A majority of herbaceous land plants have never been cultivated for food or fiber. In warm climates, wild grasses such as *Sorghum halepense* (Johnson grass), *Arundo donax* (Japanese cane), and *Bambusa* species are borderline cases where occasional use has been made of their high productivity of dry matter. In cooler climates, self-seeding plants such as reed canary grass, cattail, wild oats, and orchard grass may be viewed with mixed feeling by landowners unable to cultivate more valuable food or forage crops. Plants such as ragweed, redroot pigweed, and lambsquarters are recognized for their persistent growth habits while otherwise regarded as common pests. However, the value of such species could rise dramatically as biomass assumes a role as a nonfossil domestic energy resource.

Prior Studies on Herbaceous Plants as Energy Sources

Aside from sugarcane and "allied" tropical grasses (6,7,13,23-25), relatively little attention has been given to herbaceous land plants specifically as sources of fuels and chemical feedstocks. Studies were initiated recently at Battelle-Columbus Laboratories on common grasses and weeds as potential substitutes for fossil energy (26). Plants showing promise as boiler fuels include perennial ryegrass, reed canarygrass, sudangrass, orchardgrass, bromegrass, Kentucky 31 fescue, lambsquarters, and others. A range of species have indicated some potential as sources of oil, fats, protein, dyes, alkaloids, and rubber. Such plants include giant ragweed, alfalfa, jimsonweed, crambe, redroot pigweed, dogban, milkweed, and pokeweed.

Recently, a government-sponsored program was initiated to screen herbaceous plants to close the information gap in this area of biomass energy development (27). This program has two phases: First, to identify promising species for whole-plant biomass production in at least six different regions of the U.S., and second, to perform field evaluations on at least 20 species per region, with a view toward identifying those most suitable for cropping on terrestrial energy plantations. Arthur D. Little, Inc., conducted Phase I (2).

Six regions were designated on the basis of climatic characteristics, land availability, and land resource data provided by the U.S. Soil Conservation Service (2). A list of 280 potential species was prepared on the basis of published literature and personal interviews. These were screened in accordance with botanical and economic characteristics, with emphasis on previously uncultivated species. Certain agricultural plants were also considered.

Factors such as yield potential, cultural requirements, tolerances to physiological stress, production costs, and land availability were considered in ranking the candidate species of each region (2). Plants with yields less than 2.2 tons/acre (5 metric tons/hectare) were eliminated. For the potential energy crop species, comparisons were drawn with six categories of economic plants, including tall and short broadleaves, tall and short grasses, legumes, and tubers. Some 70 species were recommended for consideration in the program's second phase (field screening). Some of these plants (redroot pigweed, lambsquarters, Colorado river hemp, ragweed) have no prior history as cultivated crops and their cultural needs remain obscure. Other species (Bermuda grass, Kenaf, reed canary grass, sudan grass) have been improved and cultivated for decades (2).

BOTANICAL AND AGRONOMIC CONSIDERATIONS

This initial program to evaluate herbaceous land plants will help to clarify their value as a renewable energy source. However, an extensive research effort is needed to complete this task even as it applies to existing plant forms already managed as agricultural crops. A continuing effort will be needed over a period of several decades in the areas of new species evaluation, genetic improvement, herbaceous plant cropping on marginal lands, and crop tailoring to changing energy needs. The remainder of this paper offers some general guidelines and considerations for dealing with the vast pool of existing herbaceous land plants.

Botanical Considerations

Photosynthesis: Photosynthesis is the process by which the radiant energy of sunlight is converted to chemical energy by plants. Its overall reaction can be stated simply as:

$$CO_2 + H_2O \xrightarrow[\text{Green Plants}]{\text{Sunlight}} (CH_2O) + O_2$$

The amount of energy retained in the photosynthate (CH_2O) is about 468 kJ/mole.

Although not an efficient process, it is the only system of solar energy conversion on earth that has operated at any appreciable magnitude and with any appreciable economy for any appreciable period of time. An estimated 1350 J/m^2 arrives at the earth's upper atmosphere in the form of solar radiation, but only about half penetrates to the earth's surface ([28]). A theoretical 8 percent of this radiation could be converted photosynthetically; however, a maximum conversion efficiency of only 4 percent has been attained and this under conditions of low light intensity ([29]). Agricultural plants average perhaps 0.5 to 1.0 percent efficiency. Land plants on a world-wide basis probably average less than 0.3 percent efficiency. Nonetheless, the earth's plants store annually about 10 times more energy than is utilized, and some 200 times more than is consumed annually as food ([30]).

Photosynthesis consists of two phases: (a) Energy capture, yielding chemical energy and reducing power; and (b), the reduction or "assimilation" of atmospheric CO_2. The carbon reduction phase is accomplished by three distinct pathways (C_3, C_4, and CAM). Each pathway is found among the world's herbaceous land plants, but the C_3 pathway is the most widely distributed. CAM plants, which assimilate carbon at night, are relatively less important even though their utilization of water is generally more efficient than for C_3 species. The C_4 pathway was at first thought to reside only in sugarcane and related tropical grasses ([31-33]). It was soon found in other plants such as *Zea, Sorghum,* and *Amaranthus* ([34-38]). The C_4 species constitute a kind of apex in photosynthetic proficiency, aided to some extent by attributes such as a low CO_2 compensation point, a "lack" of photorespiration, and a capability to utilize both lower and higher light intensities better than C_3 species ([5,9,39-41]).

An important aspect of photosynthetic energy conversion often overlooked in higher plants is their "spectral proficiency", that is, their ability to convert different regions of the sun's spectral energy distribution. When photosynthesis by a given leaf is measured at different wavelengths of equal quantum flux, say from 400 nm in the blue-violet to 720 nm in the far-red, a photosynthetic action spectrum is attained which tells us much about the leaf's ability to "harvest" the entire package of visible light energy received from the sun. With sufficient replications, an action spectrum characteristic of the species is derived, a kind of spectral finger print complete with peaks and depressions typifying that species. Ironically, more than 60 percent of incoming solar energy is received at wavelengths shorter than 550 nm, while (apparently) most plants are photosynthetically active at wavelengths longer than 600 nm. There is some evidence that *Saccharum* and a few other species have major photosynthesis activity in the blue-violet to blue-green region (40,41). Photosynthetic action spectra have been determined for approximately 30 agricultural plants The vast majority of herbaceous land plants have not been examined in this context.

Photosynthesis in an Energy Crop Perspective: A plant physiologist or biochemist measuring photosynthesis in the laboratory usually determines the quantity of CO_2 assimilated per unit of leaf area in an hour or some other convenient time interval (mg CO_2/cm^2 -hr^{-1}). It does not necessarily follow that superior assimilation rates noted under these conditions will translate to high photosynthate yields in the field. A more convenient measure of photosynthetic potential in biomass-candidate species is the quantity of dry matter produced per square meter of leaf surface per day (g DM/m^2 -day). A majority of herbaceous land plants would produce on the order of 2-8 grams of oven-dry material per square meter per day, *during the peak of their growth or tissue-expansion phase*. A yield of 15 g/m^2 -day would be quite good and would typify some C_4 pathway species. Potential maximum yield estimates have been placed at 34 to 39 g/m^2 -day for C_3 plants and 50-54 g/m^2 -day for C_4 plants (46).

To any energy planter, the most meaningful measure of solar energy conversion to biomass is the number of kilograms of dry matter produced per hectare per year (or tons per acre per year). While photosynthetic processes per se remain an important factor, equally important are all other processes and constraints of plant growth and development which come into play as photosynthate is elaborated to harvestable biomass. Each of these factors finds expression in the energy planter's gross yield of biomass. Annual dry matter yields in the order of 22,500 kg/ha (10 tons/acre) are common for a few species but the majority of herbaceous land plants probably yield less than 4500 kg/ha (2 tons/acre).

The reckoning of dry matter yields on an annual basis rather than an hourly or a daily basis might seem inappropriate to non-woody species whose growth period lasts only a few weeks or months. However, it is correct to do so since many of the energy planter's expenses (including land rentals, taxes, equipment depreciation, and land maintenance) are incurred on an annual basis (9,47,48). Moreover, some herbaceous plant species do produce dry matter continually throughout the year and others could do so if managed as energy crops. A plant such as sugarcane propagated as a 12-month sugar crop can yield dry matter at the rate of 10-12 g/m^2-day, or about 10 tons/acre-year. The highest dry matter yields attained to date by the author were with first-ratoon sugarcane managed for total biomass rather than sugar. These amounted to 36.6 tons/acre-year, or 26.6 g/m^2-day, over a time-course of 365 days (49).

It is safe to say that for most plants, there is no direct relationship between photosynthetic potential, as determined in the laboratory, and the total dry biomass to be harvested in the field. The principal reasons for this are a series of botanical and agronomic factors which prevent the elaboration of photosynthate to biomass at rates commensurate with the plant's carbon reduction potential. Some of these factors are fundamental constraints against growth and development essentially beyond the control of the energy planter (though sometimes controllable by the plant breeder). Other constraints are a reflection of plant management and can be eliminated through research and development of the species as an energy crop.

It is also safe to say that some non-woody land plants will be found to have good biomass potentials but little prospect of ever being managed as agricultural energy commodities. For such plants, a decisive attribute will be their ability to survive and produce some biomass with the barest minimum of production inputs (8,9). Yet even in these instances one must not over-emphasize photosynthesis rate as an energy yield indicator; there is simply too much variability in the measured rates of photosynthesis and too little correlation with measured biomass (5,33,50). An example of this was found in a series of "wild" sugarcanes *(Saccharum* species) whose photosynthesis rates varied by a factor of 10 while their biomass yields varied by a factor less than 2 (40). Variation is similarly high among the hybrid sugarcanes of commerce (33,51). In a given field of sugarcane, completely uniform as to soil series, variety, planting date, and cultural management, one can expect to find photosynthesis and growth rates that vary by a factor of 3 to 5 among randomly-selected sampling sites (5).

Reduction State of the Primary Photosynthate: To this point we have considered biomass as "elaborated photosynthate", consisting mainly of cellulose, and lignin derived from glucose or polyglucosides having the basic formula $(C_6H_{12}O_6)$. This is quantitatively the most important form of biomass for both woody and herbaceous plant species. However, as a form of stored energy it has the limitation of being only partially reduced. The presence of oxygen in the structure of plant tissues, starch, and extractable sugars limits the energy content of such materials to approximately $14\text{-}16 \times 10^6$ Btu's per dry ton. Alternately, some plant species store energy in more highly reduced compounds having progressively less oxygen in their structure. Plant materials such as isoprene polymers, sterols, oils and waxes consist mainly of carbon and hydrogen and contain on the order of $40\text{-}50 \times 10^6$ Btu's per dry ton. Calvin and others have advocated the study of "hydrocarbon plants" as superior biomass energy sources (21,22,52). Many of these species have the added advantage of good adaptability to lands that are semi-arid, roughly-contoured, and otherwise marginal for the production of more conventional food and energy crops (8,53,54).

Hydrocarbon-bearing plants include both woody and herbaceous species. Some of the better-known examples, such as the rubber tree *(Hevea brasiliensis)* and guayule *(Parthenium argentatum)*, are woody perennials, while others, such as *Euphorbia* and *Calotropis* species, are borderline cases that could be managed either as forest or agronomic energy crops. Milkweed species *(Asclepidacea)* are predominantly herbaceous, but one member found in the tropics, *Calotropis procera* (the "giant milkweed"), is a woody perennial reaching heights of 9 to 12 feet over a period of several years. In Puerto Rico it is regarded as a forest specimen (55), but as an energy crop would most likely be managed as a frequently recut forage (56).

Water Utilization Efficiency: Water will quite definitely be a decisive limiting factor in the worldwide expansion of agriculture (9,54,57,58). It is therefore important that water utilization efficiency be considered in the future screening and development of herbaceous land plants as energy resources. Three factors must be assessed from the onset: (a) Utilization efficiency in photosynthetic processes; (b) water extracting capability from the candidate species' natural terrain; and (c), the species capacity for water conservation by anatomical means.

Agronomic Considerations

The production of biomass involves the collaboration of physiological, biochemical, botanical, and agronomic factors under any set of conditions. However, for the intensive management of biomass production, particular attention must be given to field-scale behavior of plant masses in which an individual plant or crown complex loses the importance we attach to it as a botanical or horticultural entity. Several agronomic considerations critical to successful biomass production are herein discussed.

Growth Characteristics: To attain maximum biomass on a per annum basis, one would ideally select a year-round growing season and plant species capable of growing on a year-round basis. Certain tropical grasses (sugarcane, napier grass, Johnson grass, bamboo) do this very nicely if planted in the tropics. Some of their members produce well also in sub-tropical or even temperate regions, but given equal management, they will realize only part of their full yield potential when growth is constrained for several months by cool temperatures.

It is important to recognize also that growth is a 24-hour process as well as a 12-month process. The photosynthetic and tissue-expansion systems that operate each day are fully dependent on the nocturnal transport and mobilization of growth-supporting compounds. For this reason, the tropics are again favored by their warm nights for biomass production. In a similar vein, the cool nights of the southwestern arid lands are probably as restrictive for biomass as are the limited moisture supplies.

Possibly the most desirable growth characteristic of all for herbaceous species is the ability to produce new shoots continually throughout the year, year after year, from an established crown. This is a predominant characteristic of sugarcane and certain other tropical grasses both related and unrelated to *Saccharum* species. Such plants do not require the periodic dormancy and rest intervals so important to most temperate species. Nor is this compensated by the intensive flush of May-June growth by temperate plants; over the course of a year the slower-growing tropical forms will out-produce them by a factor of three or four.

A less obvious but utterly critical feature of the perennial crown is its continual underground contribution of decaying organic matter to the soil. This process proceeds concurrently with the continuous renewal of underground crown and root tissues. For this reason, the long-term harvest and removal of above ground stems, together with the burning off of "trash", does not have an adverse effect on sugarcane lands. There are soils in Puerto Rico that have produced sugarcane more or less continually for four centuries without destruction of their physical properties or nutrient-supplying capability. On the other hand, seasonal crops such as field corn and grain sorghum do not develop a perennial crown. For these plants, a good case can be made against the removal of aboveground residues from the cropping site.

Tissue Expansion vs. Maturation: A common misconception is that biomass growth involves mainly a visible increase of size, and that per acre tonnages of green matter are a reasonably accurate indicator of a plant's yield potential. It is also frequently assumed that the moisture content of plant tissues is essentially constant at around 75 percent, and that dry matter yields can be calculated rather closely from green weight data. These assumptions are not correct in any case, but are particularly erroneous with respect to herbaceous species. In virtually all such plants "growth" consist of discrete, diphasic processes of tissue expansion followed by maturation. The tissue expansion phase produces visible but succulent growth consisting mainly of water (on the order of 88-92 percent moisture). The maturation phase corresponds to physiological aging and senescence, that is, to flowering and seed production, slackening of visible growth, yellowing and loss of foliage, and hardening of the formerly succulent tissues. During this period, the dry matter content will increase by a factor of two to four in a time interval that may be shorter than that of the tissue-expansion phase. For example, the hybrid forage grass Sordan 70A more than doubles its dry matter yield in a time-span of only two weeks (23), i.e., during weeks 8 to 10 in a 10-week growth and reproduction cycle. For this reason, the optimal period of harvest must be determined with care for each candidate species. Again, as a rule of thumb, the allowing of additional time before harvest will work in favor of increased biomass yields from herbaceous plants.

For most herbaceous plants, the production of dry matter can be plotted as an S-shaped curve (Figure I). Dry matter content will not ordinarily exceed 10 to 12 percent during the period of rapid tissue expansion but will begin to rise dramatically at some point in time that is characteristic of the individual species. Dry matter will rarely increase beyond 40 percent in herbaceous plants. Attempts to hasten this rise (by withholding water) or to delay it (by use of growth stimulants) have met with limited success in tropical grasses (63). Some increase in the magnitude of dry matter accumulation has been attained over short periods of time with the plant growth regulator Polaris (63).

Harvest Frequency: Once the diphasic nature of biomass growth and maturation is recognized, the importance of harvest frequency is also underscored. The optimal period for harvest in the maturation curve of one species will differ enormously from the optimal harvest period of another, even among varieties within the same genus and species. For this reason, it is convenient to group candidate species into distinct categories based on the time interval that must elapse after planting to maximize dry matter yield (63). The management and harvest requirements of each group will also vary. On this basis, it has been convenient to organize tropical grasses into "short-, intermediate-, and long-rotation" categories (Table I).

**Table I. CATEGORIES OF TROPICAL GRASSES AND LEADING
CANDIDATE CLONES UNDER INVESTIGATION AS
RENEWABLE ENERGY SOURCES IN PUERTO RICO[a]**

Category	Harvest Interval (Months)	Candidate Clones
I. Short Rotation	2-4	Sordan 70A [b]
		Sordan 77 [b]
		Trudan 5
		Millex 23
		Bermuda Grass
		NK Hybrids
		Roma (Sorghum)
II. Intermediate Rotation	4-6	Common Napier Grass (Var. Meker)
		Napier Hybrid PI 30086 [b]
		Napier Hybrid PI 7350
		NK Hybrids
		Saccharum spontaneum:
		US 67-22-2
		US 77-70
		SES 231
		S. spont. Hybrid (Wild)
		Intergeneric Hybrids
III. Long Rotation	12-18	*Saccharum* Hybrids
		NCo 310
		PR 980 [b]
		PR 64-1791
		B 70-701
		US 67-22-2 [b]
		USDA Imports

[a] DOE Contract No. DE-AS05-78ET20071.
[b] Leading candidates for their catetory.

As illustrated in Figure II, the tissue maturation curves for typical members of each category vary greatly over a time-course of 12 months. Hence, to harvest sugarcane at the 10-week intervals favorable to Sordan 70A would yield little dry matter. Similarly, any delay of the Sordan harvest beyond 12 weeks is a waste of time and production resources. Napier grass, an "intermediate rotation" species, is more than a match for sugarcane at two- and four-months of age, and will nearly equal sugarcane yields at six months, but thereafter sugarcane will easily out-produce napier grass. In this context, a short-rotation species should be harvested four or five times per year, an intermediate-rotation species two or three times per year, and long-rotation species no more than once per year. This need for careful attention to the maturation profiles of candidate species is underscored by yield data for sugarcane and napier grass harvested at variable intervals over a time-course of 12 months (Table II).

It is also evident that, while Sordan and napier grass attain rather level plateaus for dry matter, sugarcane continues to increase in dry matter beyond 12 months (Figure II). Sucrose accumulation profiles are very similar for sugarcane. For many years, sugar planters have taken advantage of this feature by extending the cane harvest interval beyond 12 months. Hence, the Puerto Rico sugar industry harvests two crops - the "gran cultura" (14 to 16 months between harvests) as opposed to the primavera crop (10 to 12 months between harvests). In Hawaii, sugarcane is commonly harvested at two-year intervals.

Energy Crop Rotations: From Figure II, one would surmise that the energy plantation manager should plant a herbaceous species such as sugarcane and leave it there - up to 18 months if possible — before harvest. In addition to maximum fiber, he would also harvest fermentable solids as a salable by-product. This reasoning would probably be correct in a tropical ecosystem suited to *Saccharum* species and where a regional tradition exists for sugar planting. However, these circumstances do not exist in many countries having an otherwise good potential for growing biomass. For example, there is no region of the U.S. mainland suited for 12- to 18-month cropping of sugarcane, although there are vast regions there suited to some form of tropical grasses. Hence, a future energy planter in Florida, Louisiana, Southern California, or Southern Texas might seriously consider whether he should harvest a 6 to 8 month crop of sugarcane per annum or two crops of napier grass in the same time frame.

Equally important is the fact that some countries will not be able to afford a land occupation of 18 months by a single energy crop. This is especially true of densely populated, developing tropical nations having an urgent need for

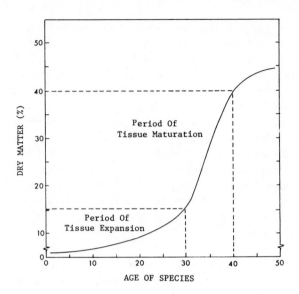

Figure 1.   A generalized representation of the maturation profile of herbaceous land plants

While no specific time-frame or plant form is depicted, the diphasic process of tissue expansion followed by maturation is typical of nonwoody plant species. With the visible growth phase essentially completed, the energy planter will gain much additional dry matter by allowing a brief additional time interval to elapse before harvest.

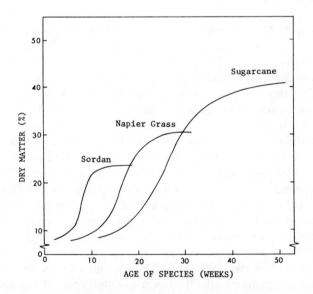

Figure 2.   Relative maturation profiles for Sordan 70A, napier grass, and sugarcane over one year. These plants are representative of the short-, intermediate-, and long-rotation cropping categories, respectively.

**Table II. DRY MATTER YIELDS OF SUGARCANE AND NAPIER GRASS
HARVESTED AT VARIABLE FREQUENCIES OVER A TIME-COURSE
OF ONE YEAR[a]**

Interval (Months)	No. of Harvests	Species	Tons DM/Acre/Year For— Plant Crop	1st Ratoon Crop
2	6	Cane[b]	6.5	3.3
		Napier[c]	12.7	11.9
4	3	Cane	11.1	11.9
		Napier	22.6	25.1
6	2	Cane	16.6	20.6
		Napier	25.6	33.0
12	1	Cane	25.5	33.6
		Napier	19.3	25.8

[a] DOE Contract No. DE-AS05-78ET20071.
[b] Computed mean of three varieties and two row spacings.
[c] Computed mean of one variety and two row spacings.

domestic food production (64). In such cases, a short-rotation species such as Sordan may be the popular choice for energy planting since it can be sown as a stop-gap between the harvest of one food crop and the planting of another. In this capacity, it would also prevent soil erosion and weed growth while acting as a scavenger for residual nutrients left over from the prior food crop.

Seasonal climate changes will also be a factor in the rotation of biomass energy species with conventional food and fiber crops. Short-rotation tropical grasses such as Sordan are ideally suited to the tropics, but they can be grown on a seasonal basis during the heat of summer in most temperate regions. Such plants could be propagated to maturity in a mid-June to mid-August time frame. In a given year, the same site could produce a cool season food crop (a *Brassica* species, spinach), or a cool season forage (ryegrass, fall barley) both preceding and following the biomass energy crop.

HARVEST AND TRANSPORTATION

Perhaps the weakest point in current production research for biomass is the lack of proven harvest equipment and methodologies for the maximized stands of biomass that each contractor strives to attain. This is most evident in woody biomass scenarios where conventional forest harvesting technology is either not applicable or simply doesn't exist in the context of silviculture energy plantations. The outlook for harvesting herbaceous land plants is considerably better but a good deal of research remains on harvest and post-harvest technology, together with equipment redesign and modification.

Mowing vs. Conditioning As Harvest Options

The vast majority of herbaceous land plants can be harvested nicely with the sickle-bar mower (assuming that land slopes and contours are otherwise suited for mechanized operations). This implement was designed more than a century ago as a replacement for the hand sickle and manual grass scythe. As a horse-drawn implement, it revolutionized the harvest of grain and forage crops. Today it is usually operated from the power take-off of Class I and II tractors. The original wooden parts have been replaced, bearings and lubrication systems have been improved, and it is no longer geared to the slow forward pace of draft animals. But it operates on basically the same principle as its horse-drawn predecessors.

There are two principal limitations of the sickle-bar mower as a harvest implement for herbaceous biomass crops: (a) It is designed to operate in relatively low-density stands of plants, and (b), its cutting process is confined to a single slice near the base of upright stems. In other words, it is a mechanized sickle for severing stems rather than a stem conditioner. This mower has a preference for dry and upright stems whose total mass does not exceed about 12 green tons per acre. It experiences real difficulty with wet and lodged materials and with plant stands in any condition whose mass exceeds 15 green tons per acre. Since its operation is based on a cutting principle, the sickle must be kept continually sharp for effective performance. Its efficiency is immediately lowered by contact with mole hills, rocks, wires, scrap metal, and durable objects of any kind encountered in the field.

In the author's experience, the modern sickle-bar mower operating in a typically dense tropical grass, such as Sordan 70A (about 20-25 green tons/acre), will experience frequent tripping of its "fail-safe" mechanism. This is a built-in feature of the implement designed to prevent its destruction when striking unseen stumps or other fixed objects at operational speed. Nonetheless, the sickle-bar mower is probably very adequate for harvesting most herbaceous land plants, that is, those plants whose standing green mass will not exceed about 12 tons per acre at any given harvest interval.

For harvesting somewhat higher densities of herbaceous material, a series of "flail" and "conditioner" designs have proven to be superior to the sickle-bar mower. Such implements do not perform on a cutting principle but rather break off the plant stem by striking it with extreme force. Sharpness of the contact blades is not a decisive feature. In fact they will perform fairly adequately even when dull from long use. These machines require high horsepower (90 to 120 hp) and high power-take-off speed (1000 rpm).

The most effective implement of this type tested to date in Puerto Rico is the M-C "rotary scythe-conditioner". The plant stems are broken off by four lines of whirling blades and are repeatedly shattered as the blades restrike the stems at 3- to 5-inch intervals. The resulting "conditioned" biomass is evenly distributed in a broad swath behind the rotary scythe. In this state, the subsequent drying and baling operations are more easily performed than with conventionally-mowed biomass, that is, with plant materials received in clumps and matts and with only one cut surface to facilitate water removal. An additional advantage of the rotary scythe-conditioner is its capacity to harvest plant densities roughly double those handled by the sickle-bar mower. A second added advantage is its ability to harvest lodged and wet materials. Such plants are harvested about as readily as those in a dry and upright condition. A third avantage is its relatively trouble-free operation. The number of parts subject to malfunction is purposely reduced to a minimum.

At this writing, the rotary scythe-conditioner has given excellent performance in plant densities amounting to about 22 green tons per acre (62). It is believed that its upper density limit will be on the order of 40 green tons per acre (65).

Plant yields considerably higher than 40 green tons per acre are anticipated for a few herbaceous species. Sugarcane yields in excess of 90 green tons per acre year were recently demonstrated in Puerto Rico (49). Most sugarcane harvesters marketed today begin to have difficulty with cane densities in the range of 50 to 60 standing green tons per acre (65). The most effective sugarcane harvester in Puerto Rico at present is the Class Model 1400. Originally developed in East Germany, the Class is a single-row, whole cane harvester which employs a powerful air blast to remove organic trash and soil from the cane at the point of harvest in the field. It has accommodated over 60 tons of green cane per acre. With modifications, it might possibly harvest 80 to 90 tons per acre (65).

Solar Drying

A characteristic difficulty with biomass is its low density relative to fossil energy and its high water content which is costly to transport to processing centers. Wherever possible, it is desirable to remove most of this water at the harvest site by solar drying. One exception to this is the use of "green" biomass for anaerobic digestion. Another exception is found in sugarcane. In this case, the whole green stalk is transported to a centralized mill for dewatering. The plant's soluble fermentable solids are recovered there from the expressed juice and sold as refined sugar or molasses.

Very adequate equipment for the solar drying of non-woody land plants can be found in the cattle forage industry. The rotary scythe-conditioner described above does much to prepare herbaceous plants for rapid drying in the sun (66,67). Ordinarily these materials would be turned over once or twice in bringing the moisture content down to about 15 percent. Three windrows would then be combined into one shortly before baling. Each of these operations can be performed with standard side-delivery forage rakes operating from the power take-off of a Class I or II tractor. When higher density biomass is to be raked (Sordan or napier grass), a heavy-duty "wheel" rake may be more suitable. These implements are also becoming standard equipment for forage-making operations.

Compaction And Baling

Solar-dried biomass is rarely transported to its processing site today in a loose state,although once this was standard practice. For economy of space in transport and storage, as well as ease of handling, such materials are first compacted and then bound with a suitable twine or wire. The standard hay "baler" today is actually a compactor. It produces conveniently-sized cubes having a controlled density range of roughly 8 to 20 pounds per cubic foot. A typical hay "bale" would weigh 60 or 70 pounds and is easily handled by one man in transport and storage procedures or in cattle-feeding operations.

A different concept in biomass baling has appeared in recent years. This is the "bulk" or "round" baler which operates as a windrow wrapper rather than a compactor. This implement produces large cylindrical bales weighing up to 1500 pounds each (68,69). Since no appreciable compaction is involved, the bale density is relatively low, on the order of 10 to 12 pounds per cubic foot. More recent modifications enable this machine to produce cube-shaped bales which are more economical of space during transport and storage. Both front- and rear-end loaders suitable for handling these bales are marketed as conventional tractor attachments (65).

There are two types of balers for sugarcane bagasse, the baling press and the briquetting press (70). The first type is a hydraulic press employing the same compaction principle used for hay. The bagasse is baled in a semi-green state and the formed cubes are tied with twine or wires to prevent them from re-expanding. Their density will range from 25 to 40 pounds per cubic foot. Bales of this type must be stacked carefully to prevent spontaneous combustion, that is, with sufficient space between them to allow air circulation. The briquetting press operates with dry bagasse having a moisture content of 8 to 15 percent. This press provides high pressures on the order of 5,000 to 15,000 psi. Under these conditions, extremely compact cubes are produced which retain their form without the use of twine or wires.

Transport And Storage

Herbaceous biomass that has been solar-dried and baled can be transported to processing or storage sites without appreciable difficulty with existing equipment. However, this can entail a significant cost. Ordinarily such materials would be loaded directly in the field on a low-bed truck. Standard bales (60-80 pounds) can be loaded manually or with mechanical loaders requiring only one laborer on the truck for final positioning of the bales. Bulk bales would be stacked two layers deep on the truck bed with tractor-mounted loaders. The same truck would transport the biomass to a final processing or storage facility without intermediate trans-shipment operations. In the case of sugarcane, the harvested whole stalks or stem billets, whatever the case may be, are hauled in carts to the adjacent mill. The same materials could be carted to an intermediate reloading point for truck delivery to more distant sugar mills.

Delivery costs will vary considerably with the individual biomass production operation. In general, a 40-ton low-bed truck with driver can be hired for about $180 per 24-hour day at current rates. Loading equipment with operators must be stationed at each end of the delivery run. In an ideal biomass production operation, i.e., one managed by a private farmer for profit, the land owner would probably own and help operate the truck and accessory equipment. An estimated delivery cost for solar-dried biomass on a 20-mile run would be $6.00 to $8.00 per ton (1980 dollars).

PRODUCTION COSTS

Published production costs for both herbaceous and woody biomass show broad variations that are both understandable and inevitable (3,6,7,48,58). A given contractor will want to present his speciality crop in the best possible light relative to the dollar inputs needed to obtain a million Btu's in biomass form. This topic has been reviewed in detail (48). It was concluded that most biomass researchers greatly underestimate the cost of biomass production, excluding from their calculations significant indirect costs, long-term repercussions on ecosystem resources, future competition for land and water, and both the cost and efficiency of biomass conversion systems.

Obtaining Correct Cost Data

A seriously misleading trend is to base the production costs of a biomass candidate on its published yield performance as a conventional food or fiber crop. Sugarcane is an appropriate example. In Puerto Rico, sugarcane managed for sucrose yields 25 to 30 green tons per acre year; as an energy crop it can yield 80 to 90 tons per acre year with only moderate increases in production costs (49). Napier grass data are similarly misleading. There is a wealth of printed matter on the yields of napier grass managed as a tropical forage crop, that is, when harvested repeatedly at five- or six-week intervals at moisture contents approaching 90 percent. As an energy crop, napier grass produces roughly two to three times more dry matter per annum at less cost than the cattle forage (49).

Production Costs For Tropical Grasses

Since June of 1977, considerable information has been gathered on production costs for sugarcane and other tropical grasses whose agriculture has been managed for maximum dry matter yield in a tropical ecosystem (62,63). A breakdown of production input charges for "energy cane" is presented in Table III. There data pertain to a privately-owned, 200-acre operation yielding 33 oven-dry tons of biomass per acre-year. Total cost, including delivery to the milling site, is $25.46 per ton or $1.70 per million Btus. Under Puerto Rico conditions, about 70 percent of this dry matter would be burned as a boiler fuel. The remainder would be extracted as fermentable solids during the cane dewatering process and later sold as constituents of high-test molasses. This is a solid credit to the insular energy cane planter owing to Puerto Rico's precarious reliance on foreign molasses as feedstock for her rum industry (71). Assuming a market price of $0.75 per gallon for high-test molasses, the fermentable solids from one such ton of energy cane would be valued at more than $45.00, or about $1500.00 per acre. Cane milling costs today in Puerto Rico are about $4.50 per ton (72).

Production costs for Sordan 70A are presented in Table IV. Although Sordan's biomass yield is lower than that of energy cane, production input costs are also lower. The final cost of an oven-dry ton of Sordan 70A is about $14.00, or $1.50 less than a ton of energy cane. In this instance, there is no sale of fermentable solids. Production costs for napier grass would be moderately lower than Sordan 70A owing to a much higher yield per acre-year for napier grass (49,62). This crop similarly has no sales for fermentable solids.

Management As A Production Cost Factor

Production costs for energy cane listed in Table III include "management" as 10 percent of the cost subtotal. This is an indefinite term covering the administrative skills expended by way of good agricultural technique to maximize biomass yield. It also reflects the morale (or profit incentive level) of the individual grower or institution in charge of production.

The management factor contribution to future biomass production scenarios can range from very good to very bad, but it will have the potential to be decisive in all production operations. Again, using sugarcane as a convenient example, it is common knowledge that little profit is to be made anywhere in the world today by planting sugar, but it is the well-managed operations that will minimize losses and offer the best prospect of survival until sugar values are again equitable. At one extreme, superior management will be found in

Table III. DRY MATTER PRODUCTION COSTS FOR FIRST-RATOON SUGARCANE MANAGED AS AN ENERGY CROP[a]

Land Area: 200 Acres
Production Interval: 12 Months
DM Yield: 33 (Oven-Dry) Short Tons/Acre: 6,600 Tons

Preliminary Cost Analysis

Item	Cost ($/Year)
1. Land Rental, at 50.00/Acre	10,000
2. Seedbed Preparation, at 15.00/Acre	3,000
3. Water (800 Acre Feet at 15.00/ft)	12,000
4. Water Application, at 48.00/Acre Year	9,600
5. Seed (For Plant Crop Plus Two Ratoon Crops 1 Ton/Acre Year at 15.00/Ton	3,000
6. Fertilizer, at 180.00/Acre	36,000
7. Pesticides, at 26.50/Acre	5,300
8. Harvest, Including Equipment Charges, Equipment Depreciation, and Labor	20,000
9. Day Labor, 1 Man Year (2016 hrs. at 3.00/hr)[b]	6,048
10. Cultivation, at 5.00/Acre	1,000
11. Land Preparation & Maintenance (Pre-& Post-Harvest)	600
12. Delivery, at 7.00/ton/20 miles of Haul	46,200
13. Subtotal:	152,748
14. Management: 10% Subtotal	15,275
15. Total Cost:	168,023
16. Total Cost/Ton (168,023 ÷ 6,600):	25.46
17. Total Cost/Million Btu (25.46 ÷ 15):	1.70

[1] DOE Contract No. DE-AS05-78ET20071.
[2] Labor which is not included in other costs

Table IV. DRY MATTER PRODUCTION COSTS FOR SORDAN 70A[a]

Land Area: 200 Acres
Production Interval: 6 Months
Sordan 70A Yield: 15 (Oven Dry) Short Tons/Acre, Total 3,000 Tons

Preliminary Cost Analysis

Item	Cost ($)
1. Land Rental, at 50.00/Acre Year	5,000
2. Water (Overhead Irrigation), 360 Acre Feet	2,160
3. Seed, at 60 lb/Acre	4,800
4. Fertilizer	10,000
5. Pesticides	4,000
6. Equipment Depreciation (6 mo.)	2,650
7. Equipment Maintenance (75% of Depreciation)	1,988
8. Equipment Operation (75% of Depreciation)	1,988
9. Diesel Fuel	2,200
10. Day Labor (90.00/Day for 140 Days)	12,600
11. Delivery, at 6.00/Ton	18,000
12. Subtotal:	65,386
13. Management (10% of Subtotal)	6,538
14. Total Cost:	71,924
15. Total Cost/Ton (71,924 ÷ 3,000):	23.97
16. Total Cost/Million Btu (23.97 ÷ 15):	1.59

[a] DOE Contract No. DE-AS05-78ET20071.

privately-owned plantations which in some countries are still basically family operations. Here the land owner has an inherent interest in his property and capital investments and possesses the skills and incentive to make a good living from agriculture. Such individuals can still be found today, for example, in the Queensland sugar industry. At the other extreme is the government-owned production operation. Historically, governments have not made good farmers. A farm manager who has little incentive to make a profit and who cannot be held accountable for a loss will ultimately have the inferior production record.

Government take-over of an agricultural commodity is sometimes viewed as a necessary intervention in a free market where important social or political considerations could not otherwise be served (73). This was the case with sugarcane in Puerto Rico where a large and otherwise unemployable labor force could no longer be sustained by private enterprise (64,79). As a consequence it now costs about 28 cents to produce a pound of sucrose in Puerto Rico, at a time when its value on the world sugar market is only about 14 cents per pound. It is fair to say that management is not the only factor contributing to high production costs — environmental quality standards have also had a negative impact on the PR sugar industry (29) — but poor management is clearly the main contributing factor.

In a well-managed production scenario for herbaceous terrestrial biomass some straight-forward steps will need to be taken to assure maximum returns from production input expenditures. These include: a) Correct land preparation, including land leveling and planning where needed; b) correct design and installation of the irrigation system; c) correct seedbed (relative to depth, density or row spacing, and season); d) reseeding of vacant space when necessary; e) correct pest control programs (including administration of control on weekends and holidays when required); f) maintenance of correct irrigation, fertilization, and cultivation programs; g) correct timing and synchronization of harvest operations; h) correct selection and use of harvest equipment; i) post-harvest maintenance of land and machinery.

For most biomass crops, the cost of these measures will accrue whether they are performed correctly or not. The decisive factor will be the skill and motivation of the operation's field managers. Good management can best be assured when production is retained in the context of privately-owned plantations that are operated for personal profit.

SUMMARY

The nature of herbaceous land plants and their potential usefulness as a future energy resource is presented in broad outline. The large number of herbaceous species found in both cool and warm climates and in both the wild and cultivated state suggests that at least a small percentage of these could become valuable sources of fuel. Extensive screening will be needed in a range of ecosystems to bring the number of candidate species to a manageable level. Both botanical and agronomic features to be evaluated during the screening process are briefly discussed. Some of the production and harvest operations required of herbaceous plants as agricultural commodities are also reviewed, together with partial cost analyses for the production operations. Management of the energy crop is seen to be the decisive cost input. This factor will be optimized in privately-owned operations motivated by a strong profit incentive.

REFERENCES

1. Schery, R.W. "Plants for Man", 2nd Ed.; Prentice-Hall, Inc.: Englewood Cliffs, N.J., 1972.

2. Saterson, K. 3rd Annual Biomass Energy Systems Conference, Colorado School of Mines, Golden, Colorado, 1979.

3. Alich, J.A. "Crop, Forestry, and Manure Residues: An Energy Resource"; Stanford Research Institute: Menlo Park, Calif., 1976.

4. Marzola, D.L.; Bartholomew, D.P. *Science* **1979,** (205), 555-9.

5. Alexander, A.G. Symposium on Alternate Uses of Sugarcane for Development, San Juan, P.R., March 1979; UPR Center for Energy and Environment Research: Caparra Heights Station, San Juan, P.R..

6. Lipinsky, E.S., *et al.* "Sugar Crops as a Source of Fuels", Battelle-Columbus Division: Columbus, Ohio, 1979; Vol. I.

7. Lipinsky, E.S., *et al.* "Systems Study of Fuels From Sugarcane, Sweet Sorghum, and Sugar Beets", Battelle-Columbus Division: Columbus, Ohio, 1977; Vol. I.

8. N.A. "Underexploited Tropical Plants with Promising Economic Value", National Academy of Sciences: Washington, D.C., 1975.

9. N.A. "Agricultural Production Efficiency", National Academy of Sciences: Washington, D.C., 1975.

10. Lipinsky, E.S.; Nathan, R.A.; Sheppard, W.J.; Otis, J.L. "Fuels From Sugar Crops"; Battelle-Columbus Division: Columbus, Ohio, 1977, Vol. III.

11. Scheller, W.A. "Symposium Papers", Clean Fuels From Biomass, Symposium sponsored by the Institute of Gas Technology, Orlando, Florida, 1977; Institute of Gas Technology: Chicago, 1977.

12. Thompson, G.D. *Proc. S. A. Sugar Technol. Assoc.* **1979,** June.

13. N.A. *Science* **1976,** *195,* 564-6.

14. Scheller, W.A., Symposium on Fermentation in Cereal Processing, American Association of Cereal Chemists, New Orleans, 1976.

15. Lipinsky, E.S. *Sugar J.* **1976,** *8,* 27-30.

16. N.A. Report to U.S. Department of Energy under Contract No. ET-78-C-01-2854, Virginia, 1978.

17. Anderson, E.V. *Chem. Eng. News* **1978,** *56* (31), 8-15.

18. Park, W.; Price, G.; Salo, D. Final Report to U.S. Department of Energy under Contract No. EG-77-C-01-4101, 1978.

19. Fricke, C.R. "A General History of the Nebraska Grain Alcohol and Gasohol Program". American Institute of Aeronautics and Astronautics, Inc.: New York, 1978.

20. Calvin, M. "Hydrocarbons Via Photosynthesis". University of California: Berkeley, Calif., 1977.

21. Calvin, M. *Bio Science* **1979,** *29* (9), 533-8.

22. N.A. *Science* **1976,** *194,* 46.

23. Alexander, A.G. First Quarterly Reports to U.S. Department of Energy under Contract No. ET-78-S-05-5912, 1978, 1979

24. Thompson, G.D. *Proc. S. A. Sugar Technol. Assoc.* **1978,** June.

25. Lipinsky, E.S. "Abstract of Papers", Annual Biomass Energy Systems Conference, Colorado School of Mines, Golden, Colo., Battelle-Columbus Division: Columbus, Ohio, 1979.

26. N.A. *Solar Energy Intelligence Report* **1979,** *5* (40), 395.

27. "Request for proposal for establishment of herbaceous screening program". DOE RFP No. ET-78-R-02-0014, 1978.

28. Monteith, J.L. *J. Roy. Met. Soc.* **1962,** *88,* 508-21.

29. Moss, D.N. "Symposium Papers", Clean Fuels From Biomass and Wastes, Symposium sponsored by the Institute of Gas Technology, Orlando, Florida, 1977. Institute of Gas Technology: Chicago, 1977.

30.   Hall, D.O. *Nature* **1979,** *278,* 114-17.

31.   Hatch, M.D.; Slack, C.R. *Biochem. J.* **1966,** *101,* 103-111.

32.   Hatch, M.D.; Slack, C.R. *Arch. Biochem. Biophys.* **1967,** *120,* 224-25.

33.   Kortschak, H.P. "Proceedings", International Symposium on Photosynthesis in Sugarcane, London, June 1978.

34.   Hatch, M.D.; Slack, C.R.; Johnson, H.S. *Biochem. J.* **1967,** *102,* 417-22.

35.   Slack, C.R. *Biochem. Biophys. Res. Commun.* **1978,** *30,* 483-88.

36.   Slack, C.R.; Hatch, M.D.; Goodchild, D.J. *Biochem. J.* **1969,** *114,* 489-98.

37.   Hatch, M.D.; Kagawa, T.; Craig, S. *Aust. J. Plant Physiol.* **1975,** *2,* 111-18.

38.   Downton, W.J.A.; Treguna, E.B. *Can J. Bot.* **1967,** *46,* 207-15.

39.   Zelitch, I. "Photosynthesis, Photorespiration, and Plant Productivity". Academic Press: New York, 1971; p 347.

40.   Alexander, A.G.; Biddulph, O. *Proc. Int. Soc. Sugar Cane Technol.* **1975,** *15,* 966-83 (South Africa).

41.   Alexander, A.G.; Biddulph, O. *Proc. Int. Soc. Sugar Cane Technol.* **1977,** *16* (Brazil).

42.   Hoover, W.N. *Smithson. Misc. Coll.* **1937,** *95,* 1-13.

43.   Balegh, S.E.; Biddulph, O. *Plant Physiol.* **1970,** *46,* 1-5.

44.   McCree, K.J. *Agr. Meteorol.* **1972,** *9,* 191-216.

45.   Bulley, N.R.; Nelson, C.D.; Treguna, E.B.; *Plant Physiol.* **1969,** *44,* 678-84.

46.   Monteith, J.L. *Exptl. Agr.* **1978,** *14,* 1-5.

47.   Jewell, W.J., Ed. "Proceedings", 1975 Cornell Agricultural Waste Management Conference. Ann Arbor Science Publishers, Inc.: Ann Arbor, Michigan, 1977; pp 5-142.

48. Zeimetz, K.A. Washington, D.C., 1979, USDA Economics, Statistics, and Cooperative Service, Agricultural Economic Report No. 425.

49. Alexander, A.G.; Garcia, M.; Ramirex, G.; Velez, A.; Chu, T.L.; Santiago, V.; Allison, W. First and Second Quarterly Report to U.S. Department of Energy under Contract No. DE-AS05-78ET20071, 1979.

50. Burr, G.O. Taiwan, 1969, Taiwan Sugar Exporting Station Research Report No. 3.

51. Alexander, A.G. "Sugarcane Physiology: A Study of the *Saccharum* Source-to-Sink System". Elsevier: New York, 1973; Chapter 1.

52. Calvin, M. Technical Congress for the Investigation and Conservation of Energy Resources, San Juan, Puerto Rico, 1979.

53. N.A. "Guayule: An Alternative Source of Natural Rubber". U.S. National Academy of Sciences: Washington, D.C., 1977.

54. "Plant Environment and Efficient Water Use". W.H. Pierre, Ed. *Am. Soc. Argon.* and *Soil Sci. Soc. Am.:* Madison, Wisc., 1965.

55. Wadsworth, F. U.S. Department of Agriculture, Institute of Tropical Forestry, Rio Piedras, Puerto Rico, personal communication, 1977.

56. Calvin, M., University of California, Berkeley, personal communication, 1977.

57. Doering, O.C. "Symposium Papers", Clean Fuels From Biomass and Wastes, Symposium Sponsored by the Institute of Gas Technology, Orlando, Florida, 1979. Institute of Gas Technology: Chicago, 1979.

58. Alich, J.A.; Inman, R.E. Palo Alto, Calif., Stanford Research Institute, Final Report for NSF Grant 38723, 1974.

59. Alexander, A.G. "Sugarcane Physiology: A Study of the *Saccharum* Source-to-Sink System". Elsevier: New York, 1973; Chapter 4.

60. Van Dillewijn, C. "Botany of Sugarcane", The Chronica Botanica Co.: Waltham Mass., 1952.

61. Humbert, R.P. "The Growing of Sugarcane". Elsevier: New York, 1968.

62. Alexander, A.G. Second Annual Report to U.S. Department of Energy under Contract No. ET-78-S-05-5912, 1979.

63. Alexander, A.G. First Annual Report to U.S. Department of Energy under Contract No. EG-77-G-05-5912, 1978.

64. Alexander, A.G. International Conference on Bioresources for Development, Houston Texas, 1979.

65. Allison, W. University of Puerto Rico, personal communication, 1979.

66. N.A. "M-C Rotary Scythe-Conditioner — The Non-Stop Mower Conditioner," Commercial brochure, Mathews Company, 500 Industrial Ave., Crystal Lake, Ill., 1979.

67. Allison, W. U.S. Department of Energy Biomass Harvesting Workshop, Baton Rouge, La., 1979.

68. N.A. "New Holland Round Balers", Commercial Brochures, Sperry New Holland, a division of Sperry Rand Corporation, New Holland, Penn.

69. N.A. "Operator's Manual. NH Round Baler Model 851." Sperry New Holland Co., New Holland, Penn., 1978.

70. Hugot, E. "Handbook of Cane Sugar Engineering", 2nd Ed.; Elsevier: New York, 1972; pp 866-68.

71. Yordan, C.L. Puerto Rico Rum Producer's Association, personal communication, 1979.

72. Rodriguez, C. Manager, Aguirre Sugar Factory, Salinas, P.R., personal communication, 1979.

73. Rask, N. "Food or Fuel? Implications of using agricultural resources for alcohol production." Discussion draft of report to the Puerto Rico Energy Office, October, 1979.

74. Romaguera, M.A. "Proceedings", Symposium on Alternative Uses of Sugarcane for Development, San Juan, P.R., March 26 and 27, 1979.

RECEIVED JULY 1, 1980.

Biomass Production by Freshwater and Marine Macrophytes

W. J. NORTH, V. A. GERARD, and J. S. KUWABARA

W. M. Keck Engineering Laboratories, California Institute of Technology, Pasadena, CA 91125

Biomass plantations for energy production in coastal and oceanic settings have several inherent attractions. Water requirements for aquatic plants may pose no serious limitations. Algal tissues do not contain high proportions of refractory materials such as lignin and cellulose (which might complicate processes for conversion to certain fuels). Many algal species show little or no seasonal changes in potential for growth and presumably can be maintained indefinitely. Photosynthetic conversion efficiencies are good. Space is abundant in the oceanic environment and environmental energy in waves and currents might be utilized for tasks such as obtaining and dispersing plant nutrients.

Research on aquatic macrophytes as producers of biomass has been undertaken at Woods Hole Oceanographic Institution (WHOI) on the east coast and on the west coast by a group of collaborators in a joint effort known as the Marine Biomass Project. Studies at WHOI have focused on estuarine and coastal situations with some attention recently to freshwater plants. The Marine Farm Project has primarily been concerned with oceanic biomass production.

A group at WHOI led by John H. Ryther has undertaken a wide variety of studies concerning aquatic macrophytes including nutrient uptake, growth,

0097-6156/81/0144-0077$05.50/0
© 1981 American Chemical Society

yields, and environmental factors affecting yields. Joel C. Goldman of WHOI has surveyed aquatic biomass production systems on a worldwide basis and is currently examining the role of carbon as a potential limiting nutrient in biomass culturing. The Marine Farm Project is presently attempting to grow giant kelp in offshore waters off southern California. Other work related to aquatic biomass production includes an investigation at the University of California, Berkeley, of microalgae in ponds. This paper will emphasize discussion of the kelp production phases of the Marine Farm Project because of the authors' direct involvement therein. We will also briefly summarize activities by the groups at WHOI.

MACROCYSTIS BIOLOGY

Two species of *Macrocystis* occur along the west coast of the United States from Baja, California to the Gulf of Alaska. *M. pyrifera* prefers temperate waters (ca. 5° to 25°C) and requires some protection from severe waves and storms from central California northwards. Adult plants typically occur in the depth range 8 to 20-30 m. This species does not usually occur much below 20 m in turbid water. *M. integrifolia* occurs along the northern portion of the range, but is not included in this paper.

The *Macrocystis* life cycle involves a heteromorphic alternation of generations between microscopic-size haploid gametophytes and macroscopic diploid sporophytes (Figure I). Our primary concern is with the large sporophyte. The adult sporophyte is anchored to the bottom by the perennial holdfast organ. Unlike true roots, holdfasts are not specialized for accumulating minerals. *Macrocystis* and indeed most seaweeds accumulate their dissolved micronutrients across all exposed surfaces. A stemlike primary stipe emerges from the holdfast apex and soon divides into a complex branching pattern. Among the first branches are blades that produce spores, and in North Pacific material, these are termed sporophylls. The basal branches support anywhere from one to hundreds of fronds. The mature *Macrocystis* frond consists of a long vinelike stipe subtending gas flotation bulbs (pneumatocysts) that in turn support leaflike blades. An older frond may display about 200 or more blades and pneumatocysts, dispersed in a regular pattern along the stipe length. The uppermost blade is meristematic and continually produces new blades, pneumatocysts, and more stipe. Basal meristems are sources of juvenile fronds. The young fronds develop rapidly, usually reaching the surface in two to four months. The upper portions of the maturing fronds then begin contributing to the canopy. Frond lifespan is only about six months (1, 3). Consequently, senescing fronds must continually be replaced by production of young fronds growing up from beneath. Plants, as a whole, may survive many years (3).

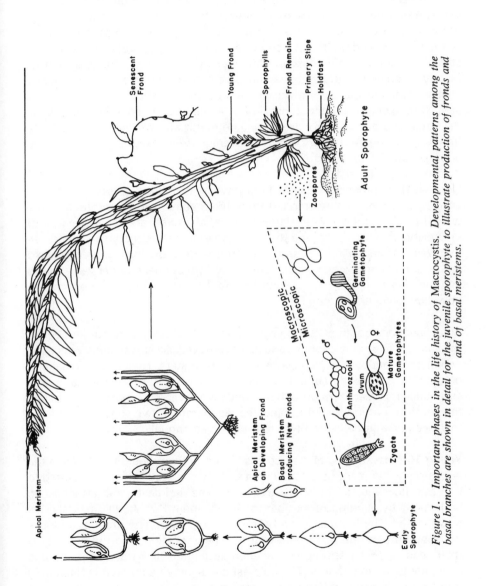

Figure 1. *Important phases in the life history of Macrocystis. Developmental patterns among the basal branches are shown in detail for the juvenile sporophyte to illustrate production of fronds and of basal meristems.*

Dense *Macrocystis* canopies are able to absorb 99 percent or more of sunlight entering the sea surface (4). Thus juvenile fronds exist in a darkened environment which has been shown to lie below the compensation level [i.e. where photosynthesis balances respiration] (5, 6). Lobban (7) and Parker (8) demonstrated existence of translocation processes in *Macrocystis*. Photosynthate produced in the *Macrocystis* canopy is translocated down the stipes to nourish juvenile fronds. Thus these young tissues are able to grow rapidly, overcoming the self-shading problem. The translocation capability enables *Macrocystis* to form very dense opulations of high biomass per unit area. The average standing crop of *Macrocystis* in southern and Baja, California is around six fronds per square meter and can range up to thirty fronds per square meter (9). The average wet weight of a frond is approximately 1 to 1.5 kg (10).

North (11) summarized results from several estimates of productivity in *Macrocystis* beds. Values ranged from 16 to about 130 metric tons of dry weight per hectare per year. Harvest yields, of course, are lower because of inefficiencies in cutting and because only upper portions of plants ae removed. The annual harvest from California waters provides an average yield in the range of one to two metric tons dry weight per hectare per year. Values may be two to four times as high for ore productive beds in areas where upwelling is well developed.

State law limits depth of cutting by commercial harvesters to four feet below the surface. Thus only canopy tissues are removed. All apical meristems lying within the canopy are also gathered by harvesters. Hence the remaining portions of cut fronds cannot significantly develop further. The canopy is essentially replaced from growth by fronds whose apical meristems lie below the depth of cutting. Usually, canopies regenerate in two to four months so that beds can be harvested two to three times annually.

Young *Macrocystis* plants may be raised from the reproductive spores in the laboratory. After plants are 10-20 cm tall, they can readily be transplanted to the sea floor or to artificial structures. The holdfasts are attached to projections by winding rubber bands around them (12). Adults may also be transplanted. The holdfast is first threaded with nylon line, then pried loose from the bottom, the plant is moved to a new location, and the holdfast is then secured to an appropriate anchorage (Figure II). Attachment to solid substrate is not mandatory. The holdfast can simply be moored by fastening it to a rope. Transplantation techniques are now used routinely by biologists from the State and from the harvesting industry to restore depleted kelp beds in southern California.

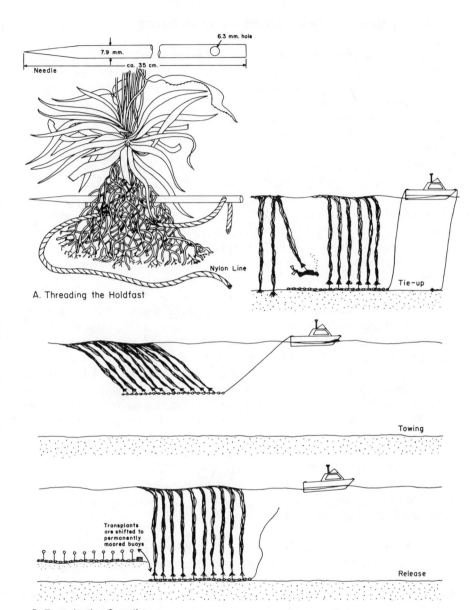

A. Threading the Holdfast

B. Transplanting Operation

Figure 2. One of several techniques in use for transplanting adult Macrocystis: *A. details of kelp needle and its use to weave nylon line between hapteral clumps in the holdfast; B. operations involved in moving transplants to new location, using chain for towing*

EARLY STUDIES BY THE MARINE FARM PROJECT

Research on oceanic farms commenced in California around 1973. They were managed by the U.S. Navy, under the direction of Howard A. Wilcox. At about the same time, a group led by Edward N. Hall at United Aircraft Research Laboratories in Connecticut was investigating methane production from organic materials and considering theoretical problems of marine biomass production. In 1976, management of the Navy project was assumed by General Electric Company. The Marine Farm Project currently exists as a group of organizations guided by, or collaborating with, the General Electric group (Global Marine Development, Inc. — engineering; Institute of Gas Technology — methane production; U.S. Department of Agriculture — kelp processing). The California Institute of Technology has been separately funded by DOE but works in close collaboration with the General Electric part of the Project. Recently, responsibility for governmental review and management was transferred from DOE to the Solar Energy Research Institute.

A review of the literature by Jackson and North (13) examined characteristics of numerous seaweed species to identify likely candidates for use on oceanic farms. Giant kelp, *Macrocystis pyrifera* was selected as a highly suitable seaweed on several counts. *Macrocystis* beds can be coppiced several times yearly by mechanical harvesting techniques. The species is highly productive. A harvesting industry utilizing *Macrocystis* has been active in southern California for almost 70 years. We thus have available a wealth of information concerning harvesting and the operations involved. Extensive research has produced techniques for transplating, predator and competitor control, culturing, and other management tools which are presently utilized in southern California (14, 15). For all these reasons, *Macrocystis* is being used as the test organism in the current studies of oceanic farming. Other seaweeds may prove to be qually suitable if the scope of the research is broadened at some future date.

The first major activity undertaken by the Marine Farm Project involved studies of *Macrocystis* transplants moored on artificial structures in oceanic environments. The largest of three such experiments consisted of a three-hectare structure designed and installed by the Naval Undersea Center off San Clemente Island, about 100 km from the mainland. The structure consisted of a grid or network of ropes, deployed 15 to 20 m beneath the sea surface by a system of cables, buoys, and anchors (16). Overall water depth ranged from about 70 to 150 m. Approximately 130 adult *Macrocystis* transplants were relocated onto the grid during summer and fall, 1974. The source of transplants was a nearby kelp bed at San Clemente Island which was also used as a control for our measurements. Growth performances by test plants and controls yielded mean elongation rates of 4.6 vs 7.6 percent per day respectively, among juvenile fronds. Nitrogen content of blade per tissues in the transplants fell to 0.7 to 0.8 percent of the dry weight compared to a range of 0.9 to 1.4 percent dry weight among the controls. The experimental plants soon displayed unusually dense accumulations of bryozoan encrustations. Presumably, the spreading rate of the bryozoan colonies exceeded expansion rate of the underlying, slowly growing blade tissues.

These general findings at the San Clemente Island farm were confirmed by additional experiments at two other sites. We concluded that low concentrations of dissolved nutrients in oceanic surface waters were apparently unable to sustain normal growth rates by kelp tissues. We also noted improvement in kelp growth during periods when natural upwelling was intense and the nutrient-rich deep water moved up into shallow depths. We further detected faster kelp growth during experiments where water artificially upwelled from depths of 30 to 45 m was introduced around the experimental plants (11). Apparently, obtaining high biomass yields from oceanic farms necessitates fertilizing operations. Costs and energetic requirements connected with dispersing commercial fertilizers on marine farms have led analysts to favor use of artificially upwelled deep water as a source of nutrients (17). There has been concern, however, that the amounts of freely available metallic ions, such as copper and zinc in deep water, might be sufficient to inhibit kelp growth. Likewise, deep water might not contain a full complement of required elements or their concentrations might not be in proper balance. We attempted to resolve these and other questions through laboratory culturing studies comparing growth in deep and surface water media. We have also attempted to determine the elemental requirements of *Macrocystis* by culturing gametophytes and juveniles sporophytes in a chemically defined artificial seawater known as Aquil.

SUMMARY OF LABORATORY FINDINGS

Culturing Work in Seawater Media

Many micronutrients in seawater occur at extremely low concentrations. Accidental contamination of laboratory ware can easily alter levels of some critical elements quite profoundly. For those who might wish to repeat our experiments, scrupulous cleanliness is mandatory in all phases of the work. Our culturing studies have utilized seawater collected from a depth range of 0 to 870 m. Most experiments, however, were conducted with water from 300 m deep, collected about 5 km offshore from our laboratory headquarters at Corona del Mar, California. Typical experiments involved batch culturing conditions using 40- l aquaria (18). Small *Macrocystis* sporophytes (wet weights of 0.5 to about 15 g) were the test organisms. The experimental design made it unlikely that nitrogen or phosphorus would be limiting when using deep water 25 to 30 μM in nitrate.

An experimental series employing nonenriched 300-m water as the medium yielded mean specific growth rates ranging from about 8 to 18 percent daily weight increases within a 13-month period of testing (Figure III). The series comprised 48 independent experiments each employing from two to nine plants. Some of the variability in results undoubtedly arose from physiological differences among the plants. There was evidence, however, that changes in composition of the deep water in part contributed to the fluctuations seen in Figure III. For example, from time to time we conducted a parallel culturing series where the 300-m water was supplemented with manganese to give 1 micromolar concentrations of Mn^{+2} in the medium. Simultaneously, background Mn concentrations in the nonenriched 300-m water were routinely determined by AAS. We found that supplementing deep water with Mn^{+2} stimulated kelp growth during periods when background Mn fell below detectable levels (Table I). Conversely, additions of Mn^{+2} could even be mildly inhibitory when the AAS analyses revealed presence of the element in deep water. We have similarly found that supplementing 300-m water with Fe^{+3} may or may not stimulate kelp growth. We have never observed a clearly stimulatory response from enriching deep water with Zn^{+2} or with Cu^{+2}. Occasionally, however, mild growth stimulation accompanied additions of copper, at appropriate concentrations, to offshore surface water. Iron and manganese concentrations have always appeared entirely adequate in surface waters although our testing has been limited.

Maximal growth by young *Macrocystis* sporophytes was obtained in a flowing system where the medium consisted of equal parts of surface and of water from 870 m deep (16). It appears that inadequacies of deep water were met by components of surface water and vice versa. Consequently, at this time, our laboratory work indicates that mixtures of the two water types should provide a near-optimal medium for fertilizing plants on oceanic farms. Most of our studies have been done at low light intensities characteristic of the sea floor in kelp beds. Recent work indicated that specific growth rates increase at high light intensities. Needs for nutrients will undoubtedly be greater to maintain such high growth rates, so that previously established nutrient resources may require reevaluation in terms of the greater needs.

Table I. EFFECT OF MANGANESE ON KELP GROWTH[a]

Dates of Testing	Mean Specific Growth Rate			Date of Sampling	Background Mn nM/ℓ
	Nonenriched Medium	Enriched Medium	Difference		
6/3/77–6/16/77	9.0 (5)	14.5 (3)	+5.5	5/5/77	ND
9/24/77–10/5/77	10.4 (6)	13.0 (3)	+2.6		
10/25/77–11/2/77	8.2 (5)	12.3 (6)	+4.1	10/31/77	ND
11/2/77–11/6/77	10.3 (3)	9.3 (3)	–1.0		
11/12/77–11/20/77	12.2 (3)	11.5 (3)	–0.7	11/15/77	ND
11/20/77–11/26/77	14.0 (3)	11.4 (3)	–2.6		
2/7/78–2/15/78	13.7 (5)	10.9 (5)	–2.8	2/6/78	3.6
				4/24/78[b]	17
5/2/78–5/8/78	10.1 (8)	7.4 (5)	–2.7	5/23/78	4.3
7/16/78–7/23/78	13.6 (5)	13.3 (5)	–0.3	7/6/78	3.6
7/23/78–7/30/78	13.8 (5)	12.0 (5)	–1.8	8/4/78	4.0
				9/19/78	1.6
10/22/78–10/29/78	13.4 (6)	13.3 (6)	–0.1	10/17/78	ND
11/5/78–11/12/78	10.3 (6)	15.0 (8)	+4.7		

[a] Temporal relations between background concentration of Mn in seawater from 300 m deep and the effect on kelp growth when 300-m water was enriched with Mn at one μM. This effect was taken as the difference between the mean specific growth rate shown by plants in Mn-enriched water minus the mean rate in nonenriched 300-m water. Background Mn was determined by atomic absorption spectroscopy. ND = Not detected (i.e. concentration below one nM). Enrichment with Mn tended to stimulate growth (i.e. positive values for the differences between experimental and control growth rates) when background Mn was low. Enrichment with Mn inhibited growth when background Mn was high. Numbers of juvenile sporophytes involved given in parentheses.

[b] Approximate end of a severe rainy season.

Culturing Work Using the Defined Medium Aquil

An artifical seawater named Aquil was devised by Morel and associates for culturing marine organisms (19). Because Aquil is a chemically defined seawater medium, chemical speciation can be computed using an equilibrium computer program called REDEQL2 (20). Aquil contains eleven major components in fixed amounts. These correspond to the principal inorganic constituents of seawater. Concentrations of macronutrients, such as nitrate and phosphate, and certain trace metals may be varied as desired for a given Aquil formulation.

We have recently been able to culture spores from *Macrocystis* through the entire gametophytic portion of the life cycle, to embryonic sporophytes as large as 30 to 40 cells. After a two-week culturing period, volumes of the embryonic plants were between 200 to 1000 times greater than the spores from which they arose. It seems unlikely that such large volume increases could have been entirely supported by reserves of nutritive elements stored in the spores. An alternative hypothesis seems more attractive — namely, that the formulation used contained all elements required by *Macrocystis*. Aside from the major salts, only nine nutrient elements were added (Table II). We will be gratified if further studies confirm this finding because the number of potentially limiting components affecting *Macrocystis* nutrition appears to be relatively small.

Table II. MEDIUM FOR KELP GROWTH[a]

Nutrient used	Amount Added, nM	Free Ion Conc. nM	Major Species Present, (%)	
Fe^{+3}	400	7×10^{-11}	FeEDTA	(100)
Mn^{+2}	10	1	MnEDTA	(65)
			$MnCl^+$	(23)
Co^{+2}	40	0.04	CoEDTA	(99)
Cu^{+2}	5	0.0000	CuEDTA	(99)
Zn^{+2}	250	0.12	ZnEDTA	(100)
MoO_4^{-2}	100	100	— —	— —
$EDTA^{-2}$	6,000	0.00007	CaEDTA	(89)
			FeEDTA	(6)
NO_3^{-1}	15,000	15,000	— —	— —
PO_4^{-3}	2,000	0.3	HPO_4^{-2}	(51)
			$MgHPO_4^{-1}$	(47)
I^{-1}	100	100	— —	— —

[a] Nannomoles of nine inorganic nutrients and of EDTA added to the completely defined artificial seawater Aquil to yield a medium that sustained development by *Macrocystis* zoospores in petri-dish cultures completely through the gametophyte stage to embryonic sporophytes.

PRESENT FIELD STUDIES

Our laboratory work thus suggests that we can expect strong stimulation of growth when deep water is dispersed among plants on a farm (e.g., compare the level of average growth calculated for small sporophytes in natural kelp beds with values achieved in 300-m water, Figure 3). It is difficult, however, to translate our laboratory data into quantitive predictions of yields from adult *Macrocystis* residing in artificially upwelled water. It is also risky to interpolate from yield measurements conducted among kelp beds. Productivity in most, if not all, of southern California's kelp beds appears to be limited for substantial portions of each year by availability of nutrients. Nutrient supplies from upwelling and runoff are quite variable and difficult to define precisely. Nonetheless, reliable information concerning yields expected from adult plants completely free of nutrient limitations is central to assessing economic feasibility of the marine farm concept. The Marine Farm Project presently is in the early stages of a large-scale field experiment intended to provide knowledge in this critical area (Figure IV).

One of our collaborators, Global Marine Development, Inc., revised the design of a structure, conceived at the Naval Ocean Systems Center, which was designed to support about 100 adult *Macrocystis* transplants and supply them with abundant quantities of water pumped up from depths of about 450 m (Figure V). This "Test Farm" was deployed about 6 km from shore, near our laboratory headquarters, during September 1978 (Figure VI). Water depth at the site was about 550 m. In mid-December, a protective curtain was installed around the western border of the farm. This curtain was intended to reduce effects of currents, which are often greater than 0.5 kt, on the transplants and to increase retention time of the artificially upwelled deep water within the farm. The curtain was lost to storms within the following week.

An initial crop of 103 adult *Macrocystis* from local beds was transplanted to the structure during November-December 1978. Our intention was to harvest and weigh upper portions of the plants at appropriate intervals to determine yields. We measured growth rates and nutrient concentrations within the tissues and in the water. We also followed reproductive success on solid substrates of the structure, general health and appearance of the test plants, and the development of an associated community, as well as other related parameters (see Figure IV). The experiment was scheduled to last for two years; however, all transplants had been destroyed by the end of two months, primarily due to lack of protection from current.

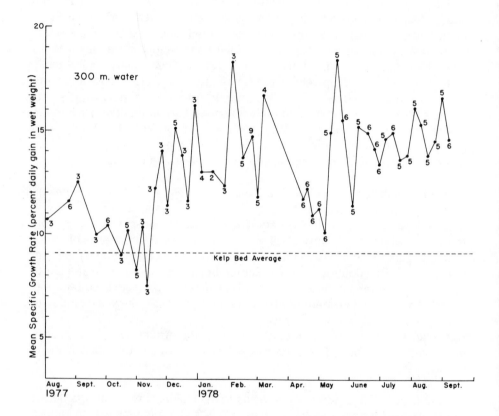

Figure 3.   Record showing variation with time of mean specific growth rates obtained from groups of juvenile Macrocystis *sporophytes cultured in seawater pumped up from depths of 300 m. The batch cultures employed 40 L aquaria with the medium being renewed every other day.*

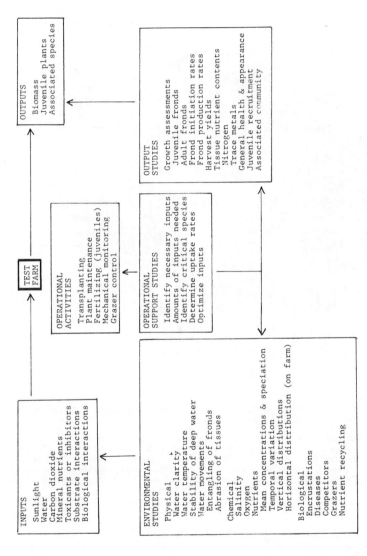

Figure 4. Relationships between fluxes of energy and materials at the Test Farm and the principal groupings that constitute operation and monitoring of the Farm by staff of the California Institute of Technology

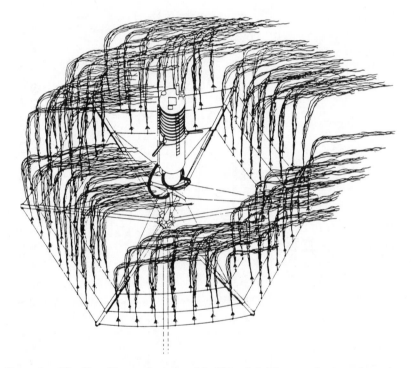

Figure 5.   The Test Farm structure with 100 adult Macrocystis *transplants indicated diagrammatically*

*A 0.61-m-diameter polyethylene pipe tending down from the Test Farm supplies 30,000
L/min of nutrient-rich water from 450 m deep to fertilize the transplants. The deep
water is discharged horizontally from three pipes 120° apart, just below the water line.
The striped cylindrical object is a buoy 17 m long that contains machinery and instrumentation. The plant holdfasts are at depths of 15–17 m. The radiating arms are about
32 m from tip to tip.*

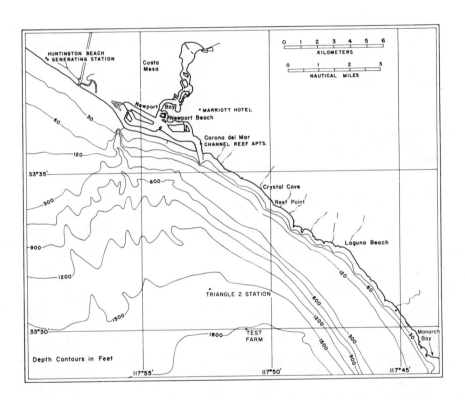

Figure 6. Chart of the southern California coastline from Huntington Beach to Monarch Bay, showing locations of the Test Farm and other geographical features described in the text

To date, investigations concerned with biological outputs from the Test Farm have encompassed two defined time periods: A. December 1978 to January 1979 when our adult transplants existed at the Farm; B. May to August 1979 when dense populations of juvenile plants appeared, presumably offspring arising from spores liberated by the adult transplants five months previously. The most important conclusions and results from our December-January monitoring were:

1. Growth rates
    Juvenile fronds: A series of seven weekly determinations between December 12 and January 29 yielded mean standard growth rates ranging from 5.4 to 7.4 percent elongation per day. These are within the normal range for natural kelp beds at this time of year but tend to lie primarily with lower portion of the range. Percent of fronds showing abnormally slow growth rates among tagged juveniles ranged from 6% to 44% of the tagged recoveries, a relatively high proportion of abnormal juveniles. Abnormally slow growth in juvenile fronds often results from severe damage to or loss of the parent adult frond, that nourishes growth of the juvenile through translocation of photosynthate.

    Adult fronds: Damage and mortality among adult fronds interferred with assessment so tha a statistically adequate evaluation of growth rate was not possible. It was established, however, that some of the tagged specimens generated reasonable rates of production of new blades.

2. Plant Mortality
    About 2/3 of the initial complement of transplants were lost between December 5 and January 5. Mortality during the next 20 days declined, as only about one third of the remaining plants disappeared. A short but violent squall on January 30 destroyed the last of the transplants. Tangling with and abrasion on various parts of the test farm structure were the sole causes of plant mortality.

3. Nitrogen contents of blade tissues
    Except for the final week of January (when all of the remaining plants had suffered significant damage), N contents remained above one percent of the dry weight. In our experience, this represents a healthy nutritional condition. Of the 82 blade samples taken, 71% were about 1.5% in N content. The highest N contents were around 2.5% and came from canopy blades during the period when the curtain was most effective in retaining the upwelled water within the farm. We concluded that, unlike our previous experimental oceanic farms, the transplants on this test farm did not suffer from inadequate nutrition.

During January 1979, we observed some small juvenile *Macrocystis* attached to several of the planting buoys. Only a month had elapsed since the transplants had been introduced. We therefore presumed that these juveniles reached the test farm as established microscopic-sized plants and did not arise from spores liberated at the test farm. Usually at least three months are needed for development of barely visible juveniles from settled kelp spores, and the time may be longer if light or nutrients are not optimal.

In late April 1979, we observed small plants developing near the ends of the test farm dispersion hoses. By May, large numbers of juveniles were appearing on most of the solid surfaces of the test farm structure down to depths as great as 30 m. Concentrations were sparse, however, below the level of the transplanting substrate (20 m). Development by most of these plants was probably stimulated not by the artificially upwelled deep water but by natural upwelling which usually is maximal during late spring. The juvenile recruits were studied intensively to gather ecological information that might be useful for encouraging and assisting kelp reproduction on this and on other oceanic farms. Several noteworthy results emerged.

1. Total plant population on the substrate arms, cables, and planting buoys was estimated to be 36,000 individuals.

2. Temporal changes in nitrogen contents of kelp blades paralleled changes in ambient nitrate concentrations (nitrate is a good measure of natural upwelling in this instance) and correlated with changes in rates of plant elongation.

3. Greatest plant mortality occurred on the smooth plastic-coated cables. Plants were probably easily dislodged by water movements from this type of substrate. High mortality rates also occurred among plants on the upwelling hoses where barnacle encrustations proliferated and created extremely abrasive surfaces. Intermediate degrees of mortality occurred on the planting buoys and substrate arms. Lowest mortality appeared among plants attached to the moderately rough surfaces provided by polyester ropes.

4. Tissue nitrogen concentration and growth was enhanced slightly by "spraying" a group of juveniles on a substrate arm, twice weekly with 1 M ammonium sulfate. Even greater enhancement occurred among plants close to bags of Osmocote pellets affixed to the side of a substrate arm. The pellets slowly released nitrogen and phosphorus into the surrounding water.

In summary, perhaps the most revealing result thus far from the test farm experiment was our failure to observe increased growth rates among the juvenile fronds during the period when the curtain was retaining the nutrient-rich deep water and hindering effects by currents. This very preliminary finding suggests that growth of juvenile fronds may be limited by the rate at which photosynthate can be translocated downward from the canopy and not from limited availability of nutrients (in this particular case). While we need more experimentation to establish this hypothesis, the possibility has important implications for optimizing biomass production. If translocation rate is important as a limiting factor in juvenile frond growth, the best strategy would involve trying to achieve a condition where availability of light becomes the principal limiting factor. Presumably this could be done by increasing frond density on the farm (i.e. placing the plants more closely together).

For the future, our collaborators at General Electric will be installing a more durable protective curtain at the periphery of the test farm in late 1979. We will then be able to resume our studies monitoring health and measuring productivity of adult kelp plants being held in the artificially upwelled deep water.

BIOMASS STUDIES AT WHOI

Studies by Ryther and co-workers of WHOI have been located for about three years at the Harbor Branch Foundation, Inc., facility in central Florida. The site has the advantage of a subtropical location with access to both freshwater and marine environments. Earlier work on *Neoagardhiella, Gracilaria, Hypnea,* and other seaweeds had been conducted directly at WHOI in Massachusetts (20). Initial phases at the Florida site included general surveys to screen the most promising candidate species in terms of ease of culturing and performance in biomass production. Of the 42 Floridanian seaweeds examined, *Gracilaria tikvahiae* showed greatest promise (22). Effects on yields of flow rates, nutrient concentrations, water temperature, solar radiation, salinity, and plant density were examined for *Gracilaria* and others (21, 23).

Yields by *Gracilaria* were determined on a weekly basis throughout the year for plants held in flowing systems enriched with ammonium or nitrate (10 to 100 μM) and with phosphate (1 to 10 μM) and essential trace metals. Cultures were exposed to ambient conditions of full sunlight and temperature. The mean annual yield for *Gracilaria* was 34.8 dry g/m^2-day (25.4 dry ash-free tons/ac-yr). Progress was made in epiphyte control by shading infested plants, by withholding nutrients for 5 to 10 days, or by use of an epiphyte-grazing snail, *Costoanarchis avara*. Preliminary attempts to raise seaweed species, including *Gracilaria*, in a semi-oceanic medium (a power plant's discharge candal) were not successful. Apparently nutrient concentrations were inadequate. Of four freshwater angiosperms evaluated, water hyacinth, *Eichhornia crassipes*, was much superior to duckweed and *Hydrilla* and well above pennywort (pennywort, however, might be useful in climates colder than tolerated by hyacinth and the others). Mean annual productivity by hyacinth was 24.2 dry g/m^2-day (range 5.3 to 34.9 g/m^2-day) or 28 dry ash-free tons/ac-yr). Yields from natural stands of hyacinths and other fresh-water macrophytes gave values less than 1/3 of those obtained from laboratory studies. Optimal culturing density for hyacinths in terms of biomass production was in the range 10 to 20 wet kg/m^2 while the range was lower for *Gracilaria, ca.* 1 to 4 kg/m^2. Hyacinth productivity estimated by nutrient uptake measurements yielded mean values about 12 percent below similar determinations by the method of weight gains. Productivity on a large plantation could probably be estimated more easily by nutrient uptake measurements than by weight changes. The presence of hyacinths increased evaporative and transpirational water losses from the culturing container by about 1.7 times above that due to simple evaporation from open water. Studies evaluated suitability as fertilizer for hyacinth culture of residues from digesters operated on hyacinth biomass. Residues supported 54 percent higher growth compared to the chemically-enriched standard medium used in routine culturing. Efficiency of utilization of nitrogen in the system hyacinth-digester residue-hyacinth was 31 percent. The digester produced 0.4 *l* of gas (60 percent methane) per gram volatile solids from hyacinths. Similar studies were progressing using *Gracilaria* as the experimental plant. Dr. Ryther's group expects to expand the operational scales for culturing *Gracilaria* and *Eichhornia*, ultimately experimenting with ponds of one-quarter acre size.

Dr. Joel Goldman is currently investigating utilization of inorganic carbon by algae to provide factual bases for ensuring that cultures never become limited by this element and for economic analysis. Studies thus far have utlized microalgae but the scope will eventually be expanded to include macrophytes. Studies include the role of carbon dioxide and of bicarbonate as carbon sources, effects of pH and of mixing, and defining culturing conditions required for the most economic and efficient means for supplying adequate carbon to mass cultures of algae (24). Tolerance to abnormally low or high pH values varied among algal species. Utilization of bicarbonate as a carbon source reduces the buffering capacity of natural waters. The pH tends to rise because hydrogen ions are assimilated and hydroxyl ions are liberated as bicarbonate is utilized. Goldman controlled pH with organic buffers in one experimental series with *Phaeodactylum tricornutum*. This marine diatom utilized bicarbonate at efficiencies of 90 to 100 percent across concentrations ranging up to more than fourfold above natural levels. Goldman concluded that bicarbonate should easily be able to fulfill carbon requirements of productive species such as *Phaeodactylum*, provided that mixing and pH control are adequate. Bicarbonate was as good a carbon source as gaseous carbon dioxide for the freshwater Chlorophyte *Chlorella vulgaris*, but not for *Scenedesmus obliquus*, under batch conditions. Goldman concluded that the rate of supply of gaseous carbon dioxide controlled its availability to the plants, rather than the concentration of carbon dioxide in the gas mixture bubbled through the medium.

ACKNOWLEDGEMENTS

Current research support from the U.S. Department of Energy under Contract E(04-3)-1275 and from the Office of Sea Grants under Grant No. 04-5-158-13 is gratefully acknowledged, as well as past support from the U.S. Navy and the National Science Foundation. Advice from Drs. Michael Barcelona, George Jackson, James Morgan, and Clair Patterson, and from Michael Burnett was invaluable. Our thanks are also due to Drs. John H. Ryther and Joel C. Goldman for helpful discussions and for supplying us with their most recent information concerning thier studies. The authors are especially grateful to Sylvia Garcia for the AAS determinations. Thanks are due to the Kerckhoff Marine Laboratory staff for assistance in all aspects of the work: Peter Allison, Brian Anderson, Barbara Barth, Randall Berthold, Elliott Crooke, Henry Fastenau, Laurence Jones, Victoria Kromer, Virginia Martini, Frank Sager, Thomas Stephan, and Mary Ann Wheeler. In part, this work is a result of research sponsored by NOAA Office of Sea Grants, Department of Commerce.

REFERENCES

1. Gerard, V. A. Ph.D. Thesis, University of California, Santa Cruz, Calif., 1976.

2. Lobban, C. S. *Phycologia* **1978** *17*, 196-212.

3. Rosenthal, R. J.; Clarke, W. D.; Dayton, P. K. *Fish. Bull.* **1974** *72*, 670-84.

4. Neushul, M. In "Biology of Giant Kelp Beds (*Macrocystis*) in California"; W. J. North, Ed.; J. Cramer, : Lehre, Germany, 1971; pp 241-54.

5. Sargent, M. C.; Lantrip, L. W. *Am. J. Bot.* **1952** *39*, 99-107.

6. Clendenning, K. A.; Sargent, M. C. In "Biology of Giant Kelp Beds (*Macrocystis*) in California"; W. J. North, Ed.; J. Cramer,: Lehre, Germany, 1971; pp 169-90.

7. Lobban, C. S. Ph.D. Thesis, Simon Fraser University, Burnaby, Canada, 1976.

8. Parker, B. C. In "Biology of Giant Kelp Beds *(Macrocystis)* in California"; W. J. North, Ed.,; J. Cramer: Lehre, Germany, 1971; pp 190-95.

9. North, W. J. "Proceedings", Symposium on Chilean Algae, Universidad Catolica, Santiago, Chile, Nov. 1978.

10. North, W. J. In "Biology of Giant Kelp Beds *(Macrocystis)* in California"; W. J. North, Ed.; J. Cramer: Lehre, Germany, 1971; pp 1-97.

11. North, W. J. "Proceedings", Fuels from Biomass Symposium, University of Illinois, Urbana-Champaign, 1977; pp 99-114.

12. McPeak, R. H.; Fastenau, H.; Bishop, D. Pasadena, Calif., 1972-73, California Institute of Technology, Kelp Habitat Improvement Project, Annual Report 91125; pp 57-73.

13. Jackson, G. A.; North, W. J., China Lake, Calif., 1973, Final Report, Contract No. N60530-73-MV176. U.S. Naval Weapons Center.

14. North, W. J., *J. Fish. Res. Board Can.* **1976** *33*, 1015-23.

15. Wilson, K. C.; Haaker, P. L.; Hanan, D. A. "The Marine Plant Biomass of the Pacific Northwest Coast"; R. W. Krauss, Ed.; Oregon State University Press: Corvallis, 1977; pp 183-202.

16. North, W. J. "Proceedings of the Symposium on Biological Conversion of Solar Energy", University of Miami, 1977; Academic Press: New York, 1977; pp 347-61.

17. Ashare, E.; Augenstein, D. C.; Sharon, A. C.; Wentworth, R. L.; Wilson, E. H.; Wise, D. L. Cambridge, Mass., 1978, DOE Report 1738R.

18. North, W. J. "Symposium Papers", Clean Fuels from Biomass and Wastes; Institute of Gas Technology: Chicago, Ill., 1977; pp 128-40.

19. Morel, F. M. M.; Rueter, J. G.; Anderson, D. M.; Guillard, R.R.L. *J. Phycol.* **1979**, *15*, 135-41.

20. McDuff, R. E.; Morel, F.M.M. Description and use of the chemical equilibrium program REDEQL2. Tech. Rpt. EQ-73-02, 1975, p 82.

21. Ryther, J. H.; Lapoints, B. E.; Stenberg, R.W.; Williams, L. D. "Proceedings" Fuels from Biomass Symposium; University of Illinois Press: Urbana, Ill., 1977; pp 83-98.

22. Ryther, J. H.; Williams, L. D.; Hanisak, M. D.; Stenberg, R. W.; DeBusk, T. A. "Proceedings," Third Annual Biomass Energy Systems Conference; SERI: Golden, Colo., in press.

23. Ryther, J. H.; Williams, L. D.; Hanisak, M. D.; Stenberg, R. W.; DeBusk, T. A. "Proceedings," Second Annual Symposium on Fuels from Biomass; Rensselaer Polytechnic Institute: Troy, N.Y. 1978; pp 947-89.

24. Goldman, J. C. Proceedings," Third Annual Biomass Energy Systems Conference; SERI: Golden, Colo., in press.

RECEIVED MAY 19, 1980.

Energy from Fresh and Brackish Water Aquatic Plants

JOHN R. BENEMANN

Ecoenergetics, Incorporated, 5691 Van Fleet Avenue, Richmond, CA 94804

The large-scale cultivation of aquatic plants and their conversion to fuels has often been suggested in recent years as a potential energy source. Large-scale systems for cultivation of microalgae (1,2), cattails (3,4), and water hyacinths (5,6) have been proposed without, however, sufficient supporting analysis. Historically, the concept of cultivating aquatic biomass for energy dates back twenty-five years when microalgae were suggested as a renewable source of methane (7). This concept was demonstrated experimentally a few years later (8) and subjected to a general analysis which, based on very favorable assumptions, concluded that the concept could be a low-cost future energy source (9). Recently, a more detailed analysis, also based on very favorable assumptions, again concluded that microalgae could be economically cultivated in large-scale systems and converted to fuels (10). A related study (11) using a similar design concept and analysis, concluded that emergent aquatic plants (e.g., water hyacinths) would be favored over microalgae because they would not be limited by the availability of an enriched carbon dioxide source. All of these analyses and proposals were based on relatively superficial considerations of the requirements for cultivation, harvesting, and conversion of these aquatic plants. This review attempts to advance the concepts of aquatic biomass energy farming based on a more detailed review of the biological data base and the technical limitations and potentials for cultivating aquatic plants. This review is based, in part, on recent reports and publications by the author and colleagues (12,13).

0097-6156/81/0144-0099$05.75/0
© 1981 American Chemical Society

A review of the aquatic plant literature (13) reveals that submerged plants, brackish water marsh plants (Spartina), small floating plants (duckweed), and blue-green algae are not as productive as emergent freshwater marsh plants (cattails, bull rushes), water hyacinths, and planktonic green algae. Thus, this paper will consider only the latter plant types. Particular applications of these plants in chemical production and utilization of marginal lands and water resources are considered. Specific conceptual systems are presented for each type of aquatic plant.

The potential of harvesting natural, unmanaged stands of marsh plants or aquatic weeds (e.g., water hyacinths) and converting the biomass to fuels is considered small by this author (12). However, management and harvesting of natural stands of aquatic plants is taking place for aquatic weed control (e.g., water hyacinths) and for wildlife management (marsh plants). Thus, this option should be considered to a greater extent in the future.

It is not possible, at present, to provide either a detailed resource base assessment (e.g., potentially available water, land, or nutrient resources), or a detailed cost analysis of aquatic plant production. Thus, this review presents general concepts of aquatic biomass farming exemplified by three systems — microalgae farming for lipid fuel and chemicals production, cattail cultivation for conversion to alcohol fuels, and growing water hyacinths for methane gas generation. Wastewater aquaculture applications are not covered in this review nor are the actual conversion processes by which aquatic biomass would be converted to fuels.

MICROALGAE FARMING FOR LIPIDS

During and after World War II, both in Germany and the U.S., the high lipid content of microalgae (up to 86% for *Chlorella*) attracted attention as a possible source of fats and oils (14-16). This led to a concerted effort in the U.S. in the late 1940's and early 1950's to develop microalgae production technology as a potential source of food. This work, which culminated in a pilot-scale project by the Arthur D. Little Co., supported by the Carnegie Institute, is reported in the book edited by Burlew entitled *Algae Cultivation from Laboratory to Pilot Plant* (17). Although not directly acknowledged, the results of this early work were not encouraging; the large plastic tube used for the pilot-scale algal culture was susceptible to leaks and overheating. Harvesting proved quite difficult, requiring expensive centrifuges. Recycling of the media appeared to give some problems. In general, costs far outweighed benefits in protein or lipids production.

Subsequent work was concentrated mainly in Japan, leading to the development of very "high technology" algal cultivation systems, some of which even grew the algae heterotrophically (on acetic acid) under sterile fermentation conditions (18). Production costs of the algal biomass produced by such systems are very high, exceeding $10,000/ton (dry) due to the use of centrifuges, drying plants, and elaborate ponding systems. The algae are used as a health food and specialty feed (e.g., for tropical fish).

In the 1960's, a number of projects were initiated for the use of microalgae in aquaculture food chains (see 19 for a review), for food production (particularly by the German and Czechoslovakia groups, see (20), and for wastewater treatment and feed production. However, at present, only one commercial production system is operating to date outside of the Far East — the *Spirulina* production plant of the Sosa Texcoco Company near Mexico City (21). Taking advantage of the naturally favorable conditions in some areas of their bicarbonate evaporation ponds, this company operates a 10 hectare *Spirulina* production pond for this filamentous blue-green alga. Harvesting is no problem, as the long filaments allow easy removal by relatively wide-mesh screens. The spray-dried product sells for about $5,000/ton, mainly to the Japanese market. Production costs are unknown.

The other major practical use of microalgae was in wastewater treatment applications (22). Microalgae are capable of providing the dissolved oxygen required in meeting the biological oxygen demand of municipal and other wastewaters. Sewage oxidation ponds have been used in the U.S. for many decades; they are simple earthen lagoons, one to two meters deep, and up to fifty acres or more in size. Several lagoons are usually operated in series to effect wastewater treatment. The microalgae culture is neither controlled nor harvested; thus, no true cultivation process is involved. Oswald in the early 1950's applied more controlled "high rate" ponds to wastewater treatment (23). These were essentially shallow (20-50 cm), mechanically mixed, and baffled ponds which allowed maintenance of a dense culture of microalgae which could more efficiently provide the oxygen required in wastewater treatment. Although these systems were studied in detail both in the U.S. (24) and more recently in Israel (25), only few pond systems of this type have been built. This is because the high algae concentration makes harvesting imperative, and microalgae harvesting was expensive.

The author, in association with W. J. Oswald, over the past four years has studied lower cost algal production and harvesting systems for application to both wastewater treatment and energy production (26-28). The research has concentrated on the problems of microalgal harvesting and species control in experimental and pilot-scale sewage high-rate pond systems. The first concept studied was to harvest the algae by micro-screens — rotating backwashed fine mesh screening devices. The critical parameter is the screen opening — a 26 micron screen size was chosen as the most cost effective. This required maintaining in the ponds colonial types of algae as average single cell algae sizes range between 2 and 20 microns. However, it was experimentally determined that relatively long detention times were required to allow maintenance of colonial green algal cultures which resulted in a significant loss of biomass productivity (about half of the total) (28). The emphasis shifted to an even lower cost method of algal harvesting — spontaneous flocculation of the algae, followed by sedimentation of the microalgae culture. The results of over two years of study, with the last year being devoted to pilot-scale (0.1 hectare) demonstration of this process, have shown that it is possible to cultivate year-round a microalgal culture that exhibits both high productivity and good harvestability (flocculation-sedimentation) (29). Data from over one year of operation is shown in Table I. Although this process remains to be demonstrated in practice, it appears that low-cost algal harvesting is feasible.

An economic analysis of microalgal biomass production must be based on a number of assumptions, only one of which is the availability (and practicality) of a low-cost harvesting process. Other assumptions must be made about the specific design and the capital cost of the pond system (e.g., lined vs. unlined), the availability and quality of water, the feasibility of recycling water, the nutrient utilization efficiency, the source and transfer efficiency of carbon (carbon dioxide), and the processing costs after the initial harvesting (defined as roughly the first 100-fold concentration). More importantly, assumptions must be made about the ability to grow certain algal species or types, preventing culture instabilities (e.g., zooplankton predation), management requirements, and productivity. The reason so many different assumptions are required is the lack of detailed and/or available information, including productivity data. The Japanese and Far East systems mentioned above are not a good guide because no first-hand technical data are available; these are commercial proprietary projects. Similarly, the Mexican *Spirulina* production project is not designed for optimal production, as it only improves on the natural situation. However, an overall production figure of 10-12 g/m^2-day averaged over 10 months has been reported (21). The best productivity data for large-scale systems (above 100 m^2) are available from high-rate sewage oxidation ponds as summarized in Table II. Even in this case, serious short comings of the data are apparent when reviewing the original literature; sewage solids, for example, are included in these production figures. However, based on available data, a biomass production rate of 40-50 t/ha-yr appears feasible and practical.

Table I. SUMMARY OF 0.1 HECTARE HIGH-RATE POND OPERATIONS AT RICHMOND, CALIFORNIA 1978-79. The two ponds were operated at variable detention times, depths, and mixing speeds, accounting for differences in productivity and harvestability (29).

Date	West Pond Total Production g/m²-day	West Pond 24-hr Imhoff Cone* % removal	West Pond Harvestable Production g/m²-day	East Pond Total Production g/m²-day	East Pond 24-hr Imhoff Cone* % removal	East Pond Harvestable Production g/m²-day
Sept 78	25.5	92	23.5	8.0	85	6.8
Oct	25.5	89	22.7	11.3	71	8.0
Nov	11.6	27	3.1	9.8	83	8.1
Dec	4.7	70	3.3	6.6	64	4.2
Jan 79	4.9	85	4.2	4.7	56	2.6
Feb	6.4	82	5.2	9.3	74	6.9
Mar	8.5	81	6.9	16.5	74	12.2
Apr	15.8	76	12.0	16.2	53	816
May	20.1	88	17.7	21.3	74	15.8
Jun	22.6	91	20.6	20.2	91	18.4
Jul	22.0	92	20.2	35.5	89	31.6
Aug	21.7	88	19.1	35.6	94	33.5
Sep	19.9	94	18.7	35.5	87	30.9
Oct	16.3	84	13.7	27.8	69	19.2

* Imhoff cone removals indicate the percentage of algal biomass that will spontaneously flocculate and settle

Table II. PRODUCTIVITIES OF MICROALGAL CULTURES IN HIGH-RATE SEWAGE PONDS

Location	m²	Scale g/m²-day	Productivity* Experiment	Duration of Time	Ref.
Richmond, Calif.	70	25.3 12.2	10 days Aug 2 mos. Nov-Dec	3 days 3 days	30
Haifa, Israel	120 150	30.2 20.2	365 days 365 days	3 days fixed 4.5 days variable	31
Melbourne, Australia	2800	14.3 17.4	30 days Mar 30 days Apr	6 days 9.3 days	32
Southern California	646	~14-18	1-2 week ave. summer	variable (semi-batch)	33
Manila, Philippines	100	~15-25	average of year-round experiments	variable	34
Richmond, Calif.	2500 1000	12 15	30 days summer 60 days Sep-Oct	5 days 2.6 days	29

*The productivities were generally not corrected for non-algal sewage solids (except for Shelef et al. 1977) and sometimes were calculated indirectly from the data presented.

Depending on assumptions, calculated production costs of microalgal biomass can amount to as little as $50/ton or up to several thousand dollars per ton. One analysis, carried out by this author, was designed to explore the lower cost limits of algal biomass production at very large scales and arrived at a production cost of $50/dry ton. This cost analysis was based on an unlined pond system consisting of very large individual ponds (100 acres), divided into long serpentine channels by baffles, mixed by paddle-wheels, and harvested by a 48-hr cycle batch settling pond. However, this analysis was not realistic nor detailed in a number of specifics. It is doubtful that any microalgal system could produce a concentrated slurry of 2-5% algae at a cost of less than $200/ton (dry weight basis), even if water and nutrients were supplied free of charge or efficiently recycled. Thus, contrary to many assertions, microalgae do not appear to be a suitable choice for energy farming, as a production cost of at least $10/10^6 Btu for "raw" biomass is foreseen. This does not detract from the potential of microalgal wastewater treatment-energy production systems, where water and nutrients are provided free of charge and credit is taken for wastewater treatment. Two independent analyses of a microalgal wastewater treatment-energy production system (based on assumptions of carbon or nitrogen as limiting nutrients) confirmed that municipal waste treatment systems could competitively produce fuel from microalgal biomass if harvesting could be carried out by a low-cost process such as microstraining (35). However, municipal wastewaters are a limited resource base; even considering energy conservation, a maximum energy contribution of about 0.1% of national energy needs can be foreseen from all wastewater aquaculture systems, microalgae being only one of these technologies (12).

Other potential contributions of microalgae for energy production are of interest. One possibility is the production of speciality chemicals, where high unit prices could defray high production costs. Such chemicals include fats, hydrocarbons, pigments, proteins, and pollysaccharides. Glycerol is a good example of such a product. It was discovered a number of years ago that the Dead Sea microalgae *Dunaliella* will produce a high fraction of its total dry weight as glycerol, as much as 50%, in response to high salt concentrations in its medium, as a method of maintaining an osmotic equilibrium (36,37). The production of *Dunaliella* has been proposed (38,39), and the development of an appropriate technology is underway in Israel (38). The concept is to co-produce protein and carotene pigments with the glycerol. Technical limitations apparently include harvesting and culture stability. Economically, the process is competing with glycerol produced during fat rendering, which sells for about $1/kg, but which could be subject to significant downward price shifts. Also, the production of this algae requires very high-strength brines and the product has a limited market.Thus, although it is the process nearest to commercialization, glycerol production from microalgae should be considered as only one of many types of microalgal chemical systems.

Under fairly optimistic assumptions and conditions, it may be possible to produce high-value liquid hydrocarbon fuels from microalgae, either by conversion of the lipids or by direct hydrocarbon production (40). In some cases, very high lipid contents have been reported in microalgae up to 86% (of dry weight) (mostly C_{16}-C_{20} fatty acids) for the unicellular algae *Chlorella* (16) and a similar amount of long-chain (C_{26}) hydrocarbons in the colonial *Botryococous* (41). However, these high concentrations are only achieved at the end of a long period of light or nitrogen limitation, which result in extremely low rates of lipid production. Whether it is possible to optimize lipid content and productivity is uncertain. About 20% lipids are present in sewage-grown microalgae (Aaronson, personal communication). However, that appears somewhat low for processing purposes. It may be possible to double this amount by strategic schedules of pond operations and nutrient additions without significantly lowering total productivity. A high productivity of *Phaeodactylum tricornutum* with about 40% lipids, using very shallow cultures, has been reported (42).

If fuel is to be produced from the algae, some type of subsidy is required even when competing against spot market prices for oil. One specific example involves the use of microalgal ponds in the evaporative disposal of brackish agricultural drainage waters whose management and disposal is a serious problem in many areas. Thus, in the Central Valley of California, elaborate discharge systems through artificial marshes are being proposed. The use for microalgal production of brackish-saline waters unsuitable for conventional agriculture is one of the most important potential applications of microalgal production systems.

In conclusion there are several, near-term applications of microalgae biomass systems in energy production. Municipal wastewater treatment systems are the most immediate ones followed by systems designed to produce speciality chemicals such as polyols (e.g., glycerol), lipids, polysaccharides and pigments. The wastewater treatment credits, and lack of alternative uses for aquatic biomass grown on sewage, or the high unit prices for some chemicals allow the relatively high production costs forecast for microalgal biomass. Such applications have, however, an aggregate potential impact on U.S. energy supplies that must be characterized as, at best, rather minor. Larger impacts involving liquid fuels production may be possible if microalgal biomass production could be combined with the management or disposal of brackish-saline agricultural wastewaters or by significant technological breakthroughs such as a continuous and spontaneous settling or flotation process for algal harvesting.

ALCOHOL FUELS FROM MARSH PLANTS

Marsh plants have been relatively little exploited by man. The immense stands of *Phragmites* (bullrush) covering about 60% of the Danube Delta, over 3 million hectares, are, perhaps, the best example of large-scale management and harvesting of an emergent marsh plant system (43). The plants are harvested on a sustainable yield basis and are used as fiber for paper manufacture, as well as some traditional uses (construction) and chemicals. In the United States, large areas of fresh brackish water marshes are cut on a more or less regular basis both in the Northern Lakes area and on the East Coast to improve the open water surface-to-marsh plants ratio optimal for migratory birds (about half and half). The concept of using this type of biomass system for energy production has been proposed, particularly in Minnesota (3,4).

Marshlands border the areas of most of the inland and coastal waters of the world and the United States. Detailed statistics on marshland areas were not reviewed by this author; however, a good estimate is that in the U.S. about 20 million hectares of marshes exist with an equal amount already drained or filled since the establishment of the U.S. Major areas with marsh lands are in the Great Lakes area, such as Minnesota with 4 million hectares, the southern states (Louisiana having 3 million hectares), the East Coast such as the Carolinas and to a lesser extent, the Sacramento Delta region on the West Coast. However, existing natural marsh lands are not likely to be used for energy production purposes to a great extent unless they can be demonstrated to be compatible with preservation of endangered plant and animal species, are conducive to wildlife management, and enhance environmental and community benefits. Such accommodations may be possible. For example, many marsh systems are essentially monocultures of specific species such as *Phragmites communis* (bullrush) or *Typha augustifolia* (cattail). Thus, one ecological objection of energy farming is overcome. Another factor that must be considered is that the cultivation of these annual plants may not allow complete harvest because that would prevent rapid regeneration without expensive replanting. This would allay the objection against clear-cutting as in tree energy farming. For maximal wildlife management, partial cutting is already undertaken as mentioned above.

The U. S. Environmental Protection Agency policy is to minimize "alterations" of quality or quantity of the natural waters that affect wetlands (44). These wetlands are recognized as sensitive ecological areas and, thus, any near-term use of significant areas must be considered unlikely. This author envisions that in the near-term, marsh plant-energy systems can be established on already-disturbed or marginal wetland areas, or where plentiful water resources allow such highly consumptive use. In the longer term (e.g., twenty years), such systems could expand into natural, non-sensitive marsh areas. Thus, eventually, a significant fraction of the large wetland resources in some states could become available for biomass production and be integrated into the higher uses of wildlife management, fisheries production, environmental protection, and recreation. Even one-tenth of all present marshlands (e.g., 2 million hectares), assuming a sustained yield of 30 t/ha-yr, which is relatively modest, could provide a significant amount of fuels, about one quad (10^{15} Btu) of raw biomass (higher heating value basis).

Before such prognosis can be made, however, the cultivation and harvesting technologies for such plants must be developed. This requires consideration of the biological characteristics of these plants. The first aspect to consider is the seasonality of these plants. They grow very rapidly in the spring time, drawing on the carbohydrate stored the previous fall in their root tubers (rhizomes). The shoots very rapidly attain a very high leaf area index, exceeding 10 in several reports, which is higher than any crop plant, even sugarcane. The vertical leaf arrangement allows gradual light attenuation and, thus, efficient light utilization. Relatively high transpiration rates allow for maximal photosynthesis rates, similar to those of tropical grasses, although the C_4 pathway of carbon dioxide fixation is usually absent. Most reports on productivity of these plants only measured the areal parts of the plants, whereas a significant fraction of the photosynthate is translocated to the roots which may contain upwards of 40% of the total biomass. Thus, the data on achievable productivity by these plants is affected by two critical problems — the seasonality of their growth and the translocation to their extensive root systems.

The root system of marsh plants evolved to tolerate the anaerobic conditions in the bottom layers of wetland areas. Two basic adaptations are found — internal air passages extending from the leaf bases to the rhizomes. Rhizomes are enlarged roots which allow for storage of carbohydrates and the extension of lateral roots and new shoots. Anaerobic roots are capable of anaerobic metabolism with ethanol (instead of the lactic acid found in animal tissues) as end product (45). It is uncertain how high a rate of ethanol production can be sustained, but it is not too far-fetched to postulate the possibility of applying genetic selections to this system to a level which would allow direct ethanol production from harvested rhizomes. Some advantages of such a system are the high solids (substrate) concentration feasible and the simplification of the process. In this context the areal part of marsh plants are lower in lignin content than terrestrial land plants, making them more suitable for enzymatic or chemical hydrolysis and subsequent use for ethanolic fermentations. Thus, on both accounts, these plants can be considered prospective sources of ethanol. Whether their fruits, which also consume a large amount of photosynthate, could be fermented is not known. Of course, combustion is a straight-forward and more efficient use of the plant for energy. However, the significance of ethyl alcohol as a liquid fuel extender makes it the preferred conversion route, even at a much higher cost or lower efficiency.

The data on productivity of cattails are summarized in Table III and are limited by the absence of total productivity data, both above and below ground. Some of the best data were collected in Minnesota as summarized in Table IV, which shows total, above, and below ground production. It should be noted that natural stands can have as high, or higher, total productivities than managed (fertilized) plots. Peat soils had somewhat lower productivities. In late summer, shoot dry weight reaches a maximum, and roots start accumulating photosynthate. Productivities of 40 t/ha-yr have been estimated for cattails in Minnesota (51).

Achievable productivity will depend on development of appropriate cultivation and harvesting technologies. The production system itself will likely be relatively simple, consisting of large (10-100 hectares) level growth areas surrounded by a low soil embankment and provided with inlet/outlet structures. The actual operations would need to be worked out: How much and when to harvest; whether only above or also below ground biomass would be harvested; how much nutrient and how to apply it; how to control possible pests; etc. Although total biomass productivity in a well-managed system would likely exceed 50 t/ha-yr, based on the data in Table II, the actual harvestable productivity is likely to be significantly less, possibly in the range of the 30 t/ha-yr. The cost of production, however, should be low if harvesting does not present too great a problem and if nutrients are available to sustain high productivities.

Typha and other similar aquatic marsh plants have nutrient concentrations as % of dry matter of about 0.5-3% N, 0.1-0.3% P, and 1.6-3.5% K (13). The actual nutrient concentration depends on the part of plant analyzed, the season (or age of plant), and, most importantly, on the nutrient supply to the plant. Nutrient limitation reduces light conversion efficiency and productivity. However, the minimal concentrations required to maintain healthy growth are not well characterized. Critical nutrient tissue levels (at which nutrient deficiency sets in) are 0.09% for P and 2.5% for K in a *Typha* hybrid (52); for nitrogen it is likely between 0.5-1.0%. For supply of such nutrient levels on a large-scale a number of sources can be considered — agricultural fertilizers, sewage and animal wastes, and recycled nutrients from a processing plant.

Table III. ABOVE-GROUND STANDING BIOMASS AND PRODUCTIVITIES OF CATTAILS (*Typha sp.*)*

Location	Season	Biomass g/m^2	Productivity g/m^2-day	Notes	Ref.
New Jersey	Summer	1380	14.9 (avg)	—	[46]
New Jersey	Summer	1565 (max)	15.4 (avg)	—	[46]
Oklahoma	April-June	1527	16.9	—	[47]
Czechoslovakia	Spring-Summer	1930-3910	15.4-30.0	—	[48]
Czechoslovakia	Summer	2972-3472	22.8-26.7	hydroponic	[48]
South Carolina	May-July	530-1132	5.8-21.5	natural site	[49]

*Adapted from Ref. 13.

Table IV. TOTAL BIOMASS PRODUCTIVITIES OF CATTAILS (*Typha*) IN MINNESOTA*

Description	Shoots g/m^2	Roots and Rhizome g/m^2	Total Productivity g/m^2-day (assuming 120 day season)	Ref.
Natural Stands	1440-1680	2650-2960	34-39	[50]
Managed Stands	810	2860	30.6	[51]
Managed Stands	780	3200	33.2	[51]
Managed Stands	1540	2670	35	[51]
Natural Stands	2320	2400	39	[3]
Natural Stands	1130	3100	35	[3]

* Adapted from Ref. 13.

Biological nitrogen fixation is well known to be associated with the blue-green algae and bacteria in marsh systems; however, its quantitative contribution to the nitrogen balance or the reduction in yield it entails are not clear from available data. Supply of nutrients must be considered as a major, though not insurmountable, problem in the optimal management of marsh systems.

In conclusion, a biomass energy system for marsh plants such as cattails can be envisioned to consist of large areas surrounded by low embankments and a water supply/drainage system which allows maintenance of optimal moisture and nutrient levels. Harvesting could be achieved by large balloon tire-mounted mowing/loading machines. Possibly, the marshes could be drained prior to harvesting. If rhizomes are also to be harvested, then a digging-cutting machine would be required and the plants would be harvested in strips to allow rapid regeneration without replanting. The harvested biomass could be used as a fuel or, preferably, for ethanol production. A variety of different products (fibers, feeds, fuels) could be produced from these plants. Large areas may be available to this type of agriculture in combination with wildlife management in Minnesota, South Carolina, and other areas of the country. The tolerance of some plants such as *T. angustifolic* to brackish water suggests their applications in coastal or agricultural drainage areas. The establishment of 64,000 acres of managed marsh lands for wildlife enhancement has been proposed as a method for handling the agricultural drainage waters of the California Central Valley (53). Thus, the development of marsh plant biomass energy systems should be given a higher priority than in the past.

WATER HYACINTH AS SOURCES OF METHANE

Water hyacinth *(Eichhornia crassipes)* is a floating aquatic plant that originated in South America and dispersed throughout the world during the past 100 years by human actions. It is probably the best known and most widespread of the aquatic weeds, choking navigation and irrigation channels and transpiring precious water, causing large economic losses. The biological characteristics that make this plant such a successful weed — rapid vegetative propagation and high productivity — also make it of interest in aquatic energy farming. Its geographic dispersal is limited primarily by its sensitivity to cold; it is killed by freezing weather and does not grow below about 15°C, although it survives, with optimal water temperatures being about 28-30°C. The plant also does not tolerate high salinity above 10% of sea water. Water hyacinths have small, round bulbs that serve as floats and produce beautiful flowers.

When crowded, the plants can exceed 1 m in height (particularly in the tropics). The vertical growth of this plant when the water surface is covered is an important factor for high productivity. As the leaves grow upward, they increase the leaf area index and, thus, the capture of solar energy becomes more efficient. Figure I summarizes the data collected by Ryther's group (54) in recent studies of the productivities of water hyacinth *Hydrilla verticillata* (a submerged plant), and *Lemna minor* (duckweed, a small floating plant). Duckweed is restricted to a two-dimensional growth pattern on the surface of the water, resulting in much lower yields than water hyacinth. *Hydrilla* is adapted to relatively low light intensities usually encountered below the water surface and also does not do well on a productivity basis. The data of Ryther agrees very closely with many literature reports, and forms the basis for excluding the submerged or small floating plants from consideration in biomass energy systems.

Table V summarizes the data on water hyacinth productivity. It should be noted that water hyacinth exhibits strong seasonality even in areas, such as Southern Florida or Louisiana, where winter temperatures are tolerated by the plant. Thus, extreme productivities of 5 and 50 g/m^2-day can be expected, even in favored areas of the U.S. The optimal standing biomass in summer will be higher than in winter, allowing some smoothing out of the variations in monthly harvest (as biomass is drawn down in winter and built up in summer). Thus, water hyacinths may, in favorable areas, exhibit less variability in harvested production than microalgae or marsh plants. This is important as it affects the capital investments in conversion plants.

The conversion of water hyacinth to methane gas by anaerobic digestion has been studied by a number of researchers (54, 61-62). The basic findings are that water hyacinth is a good substrate for anaerobic digestion, with gas production rates and yields similar to those of primary sewage sludge. Both the nutrient content (>2-3% N, 0.3-0.6% P of dry weight) and solids content of the freshly harvested plants (95% water, 5% solids) are very suitable for this conversion process.

Three different approaches to water hyacinth utilization as an energy crop can be foreseen — harvesting of natural weed populations, biomass production during wastewater treatment, and large-scale cultivation for its biomass energy value. The first option is limited by the sporadic nature of such weed harvesting operations and the cost of transporting the biomass to a central conversion facility. Opportunities may, however, exist in some tropical locations where the water hyacinth problem is severe (e.g., Brazil or Sudan). Use of water hyacinth in wastewater treatment has been studied for several years (see 12 for a review) and is being tested at the pilot or demonstration

Table V. STANDING BIOMASS AND PRODUCTIVITY OF *Eichhornia crassipes*

Location	Season	Biomass g/m^2	Productivity g/m^2-day	Comments	Ref.
Louisiana	Summer	1478	12.7-14.6	Natural Population	55
Florida	May	—	10.3	Natural Population (max. productivity)	56
Florida	Yearly	Variable	24.5 (5.4-52)	Managed Cultivation	54
Florida	Yearly	Variable	5-54	Fertilized Pond	57
Alabama	August	2130	17.7	Fertilized Pond	58
Alabama	September	—	27.6	Fertilized Pond	59
Mississippi	Apr-Jun	1090	87.5	2° Sewage Pond	60

plant scale in several locations around the U.S. These plants can be used for providing a cheap cover for ponds, thus allowing the settling of suspended solids and preventing algal growth. The nutrient absorption capacities of the plant can be used for the removal of nitrogen and phosphorus and heavy metals from wastewaters. In addition, the roots of the plants may act as suspended solids filters and may absorb some organic compounds directly. In general, water hyacinth has significant potential in wastewater treatment and the biomass produced during such a process would be quite suitable for the production of methane. However, in terms of total fuel production potential, such systems are limited to a small fraction of per-capita energy consumption, at least in industrialized countries. Large-scale energy farming of water hyacinths would be attractive where water supply is not a critical factor (water hyacinths transpire at least twice as much water as evaporates from free water surfaces). A recent report on aquatic plant biomass farming selected water hyacinth as the most promising plant, with production costs of about $30/dry ton (11). However, no detailed analysis was undertaken and the economic feasibility of water hyacinth production as an energy source is highly uncertain. One key problem is harvesting; a sufficiently low-cost process must still be developed, although that appears feasible. Nutrient recycling from an anaerobic digester effluent is possible (54).

In summary, water hyacinth is among the most productive of plants and deserves serious consideration in energy farming. However, limitations of geography due to temperature sensitivity, high water use, undeveloped harvesting technology, and uncertain economics precludes firm conclusions about the feasibility of these plants as energy crops at this time. Their use in wastewater treatment is of much more immediate value. Other floating aquatic plants should also be investigated such as the penny wort *(Hydrocotyle umbellata)* which may allow a greater geographical range (54).

SUMMARY

Aquatic plants can achieve relatively high biomass productivities when compared to terrestrial plants because they need not be water-stressed and can be optimally supplied with nutrients. Based on literature reports, productivites in southern U.S. regions of about 40-60 t/ha-yr (dry weight basis) can be predicted for green algae or marsh plants and about 80 t/ha-yr for water hyacinth. Higher productivities may be possible in exceptionally favorable locations by assuming development of advanced cultivation technologies and genetic selection of improved strains. The lack of established cultivation systems and low-cost harvesting processes imposes great uncertainties on the cost of biomass production by aquatic plants. Three potentially practical aquatic biomass energy systems are chemicals production from microalgae, alcohol production from marsh plants, and methane production from water hyacinths.

At present, aquatic plants are not being used commercially as a fuel source any place in the world. Furthermore, no large cultivation systems exist, although a few natural communities of marsh plants are subject to management and aquatic weeds are often harvested mechanically. The production and conversion technologies for aquatic plants are either conceptual or rudimentary. Thus, most of the discussions of this topic must be speculative extrapolations of the limited existing data. Nevertheless, it is clear that aquatic plants have potentially high biomass productivities and, specifically for the case of microalgae, could produce a high-quality, high-value biomass suitable for conversion to fuels and extraction of other products. Table VI gives a list of the relative advantages and disadvantages of aquatic plant energy systems in comparison with the concepts of terrestrial tree or herbaceous plant energy farming. Aquatic biomass energy systems would not compete directly with such terrestrial systems; they would be located in areas where such systems would not be feasible. Thus, in general, aquatic plants should complement other biomass energy systems.

Three favorable aspects of aquatic plant biomass systems should be stressed — the relative short-term research and development effort that will be required to determine the practical feasibility of such systems, the continuous production nature of such systems, and the relative independence of aquatic biomass systems from soil characteristics and weather fluctuations. The fast generation times of most aquatic plants allows rapid data acquisition, as compared to even short-rotation trees. This allows the rapid development of the necessary data on which to base decisions of the economic feasibility of such systems. Thus, aquatic biomass systems should not necessarily be considered as a long-range option. In case of microalgae and water hyacinth, a continuous, hydraulic production system can be designed. This allows better utilization of capital investments than in conventional agriculture which is essentially a batch operation. Finally, the independence of such systems from changes in the weather and the quality of the soil is a significant advantage in an era when climatic changes and soil erosion loom as severe future problems. Development of aquatic plant systems for waste treatment and food-feed-fiber-fuel production may be a prudent investment with a large potential return.

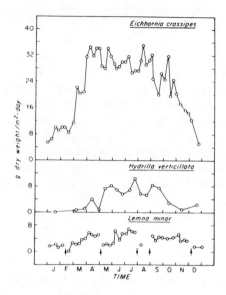

Figure 1.   *Mean daily yields of* Eichhornia crassipes, Hydrilla verticillata, *and* Lemna minor *throughout the year at Fort Pierce, Florida*

Table VI. RELATIVE ADVANTAGES AND DISADVANTAGES OF AQUATIC PLANT BIOMASS SYSTEMS RELATIVE TO TERRESTRIA PLANTS

Advantages

1. Continuous hydraulics production system (microalgae, floating plar
2. Independent of soil characteristics (allows use of marginal land)
3. Short generation times
4. Plants are not water stressed, resulting in high productivities
5. Use of brackish and wastewaters possible
6. Low lignin content, potential for high-value chemicals
7. Independent of short-term weather/climatic effects

Disadvantages

1. Low standing biomass, requires frequent harvests (microalgae, floa plants)
2. High water consumption
3. Undeveloped cultivation systems, uncertain economics
4. Harvesting systems remain major problem
5. High capital investments required
6. High nutrient contents
7. Low overall resource base

REFERENCES

1. Oswald, W.J. "Symposium Papers", Clean Fuels From Biomass, Sewage, Urban Refuse and Agricultural Wastes, Symposium sponsored by the Institute of Gas Technology, Orlando, Fla., Jan 1976; Institute of Gas Technology: Chicago, 1976; pp 311-24.

2. Benemann, J.R., *et al.* "Symposium Papers", Clean Fuels From Biomass and Wastes, Symposium sponsored by the Institute of Gas Technology, Orlando, Fla., Jan 1977; Institute of Gas Technology: Chicago, 1977; pp 101-126.

3. Andrews, N.J.; Pratt, D.C. *J. Minn. Acad. Sci.* **1978**, *44*, 5-9.

4. Minnesota Energy Agency. *The Minnesota Alternative Research and Development Policy Formulation Project Subcommittee 5 Agricultural and Wetland Sources.* St. Paul, Minn., 1978.

5. Lecuyer, R.P.; Marten, J.H. "Symposium Papers", Clean Fuels From Biomass, Sewage, Urban Refuse and Agricultural Wastes, Symposium sponsored the Institute of Gas Technology, Orlando, Fla., Jan 1976; Institute of Gas Technology: Chicago, 1976.

6. Brown, J. AICHE Symp. Ser. *181* **1978**, *4*, 13-20.

7. Meier, R.L. In *Solar Energy Research* F. Daniels, J.A. Duffie, Eds., University of Wisconsin Press: Madison, Wisc., 1977; pp 179-89.

8. Golueke, C.G.; Oswald, W.J. *Adv. in Appl. Microbiol.* **1960**, *2*, 223-62.

9. Oswald, W.J.; Golueke, C.G. *Appl. Microbiol.* **1959**, *7.*

10. Benemann, J.R.; Persoff, P.; Oswald, W.J. "Cost Analysis of Microalgal Biomass"; C.S.O. International, Inc.: Concord, Calif., 1978.

11. Ashare, E., *et al.* "Cost Analysis of Aquatic Biomass Systems"; Dynatech R/D Company: Cambridge, Mass., 1978.

12. Benemann, J.R. *Energy from Aquaculture Biomass Systems: Fresh and Brackish Water Aquatic Plants;* Office of Technology Assessment, in press.

13. Murry, M.A.; Benemann, J.R. In "Biosolar Resources" O. Zaborsky, Ed.; "Chemical Rubber Handbook on Biosolar Resources", in press.

14. Von Witsch. *Arch. Microbiol.* **1948,** *14,* 128-41.

15. Ketchum, B.H.; Redfield, A.C., *J. Cell. Comp. Physiol.* **1949,** *33,* 281-99.

16. Sphoer, H.A.; Milner, H.W. *Plant Physiol.* **1949,** *24,* 120-49.

17. Burlew, J.S., Ed. "Algal Culture from Laboratory to Pilot Plant", Carnegie Institute of Washington: Washington, D.C. 1953.

18. Tamiya, H. *Amer. Rev. Plant Physiology* **1957,** *8,* 309.

19. Goldman, J.C.; Ryther, J.H.; Waaland, R.; Wilson, E.H. Oct 1977, U.S Department of Energy Report No. DST-4000-77/1.

20. Soeder, C.J. In "Microbial Energy Conversion", H.G. Schlegel; J. Barnea, Eds. Erich Goltz KG: Gottingen, 1976.

21. Durand-Chaste, M.H. "Proceedings", Conference on Production and Use of Microalgal Biomass, Acre, Israel, Sept 17-22, 1978.

22. Gloyna, E.F.; Melina, Jr., J.; Davis, E.M. Eds. "Ponds as a Wastewater Treatment Alternative"; Center for Water Resources, University of Texas: Austin, 1976.

23. Oswald, W.J. "Developments in Industrial Microbiology", 1968; pp 112-19.

24. Oswald, W.J. "Proceedings of the 6th International Conference on Water Pollution Research";Pergamon Press, 1972.

25. Shelef, G., *et al.* Dec 1976 and Dec 1977 Technion-Israel Institute of Technology, Annual Reports.

26. Benemann, J.R., *et al.* Berkeley, Calif., Nov 1977, San. Engr. Res. Lab. Final Report No. 77-5.

27. Benemann, J.R. *et al.* "Fertilizer Production with Nitrogen-Fixing Heterocystous Blue-green Algae", San. Engr. Resch. Lab.: Berkeley, Calif., 1977.

28. Benemann, J.R., *et al.* "An Integrated System for the Conversion of Solar Energy with Sewage-grown Microalgae", U.S. Department of Energy SAN-0034-T2, 1979, Ed II.

29. Benemann, J.R., *et al.* Berkeley, Calif., Dec 1978, San. Engr. Res. Lab. Final Report No. NTIS SAN-0034-1.

30. Oswald, W.J., *et al.* J. Water Poll. Control Fed. **1957,** *29,* 437-51.

31. Shelef, G., *et al.* in "Workshop on Microalgae for Food"; Munich, 1977.

32. Caldwell-Connel Engineers, "Algae Harvesting from Sewage"; Australia Government Publication Service, 1976.

33. North American Aviation, Inc. *A Study of the Use of Biomass Systems in Water Renovation,* 1967 Final Report.

34. Oswald, W.J. *W.H.O. Cronicle* **1978,** *32,* 348-50.

35. Fraser, M., *et al.* "The Photosynthesis Energy Factory: Analysis, Synthesis, and Demonstration"; InterTechnology/Solar Corp.: Warrenton, Va., 1977.

36. Ben Amotz, A.; Avron, M. *Plant Physiol.* **1973,** *51,* 875-78.

37. Borowitzka, L.J.; Brown, A.D. *Arch. Microbiology* **1974,** *96,* 37-52.

38. Ben Amotz, A.; Avron, A. "Proceedings", Symposium Production and Use of Microalgal Biomass, Acre, Israel, Sept 17-22, 1978.

39. Williams, L. *Biotechnol. Bioeng. Symp. 8.*

40. Dubinsky, A.; Berner, T.; Aaronson, S. *Biotech. Symp.* **1978,** *8,* 51-68.

41. Brown, A.C.; Knights, B.A.; Conway, E. *Phytochem.* **1969,** *8,* 543-47.

42. Raymond, L. "Abstracts of Papers", Solar Energy Division, Amer. Soc. of Mechanical Engineers, San Diego, Calif., March 12-15, 1979.

43. Rodewald-Rudesco, L. "Das Schilfrohr. Die Binnengewaesser"; Vol. 18; E. Schweizer-bart'sche Verlagsbuckhandlmy: Yugoslavia, 1974, Vol. 18.

44. "Protection of the Nation's Wetlands: Policy Statement", *U.S. Environmental Protection Agency* **1973**, 84FR10834.

45. Hutchinson, C.E. "A Treatise on Limnology"; John Wiley and Sons: New York, 1975, Vol. IV.

46. Jervis, R. *Bull. of the Torrey Bot. Club* **1969**, *96,* 209.

47. Penfound, W.T. *Limnol. and Oceanog.* **1956,** *192.*

48. Dykyjova, D. *Photosynthetica* **1971,** *5,* 329.

49. Boyd, C.E. *Bull. of the Torrey Bot. Club* **1971,** *98,* 144.

50. Brey, J.A.; Lawrence, D.B.; Pearson, L.C. *Oikos* **1959,** *10.*

51. Moss, D.N. "Symposium Papers", Clean Fuels from Biomass and Wastes, Symposium sponsored by the Institute of Gas Technology, Orlando, Fla., Jan 1977; Institute of Gas Technology: Chicago, 1977; pp 63-78.

52. Bonnewell, V.; Pratt, D.C. *J. Minn. Acad. Sci.* **1978,** *44,* 18.

53. Interagency Agricultural Drainage Commission. *Preliminary Report on the San Joaquin Agricultural Drainage Plan;* California State Water Resources Control Board, July 1979.

54. Ryther, J.H. *et al.* "Proceedings", 3rd Annual Biomass Energy Systems Conference, Solar Energy Research Institute, Golden, Colo.; p 13.

55. Penfound, W.T.; Earle, T.T. *Ecol. Monog.* **1948,** *18,* 447.

56. Odum, H.T. *Florida Ecological Monographs* **1957,** *27,* 55-112.

57. Yount, J.L.; Crossman, R.A. *J. Water Pollut. Control Fed.* **1970,** *42,* 173.

58. Boyd, C.E.; Scarsbrook, E. *Aquatic Botany* **1975,** *1,* 253.

59. Boyd, C.E. *Econ. Bot.* **1976,** *30,* 51.

60. Wolverton, B.C.; McDonald, R.C. "Compiled Data on the Vascular Aquatic Plant Program 1975-1977"; National Aeronautics and Space Administration NSTL Station: Mississippi, 1978.

61. Chin, K.K.; Goh, T.N. "Symposium Papers", Energy from Biomass and Wastes, Symposium sponsored by the Institute of Gas Technology, Washington, D.C., Aug 1978; Institute of Gas Technology, Chicago, 1978; p 215.

62. Klass, D.; Ghosh, S. "Symposium Papers", Fuels From Biomass, National Meeting, American Chemical Society, San Francisco, Calif., Aug 25-29, 1980; Am. Chem. Soc., Fuel Chem. Div.: Washington, D.C., 1980.

RECEIVED JUNE 18, 1980.

LIQUID FUELS

6

Multi-Use Crops and Botanochemical Production

M. O. BAGBY, R. A. BUCHANAN, and F. H. OTEY

Northern Regional Research Center, SEA/AR, USDA,
1815 North University Avenue, Peoria, IL 61604

Scripture tells that early man recognized and used petroleum in the form of asphalts, tars, and oils. Coal gained acceptance in the 16th century, and its use continued to expand into the 20th century. However, before the petroleum industry began to take shape in the mid-19th century and virtually exploded onto the industrial scene with invention of the internal combustion engine, man relied principally on renewable materials for his energy, oil, and hydrocarbon resources. Today, with the decline in readily removable petroleum and rising costs of liquid fuels and chemical feedstocks, man is developing a renewed awareness of the potential value of underutilized and diverse plant species.

Numerous investigators have advanced ideas for development of energy and chemical resources from plants. The energy plantation has been analyzed in detail by Szego and Kemp (1) and Goldstein (2). Concepts of "petrol planta-tions" have been described by Calvin (3,4), and companies such as Diamond Shamrock (5) and Goodyear (6) are investigating their potential. "Integrated adaptive agricultural systems" have been discussed by Lipinsky (7), and Buchanan and Otey have advanced their "multi-use botanochemical systems" (8). In all of these proposals, entire above-ground material is to be harvested and used. And, in the most advanced schemes, fuel and chemical feedstock production is integrated with production of food and feed.

This chapter not subject to U.S. copyright.
Published 1981 American Chemical Society

If fuel is to become a major farm product, new agricultural practices and systems should maximize energy product per unit energy input. Low-energy crop production and energy-efficient processing methods and handling techniques must be components for future agricultural scenarios. Such a scheme has been proposed by Buchanan and Otey (8).

BOTANOCHEMICAL SCREENING

The plant kingdom provides a reservoir of 250,000 to 300,000 known plant species. Fewer than 0.1% have enjoyed any significant commercial recognition in the world. From this wealth of plant resources, we anticipate that new sources of energy-producing plants could be identified and exploited.

Oil- and hydrocarbon-producing plants are especially attractive as future energy and chemical resources. Plants already supply several products competitive with synthetic petrochemicals. These products include tall oil, naval stores, seed oils, and plant oils. For this discussion, we refer to such products collectively as oils and hydrocarbons.

For many years, the U.S. Department of Agriculture has actively pursued a multi-disciplined approach to identify and establish new crops as renewable resources (9). Patterned after the Department's program to identify annually renewable fibrous plants that could be cultivated for papermaking (10), an analytical screening program was instituted in 1974 to identify and evaluate species as sources of multi-use oil- and hydrocarbon-producing crops for food material and energy production (11,12). The multi-use concept requires plant breeders and agronomists to deal with a variety of new crops, each yielding several different products of varying economic value. In screening plant species as potential corps, a rating system was employed that emphasized potential economy of plant production, total biomass yield, and oil and hydrocarbon content (8). Subsequently, all candidates were ranked by this rating system. It should be emphasized that vigorous perennials were given preference over annuals, with the concept that seed-bed preparation would be infrequent for perennials.

Data for over 300 species have been accumulated, and about 40 species have been identified that have sufficient potential to merit further consideration. Nearly all of these species are being further investigated by USDA plant scientists; meanwhile, the screening program continues.

In the original scenario, potential rubber crops were considered. Since then, it was decided to develop guayule *(Parthenium argentatum)* as a domestic source for natural rubber (13). The U.S. rubber market can potentially be supplied by guayule grown in the southwest. Thus, barring discovery of an exceptional candidate, that decision preempts development of other rubber crops. However, several potential botanochemical crop species produce low-molecular-weight soluble rubbers (14) that would be valuable as a hydrocarbon component of whole-plant oils.

Our analytical procedure consists of stepwise acetone extraction followed by cyclohexane. Subsequently, the acetone-soluble fraction is partioned between hexane/aqueous ethanol (12,15), and the soluble components are freed of solvents and determined gravimetrically. For lack of specific nomenclature, the botanochemicals isolated by this technique have been referred to as "whole plant oil," "polyphenol," and "polymeric hydrocarbon." Actually, components from these extracts need to be further characterized. However, petroleum refinery processes may be sufficiently insensitive to allow use of carbon-hydrogen rich compounds represented by a broad spectrum of structures. For example, consider the diverse chemicals ranging from methanol to natural rubber which have been converted to gasoline (16). Thus, chemical species may be important if chemical intermediates are being generated but may be nonconsequential for production of fuels, solvents, carbon black, and other basic chemicals.

PROMISING SPECIES FOR WHOLE PLANT USE

Plant families from which more than one promising species has been identified thus far are Anacardiaceae, Asclepiadaceae, Coprifoliaceae, Compositae, Euphorbiaceae, and Labiatae. However, representative species from ten other families have been identified as having sufficient potential to merit further consideration. Undoubtedly, the sample base is too incomplete to establish any trends.

Crop ratings and proximate composition of promising species are summarized in Table I. Species containing the greater amount of oil, all exceeding 6%, were *Ambrosia trifida, Campanula americana, Asclepias hirtella,* and the three Euphorbiaceae. Those with the most abundant polymeric hydrocarbon, exceeding 2%, were *Asclepias subulata, A. tuberosa, Crystostegia grandiflora, Cacalia atriplicifolia, Parthenium argentatum, Elaegnus multiflora, and Agropyron repens.* Seven species had more than 15% of the polar fraction labelled polyphenol, i.e., *Acer saccharinum, Rhus glabra, Lonicera tatarica, Elaeagnus multiflora, Xylococcus bicolor, Teucrium canadensis,* and *Prunus americanus.* And one species, *Vernonia altissima,* had more than 20% of apparent protein.

Protein contents are calculated from Kjeldahl nitrogen values by applying the factor 6.25 as the nitrogen equivalent for protein. However, the nature of the nitrogenous components must be elaborated by further research. In a discussion of leafy plants as protein resources, Kohler *et al.* remind us that in addition to desirable nutrients, most plants contain compounds which, if not appropriately dealt with, may be deleterious to animals (17). They also discuss several processing schemes which may provide insights into processing plants identified during the screening program.

PLANT OIL AND HYDROCARBON-PRODUCING SPECIES

During World War II, *Parthenium argentatum, Cryptostegia grandiflora, Crysothamnus nauseosus,* and *Solidago leavenworthii* were given considerable attention as possible domestic rubber sources (18). *P. argentatum* is undergoing vigorous reinvestigation (19). *Taraxacum kok-sagbyz* (Russian dandelion), although not listed in Table I principally because of botanical characteristics, was also a strong candidate as a potential rubber crop during the 1940's (20). See Table II for comparisons of *Hevea brasiliensis* rubber molecular weights with those of species identified by Swanson, Buchanan, and Otey (14).

Relatively few plant species have been proposed as potential U.S. oil and hydrocarbon crops. As previously stated, there is considerable current interest in guayule for production of natural rubber. Rubber is a pure hydrocarbon easily depolymerized into isoprene or reformed into gasoline (16). For this use, low molecular weight may be an advantage; the polyisoprene probably is best extracted as a hydrocarbon component of a whole-plant oil. A milkweed, *Asclepias speciosa,* is being grown experimentally in Utah. USDA agronomists are studying the common milkweed, *Asclepias syriaca,* and a few rubber- and oil-producing species in other plant families.

Gutta has been found in several Gramineae species (21). Although these species are low in combined oil and hydrocarbon content, *Elymus canadensis* is being grown in small plots to test its response to plant growth stimulants, and the genetic variability of *Agropyron repens* is being evaluated.

A few Euphorbiaceae have been suggested as crops for production of a whole-plant oil low in polyisoprene content. Calvin has drawn particular attention to *Euphorbia lathyrus* and *Euphorbia tirucalli,* species that are accustomed to arid lands, and has suggested that Asclepiadaceae and Euphorbiaceae deserve increased attention because they generally contain oil- and hydrocarbon-rich latexes. Hexane-extractable material from *E. lathyrus,* representing 4-5% of plant dry weight, has been reported to have a heating value of ∼ 18,000 Btu/lb (22). This material consists almost entirely

Table I. COMPOSITION AND CROP RATING OF REPRESENTATIVE BOTANOCHEMICAL-PRODUCING SPECIES[a]

Family-Genus-Species	Common Name	Crude Protein, %	Polyphenol Fraction, %	Oil Fraction, %	Polymeric Hydrocarbons %	Crop Type[b]	Rating
Aceraceae							
Acer saccharinum	Silver maple	16.3	19.8	2.4	0.39	—	9
Anacardiaceae							
Rhus glabra	Smooth sumac	7.1	20.2	5.9	0.21	W	10
Rhus laurina	Laurel sumac	8.1	10.5	5.5	1.44	W	8
Apocynaceae							
Apocynum androsaemitolium	Spreading dogbane	17.0	7.8	3.0	0.50	R,W	10
Asclepiadaceae							
Asclepias hirtella	Green milkweed	14.2	4.4	7.7	0.49	R	10
Asclepias incarnata	Swamp milkweed	11.0	11.5	3.0	1.87	R	10
Asclepias subulata	Desert milkweed	—	—	11.4[c,d]	2.95[c]	R	8
Asclepias syriaca	Common milkweed	12.3	8.0	4.8	1.54	R	9
Asclepias tuberosa	Butterfly-weed	8.1	12.3	3.1	2.84	R	9
Cryptostegia grandiflora	Madagascar rubber vine	—	—	6.7[c,d]	2.19[c]	R	10
Caprifoliaceae							
Lonicera tatarica	Red tatarion honeysuckle	10.2	15.8	3.4	1.77	R	9
Sambucus canadensis	Common elder	6.5	6.6	2.2	0.52	R	10
Symphoricarpos orbiculatus	Coral berry	5.9	11.1	2.3	0.81	R	10
Triosteum perfoliatum	Tinker's weed	7.1	14.2	2.6	1.43	R	10
Campanulaceae							
Campanula americana	Tall bellflower	9.7	6.2	6.5	0.99	R,W	10

Compositae							
Ambrosia trifida	Giant ragweed	11.4	4.4	8.3	0.60	R	10
Cacalia atriplicifolia	Pale Indian-plantain	11.7	9.4	3.4	3.46	R	8
Chrysothamnus nauseosus	Rabbitbrush	—	—	11.5[c,d]	1.67[c]	R	8
Cirsium discolor	Field thistle	5.9	3.9	5.8	0.40	R.W	10
Eupatorium atissimum	Tall boneset	8.6	10.8	5.9	0.56	R.W	10
Helianthus							
grosseserratus	Cut-leaf sunflower	8.8	9.2	2.3	0.76	R	10
Parthenium agentatum	Guayule	18.1	7.7	4.4	4.98	R	10
Rudbeckia subtomentosa	Sweet coneflower	5.9[e]	7.8[e]	2.4[e]	1.22[e]	R	10
Silphium integrifolium	Rosin weed	6.2	7.0	2.8	0.79	R	10
Silphium laciniatum	Compass plant	9.8	8.1	3.3	0.75	R	10
Silphium terbinthinaceum	Prairie dock	4.5	6.3	2.8	0.94	R	10
Solidago graminifolia	Grass-leaved goldenrod	5.6	13.4	2.6	1.51	R	10
Solidago leavenworthii	Edison's goldenrod	12.9	8.9	5.0	1.52	R	8
Solidago ohioensis	Ohio goldenrod	5.8	8.6	2.5	0.54	R	10
Solidago rigida	Stiff goldenrod	4.9	6.8	2.4	1.48	R	10
Sonchus arvensis	Sow thistle	9.3[e]	11.0[e]	5.3[e]	0.72[e]	R.W	10
Veronia altissima	Tall ironweed	21.6	6.9	3.1	0.38	—	10
Veronica fasciculata	Ironweed	11.4	8.4	5.4	0.39	R	10
Elaeagnaceae							
Elaegnus multiflora	Cherry elaeagnus	11.9	18.9	2.3	2.03	R	10
Ericaceae							
Xylococus bicolor	—	7.2	18.6	5.4	1.08	R	10
Elymus canadensis	Canada wildrye	7.0	5.5	1.7	1.35	G	11

Table I. COMPOSITION AND CROP RATING OF REPRESENTATIVE BOTANOCHEMICAL-PRODUCING SPECIES[a]
(Continued)

Family-Genus-Species	Common Name	Crude Protein, %	Polyphenol Fraction, %	Oil Fraction, %	Polymeric Hydrocarbons %	Crop Type[b]	Rating
Euphorbiaceae							
Euphorbia dentata	Cut-leaf spurge	19.4	4.1	11.2	0.20	—	—
Euphorbia lathyris	Mole plant	12.7	7.6	9.9	0.40	—	9
Euphorbia pulcherrima	Poinsettia	16.4	6.4	6.3	0.66	W	10
Gramineae							
Agropyron repens	Quackgrass	12.2	4.6	2.4	1.95	G	10
Labiatae							
Pycnanthemum incanum	Mountain mint	13.3	8.0	2.2	1.36	R	10
Teucrium canadensis	American germander	14.3	16.7	2.7	1.44	R	10
Lauraceae							
Sassafras albidum	Sassafras	8.9	14.4	5.7	0.23	W	10
Phytolaccaceae							
Phytolacca americana	Pokeweed	15.5[e]	5.9[e]	3.4[e]	0.17[e]	—	10
Rhamnaceae							
Ceanothus americanus	New Jersey tea	12.4	12.8	3.4	0.67	W	10
Rosaceae							
Prunus americanus	Wild plum	17.3	18.5	4.6	0.20	—	10

a Values are moisture and ash free; crude protein is calculated from Kjeldahl nitrogen with the factor 6.25.

b Identified by infrared G = gutta, R = rubber, W = wax.

c Literature values.

d Also contains Polyphenol fraction.

e Values are moisture free but are not corrected for ash.

Table II. MOLECULAR WEIGHT OF NATURAL RUBBERS
RELATIVE TO *HEVEA BRASILIENSIS* RUBBER

Species	Common Name	Ratio
Hevea brasiliensis Mull. arg.	Rubber tree	1.00
Parthenium argentatum A. Gray	Guayule	0.98
Pycnanthemum incanum (L.) Michx.	Mountain mint	0.38
Lamiastrum galeobdolon (L.) Ehrend. and Polatsch.	Yellow archangel	0.32
Monarda fistulosa L.	Wild bergamont	0.32
Veronia fasciculata Michx.	Ironweed	0.32
Symphoricarpos orbiculatus Moench	Coral berry	0.28
Sonchus arvensis L.	Sow thistle	0.25
Xylococcus bicolor Nutt.	Two-color woodberry	0.25
Melissa officinalis L.	Balm	0.24
Silphium integrifolium Michx.	Rosinweed	0.22
Helianthus hirsutus Raf.	Hirsute sunflower	0.21
Cirsium vulgare (Savy) Ten.	Bull thistle	0.20
Cacalia atriplicifolia L.	Pale Indian plantain	0.20
Euphorbia glyptosperma Engelm.	Ridgeseed Euphorbia	0.20
Monarda didyma L.	Oswega tea	0.20
Lonicera tatarica	Tartarian honeysuckle	0.19
Triosteum perfoliatum L.	Tinker's weed	0.18
Solidago altissima L.	Tall goldenrod	0.18
Cirsium discolor (Muhl.) Spreng.	Field thistle	0.18
Solidago graminifolia (L.) Salisb.	Grass-leafed goldenrod	0.18
Apocynum cannabinum L.	Indian hemp	0.16
Polymnia canadensis L.	Leafy cup	0.16
Gnalphalium obtusifolium L.	Fragrant cudweed	0.16
Silphium terebinthinaceum Jacq	Prairie dock	0.15
Euphorbia pulcherrima	Poinsettia	0.15
Asclepias incarnata L.	Swamp milkweed	0.14
Grindelia squarrosa (Pursh.) Duval.	Tarweed	0.13
Veronia altissima Nutt.	Ironweed	0.13
Solidago rigida L.	Stiff goldenrod	0.12
Euphorbia corollata L.	Flowering spurge	0.12
Helianthus grossesserratus Martens	Sawtooth sunflower	0.12
Elaegnus multiflora Thunb.	Cherry Elaegnus	0.12
Rudbeckia laciniata L.	Sweet coneflower	0.12
Pycnathemum virginianum (L.) Durand & Jackson	Mountain mint	0.11
Campsis radicans (L.) Seem. ex Bur.	Trumpet creeper	0.11
Chenopodium album L.	Lambsquarter	0.11
Monarda punctata L.	Horsemint	0.11
Apocynum androsaemifolium L.	Spreading dogbane	0.11
Asclepias tuberosa L.	Butterfly weed	0.10
Nepeta cataria L.	Catnip	0.10
Teucrium canadense L.	American germander	0.10
Solidago ohioensis Riddell	Ohio goldenrod	0.10
Artemisia vulgaris L.	Common mugwort	0.10
Aster laevis L.	Smooth aster	0.10
Asclepias syriaca L.	Common milkweed	0.09
Artemisia abrotanum L.	Southernwood	0.09
Campanula americana L.	Tall bellflower	0.09
Centaurea vochinesis Bernh.	Knapweed	0.08

**Table II. MOLECULAR WEIGHT OF NATURAL RUBBERS
RELATIVE TO *HEVEA BRASILIENSIS* RUBBER**

(Continued)

Physostegia virginiana (L.) Benth.	Obedient plant	0.08
Verbena urticifolia L.	White vervain	0.08
Euphorbia cyparissias L.	Cypress spurge	0.08
Ocimum basilicum L.	Purple basil	0.08
Asclepias hirtella (Pennell Woodson)	Milkweed	0.08
Achillea millefolium L.	Yarrow	0.07
Phyla lanceolata (Michx) Greene	Frog fruit	0.07
Gaura biennis L.	Gaura	0.07

of polycyclic triterpenoids. *E. lathyris* is being further evaluated at the University of Arizona (5), and USDA is evaluating *E. pulcherrima* and several other Euphorbiaceae.

During the summer of 1979, USDA made a special effort to collect Leguminosae species. And in September 1979, Calvin drew attention to the Leguminosae *Copaifera langsdorfii* which, he observed, produces virtually pure diesel fuel (23).

POLYPHENOLS AND TANNINS

The rather simple solvent classification schemes yield complex fractions of botanochemicals. Their detailed composition depends not only on the species but also on maturity of the plant and the method of extraction (15,22). The polar fraction isolated by acetone extraction and readily soluble in 87.5% aqueous ethanol, termed "polyphenol" by Buchanan and coworkers (11,12), no doubt consists of phenolics and a wide variety of other substances. For plants of high tannin content, (e.g., *Rhus glaubra*) the polyphenol fraction might well be called tannin (24)

HARVESTING AND PROCESSING SCHEMES

Harvesting and processing technologies will need to be developed and individually tailored to each species. A multi-disciplined approach is essential to capture the full potential offered by the various species. While it is beyond the scope of this review to elaborate on the many facets, major areas of concern are: harvesting method and timing, handling, storage, and separation and recovery of materials. For example, see some USDA experiences with the development of promising new crops crambe and kenaf (25,26).

Processing of whole plant materials for oil and hydrocarbon has been discussed by Buchanan and Otey (8) and Nivert and coworkers (27). In the process envisioned, baled plant material is flaked in equipment common to the soybean processing industry. The flakes are subsequently extracted by the appropriate solvent, perhaps by a sequential extraction using several solvents.

LIGNOCELLULOSIC RESIDUE

In all plant materials, the major component will be the cellular lignocellulosic material. Several possible uses for this material exist. Some of the more attractive are cattle feed; fiber for pulp, paper, and board; chemical feedstock; or energy (8,11,12). Detailed evaluation of the cellular portion should provide bases for suggesting their most appropriate uses.

CONCLUSIONS

Green plants are solar-powered chemical factories that convert carbon dioxide and water into a variety of energy-rich compounds. Crops can be developed to help provide renewable sources of fuels and chemicals and at the same time to provide a continuing source of feed and food. Several candidates have been identified, which can become future crops for American Agriculture.

REFERENCES

1. Szego, G. C.; Kemp, C. C. *Chemtech* **1973,** 275.

2. Goldstein, I.S. *Chem. Eng. News* **1976,** *54,* 4.

3. Calvin, M. *Rubber World* **1974,** *170,* 16.

4. Calvin, M. *BioScience* **1979,** *29,* 533.

5. Diamond Shamrock Co. *Chem. Eng. News* **1978,** *56,* 21.

6. Goodyear Tire and Rubber Co. *Rubber World* **1979,** *179,* 21.

7. Lipinsky, E. C. *Science* **1978,** *199,* 644.

8. Buchanan, R. A.; Otey, F. H. *Biosources Dig.* **1979,** *1,* 176.

9. Wolff, I. A.; Jones, Q. *Chem. Dig.* **1958,** *17,* 4.

10. Nieschlag, H.J.; Nelson, G. H., Wolff, I. A.; Perdue, R. E., Jr. *Tappi* **1960,** *43,* 193.

11. Buchanan, R. A.; Cull, I. M.; Otey, F. H.; Russell, C. R. *Econ. Bot.* **1978,** *32,* 131.

12. Buchanan, R. A.; Cull, I. M.; Otey, F. H.; Russell, C. R. *Econ. Bot.* **1978,** *32,* 146.

13. 95th U.S. Congress, 1978, Public Law 95-592.

14. Swanson, C. L.; Buchanan, R. A.; Otey, F. H. *J. Appl. Polym. Sci.* **1979,** *23,* 743.

15.   Buchanan, R. A.; Otey, F. H.; Russell, C. R.; Cull, I. M. *J. Am. Oil Chem. Soc.* **1978,** *55,* 657.

16.   Weisz, P. B.; Haag, W. O.; Rodewald, P. G. *Science* **1979,** *206,* 57.

17.   Milner, M.; Scrimshaw, N. S.; Wang, D. I. C., Ed. "Protein Resources and Technology: Status and Research Needs"; Avi Publishing Co.: Westport, Conn., 1978; p 543.

18.   Trumbull, H. L. *Ind. Eng. Chem.* **1942,** *34,* 1328.

19.   McGinnies, W. G.; Haase, E. F., Ed. "An International Conference on the Utilization of Guayule," Proceedings of Meeting held at Tucson, Arizona, November 1975.

20.   Whaley, W. K.; Bowen, J. S., "Russian Dandelion, An Emergency Source of Natural Rubber," U.S. Department of Agriculture No. 618, 1947.

21.   Buchanan, R. A.; Swanson, C. L.; Weisleder, D.; Cull, I. M. *Phytochemistry* **1979** *18,* 1069.

22.   Nemethy, E. K.; Otvos, J. W.; Calvin, M. *J. Am. Oil chem. Soc.* **1979,** *56,* 957.

23.   Calvin, M. (Thomas H. Maugh II) *Science* **1979,** *206,* 436.

24.   Clarke, I. D.; Rogers, J. S.; Sievers, A. F.; Hopp, H. "Tannin Content and Other Characteristics of Native Sumac in Relation to Its Value as a Commercial Source of Tannin"; U.S. Department of Agriculture, No. 986, 1949.

25.   Princen, L. H. "Potential Wealth in New Crops: Research and Development", in "Crops Resources", Academic Press: New York, 1977.

26.   Bagby, M. O. "Kenaf: A Practical Fiber Resource", Atlanta, Ga., TAPPI Press Reports Nonwood Plant Fiber Pulping Progress Report No. 8, 1977.

27.   Nivert, J. J.; Glymph, E. M.; Snyder, C. E. "Preliminary Economic Analysis of Guayule Rubber Production", 2nd International Guayule Conference, Saltillo, Coahuila, Mexico, Aug. 1977.

RECEIVED JULY 1, 1980.

Effects of Reaction Conditions on the Aqueous Thermochemical Conversion of Biomass to Oil

P. M. MOLTON, R. K. MILLER, J. A. RUSSELL, and J. M. DONOVAN

Batelle Pacific Northwest, P.O. Box 999, Richland, WA 99352

Thermochemical liquefaction of biomass is basically a simple process whereby it is heated with alkali under pressure at temperatures up to 400°C. This simple procedure converts the biomass to a mixture of gas (2-10%), char (5-40%), and oil (up to 40%), on a weight basis. It is one of several methods available for conversion of biomass to potential liquid fuels, the others being direct heating of dry matter (destructive distillation, pyrolysis) (1), fermentation (or anaerobic digestion) (2), and gasification (partial oxidation) (3) followed by liquefaction to methanol. There are variants on all of these processes.

The most interesting variant on the basic thermochemical liquefaction process involves the addition of an overpressure of carbon monoxide and hydrogen to the reaction, which is also performed in a non-aqueous solvent (anthracene oil or recycled product oil). Yields of oil up to 70% of the weight of the Douglas fir wood feedstock have been reported in an investigation by Elliott (4-8), Elliott and Walkup (9) and Elliott and Giacoletto (10). This process variant (also known as the Albany, PERC, or CO-Steam Process) is described in more detail in the Results and Discussion section.

The thermochemical liquefaction process and its variants are of interest because they appear to have several advantages over the other methods. They do not require preliminary drying of the feedstock; they operate at a

0097-6156/81/0144-0137$06.50/0
© 1981 American Chemical Society

relatively low temperature; and they convert all of the biomass, leaving only a relatively small unusable residue which is as little as 5% of the feedstock. In addition, the crude liquid product separates spontaneously from the aqueous phase after the reaction.

Historically, the aqueous degradation of biomass in the form of newsprint or cotton was examined by Berl and Schmidt (11-13) and Berl, Schmidt, and Koch (14), and by others who were primarily interested in the mechanism of geological formation of coal from plant material (e.g., 15). The alkaline reaction of cellulose was further investigated by a series of authors, notably Samuelson (16-28) and Heinemann (29). The reactions of cellulose at temperatures up to 180°C have long been of interest in the paper pulping industry. Although conditions in pulping have some relevance to the conversion of cellulose to oil, as discussed in more detail later, they are generally milder (since dissolution of the cellulose is definitely not required). Other chemicals are also added, such as sodium sulfide in Kraft pulping, and no attempt is made to exclude oxygen from the reaction. The development of thermochemical liquefaction techniques for conversion of biomass to oil has been reviewed by Molton and Demmitt (30). The effect of alkali on cellulose has been reviewed by Richards (31) and Meller (32).

In 1924, Waterman and Kortlandt (33) observed that semicoke obtained from lignite was liquefied more rapidly if there was an overpressure of hydrogen and/or carbon monoxide. Fischer and Schrader (34) observed that sodium formate in large amounts facilitated the liquefaction of various materials including peat and cellulose at 400°C. The effect of formate or carbon monoxide on the rate of biomass liquefaction was reported in 1960 by Appell, Wender, and Miller (35) working at the Bureau of Mines in Pittsburgh. In a series of publications (36-40), these and other workers at the Bureau of Mines showed a definite effect of carbon monoxide on the alkaline liquefaction of biomass.

As a result of the Bureau of Mines work, a pilot plant was constructed at Albany, Oregon. This was designed to liquefy biomass as a slurry in recycled product oil in the presence of 5% aqueous sodium carbonate at 290-370°C. The residence time was estimated at 20 min to 1 hr in a stirred tank reactor. Initially, wood flour was used as the biomass source, after hammer-milling and pre-drying. The pilot plant started operation in 1977, although significant amounts of oil product were not obtained until several months later. Operation of the pilot plant has been dogged by problems of plugging and corrosion of pipes and mechanical difficulties.

At the time of pilot plant completion, only the original Bureau of Mines work was available for estimation of process parameters. This work was used in the design of the Albany pilot plant. Thus, the original conditions for operating the pilot plant were assumed, with no guarantee that they were optimum conditions. Since pilot plant startup, laboratory investigations have been performed in support of the pilot plant (4-10), and have included work on process optimization.

For some years we have been working on the determination of the chemistry involved in biomass liquefaction, using pure cellulose (Solka-floc) as our initial model (41-43). We believe that a comparison of our results using pure cellulose in an aqueous system, with results obtained in a closer simulation of the Albany pilot plant conditons, is useful in predicting some of the complex chemistry occurring in the pilot plant. A more direct comparison using wood substrate and recycle oil is difficult due to the great complexity and variability in the system.

In this paper, we report the results of some of our experiments on the variation of reaction parameters and product formation from pure cellulose in an aqueous system closely related to the Albany pilot plant conditons.

MATERIALS AND METHODS

Throughout this work we used pure cellulose (Solka-floc; Brown Co., NJ) as our model biomass source because of its high purity. Use of newsprint or wood would have made our results difficult to reproduce due to the heterogenity of such materials. The effect of the lignin component of wood on this liquefaction process is being investigated in a separate series of experiments. The Solka-floc that we obtained was of 95.8% purity, and contained water (3.9%), ash (0.3-0.5%) and nitrogen (0.004-0.0034%). Solka-floc of the grade used by us contains over 90% alpha-cellulose, is greater than 50% crystalline, and consists of a 3:1 mixture of fibers with degree of polymerization (DP) of 600 and 1100. The average DP is therefore about 750, comparing with Douglas fir cellulose of DP about 800. The degree of crystallinity (44) and the DP value (45,46) have both been shown to exert highly significant effects on the rate of degradation of cellulose. Anhydrous sodium carbonate (Fisher certified ACS grade) was of minimum 99% purity. All water was distilled, and all solvents were of analytical reagent grade.

Experiments to determine the effect of carbon monoxide on oil product yield were performed in 2.5 and 10 gallon autoclaves; experimental conditions are shown in Table I. Cellulose, sodium carbonate, and water were mixed in a 30% cellulose slurry, and the slurry was poured into the autoclave, which was then sealed and flushed with pure nitrogen for 5 min to exclude oxygen. The autoclave was then raised to the correct temperature (heating time 2.5 hr), maintained at this temperature for 20 to 60 min, and cooled (cooling time 3.5 hr). The carbon monoxide, where added, was added after the nitrogen flush, at 250 or 500 psig initial pressure. After cooling, the gas was vented and the autoclave opened. The aqueous phase was separated and the oil residue and char poured off and dried. The product oil of chief interest in these experiments was obtained by a 3-day Soxhlet extraction of the tar plus char fraction, using acetone as the extracting solvent. After removal of the acetone under reduced pressure, the oil yield was determined on a weight basis.

To determine the effects of alkali concentration, final reaction temperature, and reaction time on the kinetics of the oil-forming reactions, a different procedure had to be used. For these experiments, small reactors were used, of approximately 7 ml total volume. After mixing the cellulose with water and alkali, inserting the mixture into the bomb and flushing with nitrogen, the reactors were immersed in a sand bath already at the correct temperature. Heating time was thus reduced to 5 min, and cooling time to 7 min, a procedure which minimized the reactions occurring during the heating and cooling periods. However, because of the small amounts of material in these reactors, a product yield could not be obtained. Instead, the oil-forming reactions were monitored by gas chromatography on a Perkin-Elmer 900 instrument. The number and distribution of volatile products were used as an index of the degree of reaction. Further identification of products was performed on a Hewlett Packard 5992 GC/MS instrument.

A series of experiments was performed with the small reactors, based upon statistical experimental design techniques. A Box-Behnken design was used, with the experimental parameters shown in Tables II to IV for determination of the effects of reaction time, temperature, and alkali concentration.

After reaction, the aqueous and oil phases were separated and examined separately. In these experiments, for purposes of comparison of product distribution as a function of the variable experimental parameter, the GC trace of the untreated aqueous phase was used. Further work is being performed to identify the organic components of the product oils and aqueous phases to determine the reaction mechanisms involved in oil formation. Aqueous phase components were found to be similar in composition to most oil components, but are better resolved on the GC. There is a possibility that higher molecular weight oil components may change independently of lower molecular weight components, but we have no way of testing this as the separation and identification of high molecular weight materials in such a complex mixture are beyond our current analytical capabilities.

Table I. EFFECT OF CARBON MONOXIDE ON CELLULOSE
LIQUEFACTION

CO Pressure (psig)	Alkali Conc. (N)	Final Temp. (°C)	Time at Temp. (min)	Yield* (%)
0	0.3	318	20	31
0	0.3	350	20	22
0	0.3	407	60	23
250	0.3	293	20	14
250	0.3	300	60	28
250	0.3	332	60	33
0	0.6	317	20	31
0	0.6	337	30	26
0	0.6	304	60	29
0	0.6	337	60	29
0	0.6	360	60	26
0	0.6	385	60	21
250	0.6	299	20	34
250	0.6	318	20	34
250	0.6	359	60	28
500	0.6	268	20	26
0	1.18	307	20	19
0	1.18	324	20	27
250	1.18	321	20	33
500	1.18	268	20	31
500	1.18	307	20	27
0	1.79	275	20	18
250	1.79	268	20	24
500	1.79	275	20	23
0	2.37	291	20	4
0	2.37	343	20	11
500	2.38	270	20	15

* Based on original weight of cellulose added.

Table II. CONDITIONS USED FOR BOX-BEHNKEN STUDIES OF
CELLULOSE LIQUEFACTION, 250-290°C

Run	Temp. (°C)	Time (hr)	[Na_2CO_3] (N)
1	270	1.0	0.6
2	270	3.5	0.3
3	270	6.0	0.0
4	250	3.5	0.0
5	290	1.0	0.3
6	290	3.5	0.0
7	290	6.0	0.3
8	270	3.5	0.3
9	250	1.0	0.3
10	250	3.5	0.6
11	250	6.0	0.3
12	290	3.5	0.6
13	270	1.0	0.0
14	270	3.5	0.3
15	270	6.0	0.6

Table III. CONDITIONS USED FOR NON BOX-BEHNKEN KINETIC STUDIES OF CELLULOSE LIQUEFACTION, 150-230°

Run	Temp. (C°)	Time (min)	$[Na_2CO_3]$ (N)
1	210	45	0.3
2	210	60	0.3
3	170	60	0.3
4	170	45	0.3
5	150	45	0.3
6	150	60	0.3
7	190	45	0.3
8	190	60	0.3
9	230	15	0.3
10	230	30	0.3
11	190	15	0.3
12	190	30	0.3
13	230	60	0.3
14	230	45	0.3
15	210	30	0.3
16	210	15	0.3
17	170	15	0.3
18	170	30	0.3
19	150	30	0.3
20	150	15	0.3

Table IV. CONDITIONS USED FOR FIGURES OF GC TRACES OF CELLULOSE LIQUEFACTION PRODUCTS

Figure	Temp. (°C)	Time (hr)	$[Na_2CO_3]$ (N)
1 a	270	1.0	0.6
1 b	270	3.5	0.3
1 c	270	6.0	0.6
2 a	150	1.0	0.3
2 b	210	1.0	0.3
2 c	290	1.0	0.3
3 a	270	3.5	0.07
3 b	270	3.5	0.03
4 a	270	6.0	0.6
4 b	270	6.0	0.0

RESULTS AND DISCUSSION

Effect of Carbon Monoxide

The results of 27 autoclave experiments with cellulose are shown in Table I. Oil yields range from 4% to 34% of the original weight of cellulose added, with no clear pattern emerging. In part this may be due to the masking effect of the heating and cooling times for the autoclaves. Based on pure cellulose, the theoretical yield of a hydrocarbon oil (C_6H_{10}) is 51% by weight; with 8% oxygen, as is characteristic of the Albany pilot plant product, this is increased to 55%. Hence, our highest oil yield is 34/55 X 100, or 62% of theoretical. Lignin and hemicellulose are of more variable composition, but the theoretical yields of oil containing 8% oxygen are about 75% and 50% respectively. Based on this data, a crude estimate of the theoretical oil yield from wood containing 50% cellulose, 25% lignin, and 25% hemicellulose is 60%. Elliott and Walkup [9] have presented data for the Albany model system showing oil yield to be a function of carbon monoxide overpressure in the range of 250-1500 psig initial pressure, with a maximum oil yield of 70%. This indicates that a theoretical oil yield can be achieved at high CO overpressure, with the possibility of some incorporation of CO into the product, although one experiment carried out with [14]CO did not result in ^{14}C incorporation into the oil product.

Elliott [7] has reported some experiments using wood flour, lignin, and cellulose without any addition of CO. At 350°C final temperature for 1 hr, with 18% substrate slurry, the oil yields were 39, 23, and 30% respectively. The result with cellulose is in agreement with our data. The effect of carbon monoxide addition to an aqueous system with a 30% cellulose slurry was to increase oil yield by only 3-4% in our experiments, up to 500 psig initial CO pressure. This is also in agreement with Elliott's data. The CO-enhancement effect on oil yield is thus barely noticeable with cellulose at CO pressures below 500 psig.

The only other noticeable effect of variation of the experimental parameters shown (Table I) is a negative effect of high alkali concentration on oil yield, when the sodium carbonate concentration is raised to 2.37 N (12.6%) at 270-343°C. This is counter to Elliott and Giacoletto's results [10]. The effect of alkali on cellulose chain peeling and cleavage was found to be directly proportional to alkali concentration at lower temperatures (185°C) [47]. Since high alkali concentrations favor the hydride- transfer mediated Cannizzaro reaction to yield acid salts and alcohols from aldehydes and some ketones, we suggest that perhaps the lower oil yield may be due to removal of carbonyl intermediates from the reaction.

Because of the masking effect of long heating and cooling times in large autoclaves (2.5, 10 gallon), we continued our experiments in small reactors of 7 ml total volume. This permitted us to achieve short heating and cooling times, but because of the small amount of reactants, prevented us from assessing oil yields. Since we are currently interested in reaction chemistry, this was not considered an overriding disadvantage. We were able to estimate the degree of cellulose conversion from the gas chromatograph traces of the products, as shown below, but these do not represent a quantitative yield estimate.

Effect of Reaction Time

Figure I shows the GC traces obtained from direct injection of $1\mu l$ quantities of the aqueous phases from three experiments. We chose to use the aqueous phases because they contain lower molecular weight materials than the oils, and hence presumably products formed earlier in the oil generation reaction sequence, and because the aqueous phases give cleaner GC traces than the oils, which are very difficult to resolve. Only minor differences are noticeable between the traces from reactions carried out for 1, 3.5, and 6 hr at a constant temperature of 270°C. This temperature is much lower than the temperature normally used in wood liquefaction (370°), yet most of the cellulose was dissolved. The yield of char or insoluble residue was generally less than 20% of the weight of cellulose added.

Effect of Reaction Temperature

The GC traces from three experiments run at temperatures of 150, 210, and 290°C and constant time (1 hr), alkali concentration of (0.30 N), and without carbon monoxide, are shown in Figure II. At the two lower temperatures the most noticeable feature is a pair of large peaks close to the injection point. One of these has been identified as acetone. The quantity of acetone is less at the higher temperature 290°C, which incidentally is the lowest operating temperature at Albany. The remainder of the peaks are products derived in part from condensation of acetone under alkaline conditions (Aldol condensation), and increase in amount with increasing temperature, although the distribution remains relatively constant. The conclusions to be reached from these experiments are clear: *Acetone is an early major product from cellulose, is produced at a much lower temperature than generally assumed (150°C and above), and then condenses to other products in a series of reactions which are constant over the range 150-290°C.* The nature of these reactions is still being elucidated, although it appears that acetone is only one of at least three major small molecule intermediates leading to the formation of oil, the others being acrolein and acetoin. At the lower reaction

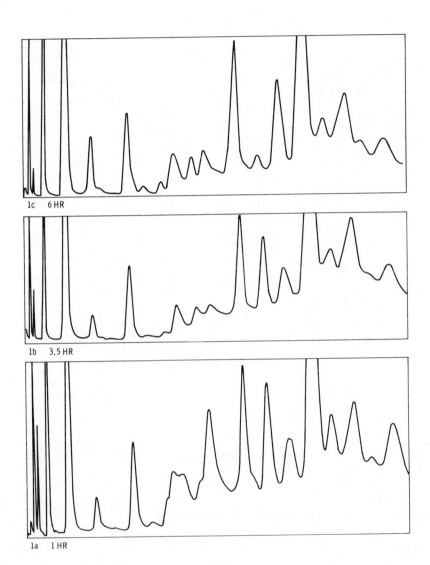

Figure 1. Effect of reaction time

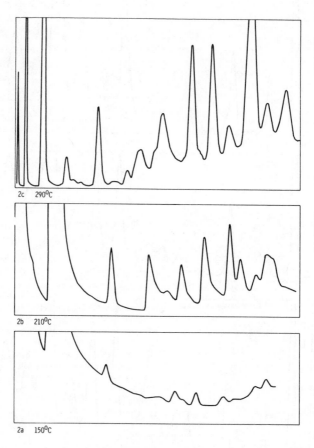

Figure 2. Effect of temperature

temperatures 150°C and 200°C, very little cellulose was dissolved (it was merely browned). The volatile products observed account for an estimated 2-3% of the total carbon in the reactor.

Effect of Alkali Concentration

A level of 5% sodium carbonate based on organic feed material is used at Albany. The aqueous stream is added to the reaction mixture as 20 wt % carbonate in water. Because the Albany system is mainly non-aqueous, the activity of the sodium carbonate is unknown. At the operating temperature of the Albany reactor, the amount of liquid water remaining has not been determined; the sodium carbonate may be dispersed in the oil vehicle. Some water is produced during the liquefaction process, and some acidic materials which will react with carbonate are produced, so the chemical form and activity of the catalyst is difficult to determine in the Albany process. Our reactions were all carried out in the aqueous system, so the sodium carbonate is readily available for reaction at the added concentration, rather than at some undetermined lower concentration.

In two experiments shown in Figure III, sodium carbonate concentrations of 0.07 and 0.30 N resulted in roughly the same total volatile organic material concentration in the aqueous phase of the product. There are some differences in relative product concentrations, but the overall GC traces are very similar. Actual identity of each component will be confirmed by GC/MS but is irrelevant here. In Figure IV, a GC trace at an alkali concentration of 0.60 N but with a reaction time of 6 hr gave a very similar product distribution to that observed after 3.5 hr (Figure III). However, a zero alkali concentration (Figure IVa) resulted in only traces of product under the same conditions.

The conclusions from this series of experiments are also clear: Alkali concentration has little effect on the liquefaction reaction over the range 0.07 to 0.6 N. The reaction proceeds to the same products at concentrations as low as 0.07 N. The presence of some alkali is necessary for the reaction to proceed, even though in all of our experiments, the pH at the conclusion of the experiment dropped to the acid side (5-6 on average). Results obtained by Lai and Sarkanen (47) showed that the rate of cellulose degradation is directly proportional to the alkali concentration.

Elliott and Giacoletto (10) have also examined the effect of sodium carbonate concentration in the oil-based liquefaction system and have shown no change in alkali effect over the range of 6-18% addition by weight. They estimate that a concentration of 3% may be practicable before an adverse effect on liquefaction is observed, while our lowest value of 0.07 N (0.37%) is almost 10x lower than this in the cellulose/water system.

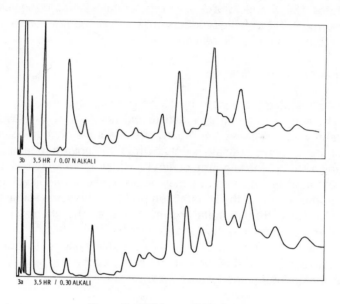

Figure 3.   Effect of alkali (3.5 h)

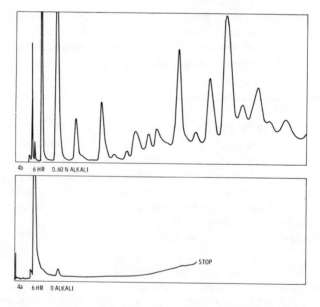

Figure 4.   Effect of zero and 0.6N alkali

A great deal of work has been done on the effect of aqueous alkali on cellulose, from the viewpoint of the pulping industry (e.g., 16-28, 48-59). The minor organic volatile products observed here at similar temperature to those used in Kraft pulping (150-180°C) eventually lead to colored product formation, which is of concern to the paper industry. The formation of acetone from cellulose has long been known. Generally, most interest has been shown in the nature of the residual cellulose after alkali treatment, not in the nature of the volatiles. From the viewpoint of determining the chemistry of oil formation from cellulose, the intermediate volatile products are all-important and the residual cellulose is of little interest.

In our view, the economics of the Albany pilot plant could be improved somewhat by using a lower sodium carbonate concentration. This would also reduce the chance of alkali stress corrosion.

Oil Composition

We are currently studying the variation in oil product composition and aqueous phase organic product composition as a function of reaction time, temperature, and alkali concentration. This work is in its preliminary stages, and should eventually lead to a detailed understanding of the chemical reactions occurring during liquefaction. At present, we have identified 78 components of the oils and aqueous phases derived from liquefaction of pure cellulose. These compounds are listed in Table V. They include a wide range of aliphatic and aromatic alcohols, phenols, hydrocarbons, substituted furans, and alicyclic compounds. Their individual concentrations are normally less than 1% of the total organic content of the product. The available evidence supports a degradation of cellulose to acetone, acrolein, and acetoin (among other products). These small molecules then re-condense under the alkaline conditions of the reaction to give the observed products. Aromatic hydrocarbon and phenol formation from these molecules under these conditions appears to involve a variant of the Aldol condensation. These conclusions were also reached by Theander, Popoff, and co-workers and reported in an earlier series of articles (60-66). Our results support their conclusions and proposed mechanisms.

For example, 1,4-addition of the acetone carbanion to acrolein would be a facile reaction under the reaction conditions used for cellulose liquefaction (both of these intermediates are formed from cellulose). The product, 5-ketohexanal, could cyclize to 3-hydroxycyclohexanone, which would then dehydrate and dehydrogenate to phenol and related aromatic products. This route to an observed product (phenol) is still speculative, but we have shown the more direct route to phenol from cyclohexanone does not occur under the

Table V. PRODUCTS IDENTIFIED FROM CELLULOSE LIQUEFACTION

OPEN CHAIN ALIPHATICS

Aliphatic Hydrocarbons	Probability	Mol. Wt.
cis- or trans-3-hexene	T*	84
3-heptene	T	98
4-octyne	T	110
2,5,-dimethyl-2,4,-hexadiene	P	110
2,5,-dimethyl-trans-3-hexene	T	112
octene	T	112
decyne	T	138
2,6,10,14-tetramethyl-heptadecane	P	296
Aliphatic Alcohols		
ethylene glycol	P	62
3-methyl-1-butanol	T	88
diethylene glycol	H	106
(2- or 3-) methyl-2,4,-petanediol	H	118
diethylene glycol, monoethyl ether	H	134
2-(2-butoxyethoxy) ethanol	H	162
Aliphatic Ketones		
4-methyl-pent-3-en-2-one	T	98
4-methyl-2-petanone	P	100
3-heptanone	P	114
4-hydroxy-4-methyl-2-pentanone	T	116
3-octadienone	T	124
2-methyl-hepten-3-one	P	126
acrolein hydrate monotrifluoroacetate†	T	170

Table V. PRODUCTS IDENTIFIED FROM CELLULOSE LIQUEFACTION
(continued)

CYCLIC COMPOUNDS

Cyclic Hydrocarbons	Probability	Mol. Wt.
1-ethyl cyclohexene	P	110
ethyl- or dimethyl-cyclohexane	T	112
methyl-n-propylcyclopentene	T	124
methyl ethyl cyclohexane	T	126
1,5,5,6-tetramethyl-1,3-cyclohexadiene	T	136
(1,1)-, (1,2)-, (1,3)-dimethylindan	P	146
1,1-dimethyl-1,2,3,4-tetrahydronaphthalene	T	160

Cyclic Alcohols		
cyclopentanol	P	86
cyclonhexanol	P	100

Cyclic Ketones		
cyclopentanone	H	84
2-methyl cyclopentanone	T	98
3-methyl cyclopentanone	P	98
cyclohexanone	H	98
2,5-dimethyl cyclopentanone	H	110
2,4-dimethyl cyclopentanone	T	112
2-ethylcyclopentanone	H	112
3-methylcyclopentanone	H	112
cyclohexene-1-enyl methyl ketone	T	124
4,4,5-trimethylcyclohex-2-enone	T	138
2-cyclopentyl-1-cyclopentanone	T	152
ethyl-2-methyl-cyclohex -1-enyl ketone	T	152

Furans		
8-butyro lactone (keto form of 2-hydroxy-4,5-dihydrofuran)	P	86
2,4-dimethyl furan	P	96
2,5-dimethyltetrahydrofuran	H	100
2-hydroxy-2-methyl-tetrahydrofuran	T	102
2,3,4-trimethyl furan	P	110
ethyl methyl furan	T	110

Table V. PRODUCTS IDENTIFIED FROM CELLULOSE LIQUEFACTION
(Continued)

Alkyl Benzenes	Probability	Mol. Wt.
propyl furan	T	110
2-iso-propyl furan	T	110
2,5-diethyl furan	T	124
butyl furan	T	124
2-methylbenzofuran	P	132
pentyl furan	T	138
heptyl furan	T	166

AROMATICS

o-, p-, or m-xylene	H	106
ethyl benzene	H	106
2-methyl styrene	P	118
1-methyl-2-ethyl benzene	H	120
1-methyl-3-ethyl benzene	H	120
n-propyl benzene	H	120
o-allyl toluene	T	132
1-methyl-2-n-propyl benzene	H	134
cyclopentyl benzene	T	146
1-methyl-4-isobutyl benzene	P	148
2-phenyl-3-methyl butane	T	148

Phenol Derivatives

phenol	H	94
o-cresol	H	108
m-cresol	H	108
p-cresol	H	108
2-ethyl phenol	T	122
o-methoxyphenol	H	124
2-phenoxyethanol	H	138
2-hydroxy-4-methyl-acetophenone	P	150
n-propyl cresol	T	150
4-t-butyl (o- or m-) cresol	P	164

POLYFUNCTIONALS

2-formyl-2,3-dihydro-pyran	P	112
2,5-dimethylterephthaldehyde	T	162

The above compounds were identified from 18 samples. Eleven samples were analyzed on a Hewlett-Packard 5980 GC/MS; 3 samples on a 50-ft. stainless steel LB550X column programmed from 70°C to 140°C at 4°/min, with a 2 min initial hold; and 8 samples on a 30-m glass capillary SP2100 column programmed from 50° at 260° at 4°/min with a 4 min initial hold. Six samples were analyzed on a Hewlett-Packard 5992 GC/MS on a 6-ft stainless steel Carbowax 20 m column, using 3 programs: 50°/2 min, 2°/min to 190°; 85°/10 min, 8°/min to 115°; and 115°/5 min, 2°/min to 150°. A 7th sample analyzed on this instrument was at 34°/2min, 2°/min to 135° on a 6-ft stainless steel OV-17 column.

* T - Tentative	P - Probable	H - Highly Probable
(10-30%)	(31-75%)	(75-99%)

† Identified in a trifluroacetylated oil sample.

conditions used for cellulose liquefaction. Similar Aldol-type condensations can account for many more of the observed products. Little else of relevance to the present discussion can be said at present regarding the mechanisms of formation of these products.

The nature of the products allows us to predict that the oil will slowly polymerize due to reaction between the phenols and furans, with side reactions from unsaturated hydrocarbons. The oil will also be rather corrosive because of the acidic nature of the phenols. We have observed both of these in practice. Oil products are stable only if kept in a freezer, and syringes used for injection of samples into the GC/MS instrument characteristically seize. Plugging of pipes together with rather serious corrosion problems have occurred in practice at Albany, and are consistent with our observations on oil composition. Under these circumstances, to suggest direct use of the oil as a fuel in conventional boiler equipment without pretreatment is unwise.

Implications for Albany

The experimental work reported here deals with optimum reaction conditions for the liquefaction of the major biomass component, cellulose, in an aqueous system. While lignin in wood may change the results qualitatively, we do not expect that transfer of our conclusions to a wood-based system will invalidate these results. However, there are some differences and our results should be used as indicators rather than quantitative measures of performance in other systems. For instance, at Albany, the reactor is a stirred tank, and consequently some of the oil has a longer residence time, while some proceeds through rapidly. This causes some difference between results at Albany and our experiments, which were performed in small autoclaves. Here we simply apply the results of our work generally to biomass liquefaction systems, with Albany as an example, without attempting to claim a 1:1 correspondence. In any event, laboratory work in support of the Albany pilot plant is proceeding separately at our laboratory (4-10), while the research reported here is basic in nature (41-43).

In Table VI, we show a comparison between the reaction conditions used by us and those used at Albany. Cellulose liquefaction in an aqueous system is improved only slightly by the addition of up to 500 psig of carbon monoxide, while use of wood and larger amounts of carbon monoxide results in improved oil yields. There are problems with the use of an oil vehicle at Albany, such as the continued polymerization of the vehicle with repeated passes through the reactor. The original intent in using recycle oil was to reduce reactor operating pressure since water at 370°C generates a very high pressure. Also, it was felt that there could be some hydrogen donor solvent

Table VI. COMPARISON OF REACTION PARAMETERS: BATTELLE-NW, AND ALBANY, OR PILOT PLANT

	Temperature, (°C)	Media	Gas	Catalyst	Residence Time, (min)	Feedstock
Albany (1977-7979)	290-370	Anthracene oil	CO	Na_2CO_3 (5%)	5-12	Douglas Fir
Albany, 1978	290-370	Heavy cycle oil	$CO:H_2$ 60:40	Na_2CO_3 (5%)	60	Douglas Fir
Albany, 1979	290-370	Water	$CO:H_2$ 60:40	Na_2CO_3 (5%)	90	Acid-hydrolyzed Douglas Fir
Battelle-NW	150-407	Water	CO, CO_2, N_2	Na_2CO_3 (0.37-12.6%)	20-360	Cellulose (Solka-floc)

effect in using recycle oil. This effect does operate to some extent, as shown by Elliott and Giacoletto (10). All of the experiments on the effect of CO overpressure above 500 psig have been performed in an oil vehicle; if the same yield-increasing effect can be shown to occur in an aqueous solvent, then many of the vehicle-associated problems at Albany could be eliminated by using water as the vehicle. This would of course affect the economics of the process, as the operating pressure would increase, but to some extent this would be alleviated by simplification of the product separation step — water and oil separate spontaneously (leaving an oil product containing 10-15% water), and there would be a lower load on the centrifuge. Polymerization and corrosion problems might still occur. The aqueous solvent, containing some of the water-soluble products, could in principle be recycled to provide additional oil.

The hydrogen donor solvent-effect should still operate with an aqueous vehicle, since product oil would dissolve most of the early reaction products; these being organic in nature, would tend to dissolve readily in oil already formed. Hydrogenation of intermediate products would thus still be expected to occur, even though water is a poor donor solvent.

Polymerization of the product oil during repeated recycling is not surprising, since many of the reaction products are reactive. Fu, Illig, and Metlin (67) observed this polymerization phenomenon during an investigation into oil derived from bovine manure; they found that the polymerization could be prevented by intermediate hydrogenation of the initial product over a cobalt molybdate catalyst. If recycle oil is found to be an essential part of the Albany process, such an intermediate hydrogenation could increase vehicle stability.

Regardless of the vehicle used, it appears that the alkali concentration used at Albany is too high. Our results with cellulose indicate that a reduction of up to 13x may be possible without interfering with product *composition* (not yield), while Elliott and Giacoletto's results (10) indicate that a 2x reduction is possible without affecting oil *yield* from wood.

It is difficult to determine an optimum time or temperature for the liquefaction process because reaction time and temperature are interactive variables. Reaction temperature is the most important factor according to our research, in agreement with work reported by Corbett and Richards (68) in terms of the amount of volatile products formed from cellulose. Inclusion of lignin and hemicellulose in a wood substrate may alter the optimum temperature, but should not alter the fact that reaction temperature is controlling. However, oil yield itself may not be a suitable parameter for measuring the efficiency of the liquefaction process. A high yield of oil may be achieved at the expense of a low heat content through incorporation of oxygen into the product. Furthermore, there may be other trade-off factors to be taken into consideration, such as the chemical composition and corrosivity of the product. Maximization of the heat content of the product as a dual function of time of reaction and maximum temperature has not yet been studied but would be very useful. According to our results with cellulose, the primary oil-forming reactions leading to volatile carbonyl compounds are complete at 230°C after 1 hr, with subsequent reactions involving intermediates already formed and leading to increased viscosity and molecular weight. Earlier work showed dissolution of cotton hydro-cellulose in 0.5 hr at 100°C in 0.5 N alkali (68-71), although the molecular weight of the hydrocellulose was not disclosed. This is much lower than the 370°C currently used at Albany, although liquefaction of the lignin component of wood may well require a much higher temperature than required for cellulose liquefaction.

Regarding the design of the process, we feel that a continuous plug flow reactor is inherently better than the stirred tank currently in use. A plug flow reactor would also give greater control over the reaction parameters. We also feel that more careful attention should be given to the relationship between reaction parameters and product yield and quality in future design of biomass liquefaction plants, since our work, and that of Elliott (4-10) demonstrates that a number of erroneous assumptions were made in designing the pilot plant at Albany.

SUMMARY

The direct thermochemical conversion of woody biomass to oil is one method of producing liquid fuel. This method is currently being examined in a pilot plant at Albany, Oregon. At Albany, wood is liquefied at a temperature of 290-370°C, a pressure of over 3000 psig, and a residence time of 20-90 min in the presence of a 60:40 mixture of carbon monoxide:hydrogen, and a catalyst consisting of 5% sodium carbonate. In our work on the fundamental chemistry of this liquefaction technique, laboratory studies using pure cellulose as a model compound showed that liquefaction occurs rapidly at 300°C and below at a pressure of 2800 psig of steam with no added gas, in less than 1 hr. Furthermore, the effect of the sodium carbonate "catalyst" on the early liquefaction reactions is independent of its concentration to a level of below 0.8%. Gas chromatographic and mass spectrometric examination of the product oil shows its composition to be almost constant over the whole range of reaction conditions used. A re-examination of reactor design of future biomass liquefaction plants appears warranted in light of these results.

158   BIOMASS AS A NONFOSSIL FUEL SOURCE

ACKNOWLEDGEMENT

We gratefully acknowledge support for this work under contract number EY-76-C-06-1830 from the U.S. Department of Energy, Division of Chemical Sciences, Office of Basic Energy Sciences.

REFERENCES

1. Molton, P. M.; Demmitt, T. F. *DOE Pacific Northwest Laboratory Report BNWL-2297*, August 1977.

2. Pohland, F. G. "Anaerobic Biological Treatment Processes," *Adv. Chem. Series 105*, American Chemical Society: Washington D.C., 1971.

3. Mudge, L. K.; Rohrmann, C. A. "In Solid Wastes and Residues Conversion by Advanced Thermal Processes," *ACS Symposium Series, 76;* American Chemical Society: Washington, D.C., 1978; 126-141.

4. Elliott, D. C. 4th Biomass Thermochemical Coordination Meeting, Battelle Memorial Institute, Columbus, Ohio, April 1978,; Battelle Pacific Northwest Laboratory: Richland, Wash., 1978; DOE Fuels from Biomass Program Report.

5. Elliott, D. C. 6th Biomass Thermochemical Conversion Contractors Meeting Jan 1979; Battelle Pacific Northwest Laboratory: Richland, Wash. 1979; DOE Fuels from Biomass Program Report.

6. Elliott, D. C. 7th Biomass Thermochemical Conversion Contractors Meeting Roanoke, VA., April 1979; Battelle Pacific Northwest Laboratory: Richland, Wash., 1979; DOE Fuels from Biomass Program Report.

7. Elliott, D. C. 8th Biomass Energy Systems, Thermochemical Conversion Contractors Meeting, July-Aug 1979, Seattle, Wash; Battelle Pacific Northwest Laboratory: Richland, Wash., DOE Fuels from Biomass Program Report.

8. Elliott, D. C. Canadian Chemical Engineering Conference, Sarnia, Ontario, Oct 1979. Bench-scale Research in Biomass Liquefaction by the CO Steam Process.

9. Elliott, D. C.; Walkup, P. C. Bench-Scale Research in Thermochemical conversion of Biomass to Liquids in Support of the Albany, Oregon Experimental Facility, Coordination Meeting: Thermochemical Conversion Systems, Wright-Malta Corp., Ballston Spa N.Y., Oct 1977. Battelle Pacific Northwest Laboratory Richland Wash., 1977.

10. Elliott, D. C.; Giacoletto, G. M. Bench-Scale Research in Biomass Liquefaction in Support of the Albany, Oregon, Experiment Facility, 3rd. Annual Biomass Energy Systems Conference, Golden, Colo., June 1979.

11. Berl, E.; Schmidt, A. *Justus Liebig's Ann. Chem.* **1928,** *461,* 192.

12. Berl, E.; Schmidt, A. *Justus Liebig's Ann. Chem.* **1932,** *493,* 97.

13. Berl, E.; Schmidt, A. *Justus Liebig's Ann. Chem.* **1932,** *496,* 283.

14. Berl, E. Schmidt, A.; Koch, H. *Z. Angew. Chem.* **1930,** *43,* 1018.

15. Bergius, F. *Naturwiss.* **1928,** *16,* 1.

16. Franzon, O.; Samuelson, O. *Svensk Papperstidn.,* **1957,** *60,* 872.

17. Christofferson, D.; Samuelson, O *Svensk Papperstidn.* **1960,** *63,* 729.

18. Alfredsson, B.; Gedda, L.; Samuelson, O. *Svensk Papperstidn.* **1961,** *64,* 694.

19. Albertson, U.; Samuelson, O. *Svensk Papperstidn.* **1962,** *65,* 1001.

20. Gunne, I.; Samuelson, O. Thede, L. *Svensk Papperstidn.* **1968,** *71,* 161.

21. Samuelson, O.; Thede, L. *Acta Chem. Scand.* **1968,** *22,* 1913.

22. Samuelson, O.; Stolpe, L. *Svensk Papperstidn.* **1969,** *72,* 662.

23. Samuelson, O.; Stolpe, L. *Svensk Papperstidn.* **1971,** *74,* 545.

24. Samuelson, O.; Sjoberg, L. *Svensk Papperstidn.* **1972,** *75,* 583.

25. Samuelson, O.; Stolpe, L. *Acta Chem. Scand.* **1973,** *27,* 3061.

26. Samuelson, O.; Stolpe, L. *Svensk Papperstidn.* **1974,** *77,* 16.

27. Samuelson, O.; Stolpe, L. *Svensk Papperstidn.* **1974,** *77,* 513.

28. Peterson, G.; Samuelson, O. *Acta Chem. Scand.* **1976,** *B30,* 27.

29. Heinemann, H. *Pet. Refiner* **1954,** *33,* 161.

30.   Molton, P. M.; Demmitt, T. F. *Polym.-Plast. Technol. Eng.* **1978,** *11,* 127.

31.   Richards, G. N. *High Polym.* **1971,** *5,* 1017.

32.   Meller, A. Holzforschung **1960,** *14,* 129.

33.   Waterman, H. I.; Kortlandt, F. *Rec. Trav. Chim.* **1924,** *43,* 691.

34.   Fishcher, F.; Schrader, H *Brennstoff.-Chem.* **1921,** *2,* 161.

35.   Appell, H. R.; Wender, I.; Miller, R. D. *Chem. Ind. (London),* **1960,** *19,* 703.

36.   Friedman, S.; Ginsberg, H.; Wender, I.; Yavorsky, P. Third Mineral Waste Utilization Symposium, IIT Research Institute, Chicago, March 14-16, 1972.

37.   Appell, H. R.; Wender, I.; Miller, R. D. Third Annual Northwest Regional Anti-Pollution Conference, University Rhode Island, July 21-23, 1970.

38.   Appell, H. R., *et al.* U.S. Bureau of Mines Rept. 7560, 1971.

39.   Wender, I.; Steffgen, F. W.; Yavorsky, P. M. "Recycling and Disposal of Solid Wastes," Yen, T. F., Ed., Ann Arbor Scientific: Ann Arbor, Mich., 1974; p 59.

40.   Appell, H. R.; Wender, I.; Miller, R. D. U. S. Bureau of Mines Technical Progress Report 25, 1970.

41.   Molton, P. M.; Miller, R. K.; Donovan, J. M.; Demmitt, T. F. Thermochemical Conversion of Biomass Residues Technical Seminar, sponsored by Solar Energy Research Institute, Golden, Colo., Nov 30 — Dec 1, 1977.

42.   Molton, P. M.; Demmitt, T. F.; Donovan, J. M.; Miller, R. K., Energy From Biomass and Wastes, symposium sponsored by the Institute of Gas Technology: Chicago, 1978, Aug 14-18, 1978.

43.   Molton, P. M.; Miller, R. K.; Donovan, J. M.; Demmitt, T. F. *Carbohydr. Res.,* **1979,** *71,* 331.

44.   Basch, A.; Lewin, M. *J. Polymer Sci., Polymer Chem. Ed.;* **1973,** *11,* 3071.

45. Pacault, A.; Sauret, G. *Compt. Rend. Acad. Sci.* **1958,** *246,* 608.

46. Broido, A.; Javier-Son, A.; Quano, A. C.; Barrall II, E. M. *J. Appl. Polymer Sci.* **1973,** *17,* 3627.

47. Lai, Y-Z.; Sarkanen, K. V. *Cellul. Chem. Technol.* **1967,** *1,* 517.

48. Kleinert, T. N. *Papier (Darmstadt)* **1969,** *23,* 135.

49. Kleinert, T. N. *Papier (Darmstadt)* **1970,** *24,* 73.

50. Kleinert, T. N. *Holzforschung* **1975,** *29,* 134.

51. Funasaka, W.; Yokokawa, C.; Kajiyama, S. *Mem. Fac. Eng. Kyoto Univ.* **1953,** *15,* 116.

52. Lai, Y-Z. Ph.D. Thesis, University of Washington, 1968.

53. Lai, Y-Z. *Carbohydr. Res.* **1973,** *28,* 154.

54. Machell, G.; Richards, G. N. *Tappi* **1958,** *41,* 12.

55. Meller, A. *Svensk Papperstidn.* **1962,** *65,* 629.

56. Shivaza, A. Ya.; Ivanov, V. I.; Lyalyushkin, A. Ya., *Izv. Akad. Nauk Kirg. SSR* **1973,** *3; Chem. Abs.* **1973,** *79,* 80512n.

57. Sihtola, H.; Lamanen, L. *Cellul. Chem. Technol.* **1969,** *3,* 3.

58. Sjostrom, E. *Tappi* **1977,** *60,* 54.

59. Young, R. A.; Sarkanen, K. V. *Carbohydr. Res.* **1977,** *59,* 193.

60. Theander, O.; Popoff, T. *J. Chem. Soc. D.* **1970,** 1576.

61. Popoff, T.; Theander, O. *Carbohydr. Res.* **1972,** *22,* 135.

62. Forsskahl, I., Popoff, T.; Theander, O. *Carbohydr. Res.* **1976,** *48,* 13.

63. Popoff, T.; Theander, O. *Acta Chem. Scand.* **1976,** *B30,* 397.

64. Popoff, T.; Theander, O. *Acta Chem. Scand.* **1976,** *B30,* 705.

65.   Olsson, K.; Pernemalm, P-A.; Popoff, T.; Theander, O. *Acta Chem. Scand.* **1977,** *B31,* 469.

66.   Popoff, T.; Theander, O.; Westerlund, E. *Acta Chem. Scand.* **1978,** *B32,* 1.

67.   Fu, Y. C.; Illig, E. G.; Metlin, S. J. *Envir. Sci. Tech.* **1974,** *8,* 737.

68.   Machell, G.; Richards, G. N. *J. Chem. Soc.* **1960,** 1924.

69.   Machell, G.; Richards, G. N. *J. Chem. Soc.* **1960,** 1932.

70.   Machell, G.; Richards, G. N. *J. Chem. Soc.* **1960,** 1938.

71.   O'Meara, D.; Richards, G. N. *J. Chem. Soc.* **1960,** 1944.

RECEIVED JUNE 18, 1980.

Liquid Hydrocarbon Fuels from Biomass

JAMES L. KUESTER

College of Engineering and Applied Sciences, Arizona State University,
Tempe, AZ 85281

Considerable attention is currently being focused on development of synthetic liquid fuels from non-petroleum feedstocks. The major feedstock candidates are coal, oil shale and biomass. Potential products include alcohol fuels, alcohol-gasoline blends and equivalents of commercial materials currently derived from petroleum (kerosine, diesel, jet fuel, high octane gasoline, etc.). Various advantages and disadvantages of coal and oil shale feedstocks vs. biomass are listed in Table I. It seems obvious that coal and oil shale will have the largest short-term impact due to the large quantities available. Initially, biomass use will concentrate on waste streams, (industrial, agricultural, urban) and surplus agricultural crops. Longer term, agricultural and forest residues (not currently collected) and energy crops will play an increasingly important role. Eventual widespread use of renewable biomass sources appears inevitable (1, 2) with the main question being primarily one of timing.

Several approaches have been proposed for the production of liquid fuels from biomass. Alcohol production via fermentation is state-of-the-art technology for specific feedstocks (grain etc.). The use of non-food sources (urban refuse, industrial wastes, etc.) is not fully developed. Processing times are on the order of days however for biological conversion. Non-biological methods fall into two categories: (1) direct liquefaction, and (2) indirect liquefaction. Both involve a thermal conversion step. Direct liquefaction

0097-6156/81/0144-0163$05.50/0
© 1981 American Chemical Society

processes (3, 4) attempt to produce a product without going through the gas phase. The product however contains an appreciable amount of oxygenated compounds thus leading to quality and stability problems. Current research in this area is aimed at eliminating the oxygen. Process conditions appear severe (3). Indirect liquefaction methods convert the biomass to a synthesis gas and then recombine components to form quality hydrocarbon fuel products free of oxygenated compounds. Three approaches have been Mobil, Naval Weapons Center (China Lake), and Arizona State University. The Mobil process (5) converts methanol to high octane gasoline via a catalytic process at mild operating conditions. The methanol is synthesized from carbon monoxide and hydrogen which can be obtained from fossil fuels or biomass via gasification. The Naval Weapons Center approach (6) thermally polymerizes olefins (primarily ethylene). The olefins are separated from the biomass gasification stream. The Arizona State University process converts the hydrogen, carbon monoxide and olefins produced in a gasification step to paraffinic fuels (e.g. diesel, jet, kerosine) via a catalytic reactor at mild operating conditions. Prior separation of the unreactive gas components (ethane, methane, carbon dioxide, etc.) is not required. If a high octane gasoline is desired, the paraffinic fuels are passed through a conventional catalytic reformer.

The present status and future projections of the Arizona State University process are described in this paper.

PROCESS DESCRIPTION

The basic conversion scheme is depicted in Figure I and the processing equipment is shown in Figure II. The system is operated continuously and is a simulator of commerical scale processing for the most part. Thus equipment and procedure development has accompanied factor and optimization studies. The system is conveniently divided into two sections: (1) gasification, and (2) liquid fuels synthesis.

Gasification. The pyrolysis reactor consists of a fluidized bed where the solids-fluidized medium is either inert (e.g. sand) or a catalyst in the 60-120 mesh range. The bed diameter is 10 in. with a length of 4 ft. The fluidized bed zone is approximately 18 in. A second identical fluidized bed serves as the solids heater and catalyst regenerator with continous transfer between the two vessels. The principle is identical to that employed in the catalytic cracking of petroleum, a process that has been successfully employed by petroleum refiners since the 1940's. The advantages of the system are efficient heat transfer, effective temperature control, continous catalyst regeneration, and the production of high quality gas free of combustion

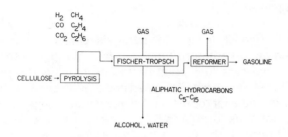

Figure 1.   Basic chemical conversion

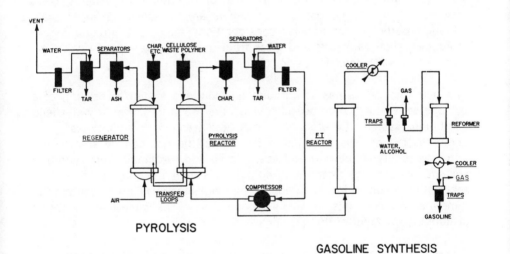

Figure 2.   Conversion system schematic

products. It is anticipated that the final configuration will include the use of a continuously regenerated cracking catalyst with steam addition to the pyrolysis reactor (to promote the water gas shift reaction). If steam addition were used without continous catalyst regneration, a steam preheater probably would replace the second fluidized bed, i.e., no solids loop transfer would be employed.

Feedstock is continously fed into the pyrolysis reactor via a screw feeder. The bed is operated at a controlled temperature in the 600-800°C range and a slight positive pressure (0-5 psig). Residence times are in the 0-5 second range. At these conditions, the feedstock is flashed to a gas. Lower temperatures promote tar formation. The maximum temperature is limited by metallurgy constraints (distributor plates, transfer loops). Work to date has been limited to dry, finely ground feedstocks free of inorganic matter. Wet feeds should not present a problem from a chemistry viewpoint, but the operational problem of reliable continuous feeding would have to be demonstrated. A pumped water-slurry feed might be preferable to a water-wet cake. Particle size is limited by the scale of the equipment to about 1/4 in. maximum. Thus, larger particles or pellets are feasible at a larger scale although particle heat transfer considerations may be a limiting factor. Inorganics such as metals and glass from urban uefuse would have to be removed from the feedstock prior to the gasification step.

The pyrolysis gas is passed through an overhead system consisting of a cyclone separator (to remove char and other solid particles), a water-feed scrubbing system (to cool the gas and remove condensibles), and compressor. From the compressor discharge, the gas can be split to the pyrolysis reactor (fluidizing and sparge gas) and the liquid fuels synthesis system. The use of a steam-fluidized pyrolysis system without pyrolysis gas recycle is also being investigated. Storage tanks are available to contain the generated pyrolysis gas and allow for operation of the liquid fuels synthesis system independent of the gasification system if desired.

The fluidized bed regenerator (or steam preheater) would be fueled with air and recycle char plus off gases from the liquid fuels synthesis reactors on a larger scale. In the research unit, propane and oxygen are used without recycle of char from the cyclone or downstream gases. Some char undoubtedly circulates in the transfer loops and coke is burned off the catalyst (if used) in the regenerator. The overhead system for the regenerator consists of a cyclone separator and scrubber with combustion gases vented to the atmosphere.

Liquid Fuels Synthesis. The first step in the liquid fuels synthesis system is a fluidized bed catalytic reactor (2 in. × 6 ft.) containing a modified Fischer Tropsch type catalyst (cobalt-alumina). Raw pyrolysis gas (without gas separation) is fed to this system at mild operating conditions (e.g., 250-300°C, 125 psig, 18 seconds residence time). The fluidized mode is employed to achieve temperature control with the significant exothermic heat of reaction. Continous regeneration is not used. The reaction is largely self-sustaining. The gas product is cooled in a condensor and two liquid phases result — a hydrocarbon phase and an alcohol-water phase. The relative amounts of these phases are dependent on the operating conditions and can vary from 100% hydrocarbons to 100% water-alcohol phase. The hydrocarbon phase is paraffinic in nature with some branched compounds, olefins and aromatics. The water-alcohol phase is essentially a binary of normal propanol and water with the alcohol content as high as 15 wt %.

The hydrocarbon phase is similar to JP-4 jet fuel. A simple distillation will isolate a kerosine-diesel fuel type fraction. A high octane gasoline is readily achieved by passing the Fischer Tropsch organic phase through a conventional catalytic reformer (2 in. × 2 ft.). Alternatively, a composite catalyst is being explored in the Fischer Tropsch step to produce a high octane product directly.

Off gases are generated from both the Fischer Tropsch and reforming steps. Quality is very high (600-2500 Btu/SCF) with a heavy concentration of low molecular weight paraffins. On a commercial scale, these gases would be recycled back to the gasification system.

Photos of the conversion system laboratory are shown in Figure III.

GASIFICATION STUDIES

Gasification studies completed or in progress are listed in Table II. The original studies were on a 4-in. diameter fluidized bed with paper chips feedstock. A temperature and feed rate factorial study with sand as the fluidized media revealed that the best conditions with regard to gas phase yields and composition were at the upper imposed constraints of temperature and feed rate (Figure IV, Table III). In order to achieve higher temperatures and feed rates, the 4-in. beds were replaced by 10-in. beds. With the original 4-in. beds, 10-60 fluidized solid particles were used. With this size particle, the temparature differential between the solids heater and pyrolysis reactor was about 300°F. Thus, the pyrolysis reactor temperature was limited to about 1500°F (since combustor temperatures above 1800°F can result in melting of metal parts). To reduce the temperature differential

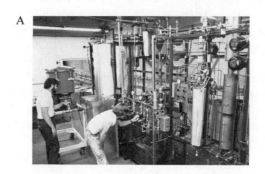

Figure 3. Conversion laboratory: a, conversion equipment; b, control room; c, analytical equipment

Table I. SYNTHETIC LIQUID FUELS FEEDSTOCK COMPARISON

	Coal, Oil Shale	Biomass
Advantages:	1. large quantities 2. low oxygen content	1. renewable 2. high hydrogen content 3. low contamination (sulfur, etc) 4. minimal enviromental problems (collection, processing)
Disadvantages:	1. nonrenewable 2. low hydrogen content	1. land use competition 2. high oxygen content 3. high contamination (sulfur, etc) 4. environmental problems (mining, processing)

Table II. GASIFICATION STUDIES

1. BASE ON 4-in. BEDS — COMPLETED
2. TEMPERATURE, FEED RATE STUDY ON 4-in. BEDS — COMPLETED
3. BASE ON 10-in. BEDS — CONTINUING
4. WATER GAS SHIFT CATALYST ON 10-in. BEDS — CONTINUING
5. CRACKING CATALYST ON 10-in. BEDS — CONTINUING
6. ONE PASS STEAM ON 10-in. BEDS — STARTING
7. MATHEMATICAL MODELS — CONTINUING
8. FEEDSTOCK CHARACTERIZATION — CONTINUING
9. ENVIRONMENTAL ASSESSMENT — CONTINUING

and allow for desired higher pyrolysis temperatures, a smaller solid particle is called for (60-120 mesh). Thus, the larger diameter beds were installed to prevent solids blowover and maintain transfer loop surge capacity for decreased bed heights. A resulting temperature differential of 100°F was achieved. A larger capacity compressor was also installed to handle the expected increase in gas phase yields. Studies are in progress on this revised system to explore the new range of temperature and feed rate.

Additional studies in progress include a survey of water gas shift and cracking catalysts, one pass steam effect (no pyrolysis recycle) and a survey of various feedstocks. Commercial water gas shift catalysts are limited to a maximum temperature of about 900°F and thus are not appropriate for fluidization at the temperatures under investigation. A fixed bed in the overhead system however with steam feed significantly altered the product composition (Table IV) as predicted:

$$CO + H_2O \rightarrow CO_2 + H_2$$

Use of cracking catalyst appears highly desirable to promote the secondary reactions involving the decomposition of tars:

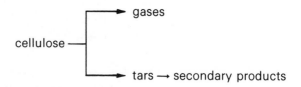

There is some indication that an improvement in gas phase composition can be achieved as well as an increase in gas phase yields and a decrease in tar residue. The use of steam is required to promote the water gas shift reaction. Thus the option exists to fluidize completely with steam and not recycle pyrolysis gas. The effect of this option is under study.

Feedstocks that have been tested on the system include a preprocessed municipal refuse (Eco-Fuel II, Combustion Equipment Associates), kelp residue (Kelco Co.), paper chips, synthetic polymer (polyethylene), and quayule bagasse (Centro de Investigacion en Quimica Aplicada, Saltillo, Coahuila, Mexico). Typical feedstock analysis for selected materials is shown in Table V. The paper chips, Eco-Fuel II and guayale bagasse appear to be similar in performance. The kelp residue was unacceptable due to the large amount of inert filter aid in the material. Polyethylene was spectacular in performance (high olefins, low oxygenated compounds). However, a commercial scale waste supply of synthetic polymer is not realistic.

Table III. PYROLYSIS GAS YIELDS
(lb gas/lb feed × 100)

Run	H_2	CO_2	C_2H_4	C_2H_6	CH_4	CO_2	TOTAL
No. 1 Base	1.11	12.9	3.38	1.11	4.91	42.0	65.4
No. 2 Base	1.42	14.6	3.15	1.21	5.57	45.3	71.2
No. 3 Base	1.15	10.8	2.86	0.98	4.68	40.9	61.3
+1,+1	1.61	12.9	5.90	0.94	6.90	54.2	82.4
+1,−1	1.36	9.4	2.21	0.32	5.32	36.7	55.4
−1,+1	0.50	10.9	1.28	0.61	2.27	30.2	45.7
−1,−1	0.78	17.1	1.27	0.84	2.69	38.6	61.3

Table IV. WATER GAS SHIFT CATALYST

Mole %	Base	Catalytic
H_2	14	46
CO	58	14
CO_2	6	23
Other	22	17

Table V. FEEDSTOCK ANALYSIS
(wt %)

	ECO Fuel II	Paper Chips	Cotton Gin Trach	Guayule Bagasse
C	38.0	41.7	49.3	40.2
H	4.9	5.7	5.7	4.7
O	32.0	52.0	41.7	47.2
N	<0.3	<0.3	<0.3	<0.3
S	0.6	0.1	1.2	1.1
Ash	24.5	0.5	2.1	6.7

A mathematical development of the gasification system is in progress. Also an environmental assessment of the configuration continues. The primary stream of concern is the pyrolysis reactor scrubber liquid discharge. The thrust here is to characterize the stream and hopefully minimize formation of any contaminates in the pyrolysis step. Alternatively, a water cleanup steam would have to be incorporated.

A typical pyrolysis gas composition range for the system (without steam or catalyst usage) is indicated in Table VI. Further manipulation of composition (olefins, carbon monoxide, hydrogen) is expected based on studies in progress.

LIQUID FUELS SYNTHESIS STUDIES

A list of liquid fuels studies completed or in progress is given in Table VII. Initial work was aimed at developing a catalyst to use in conjunction with a mixed pyrolysis gas feed stream. The expected reactions are indicated in Table VIII. A list of catalyst candidates tested is shown in Table IX. A modified Fischer Tropsch type (cobalt-alumina) was the most successful. Subsequent work on this reactor was aimed at establishing the effect of temperature, pressure, catalyst loading, and feed composition on reactor performance. Results are shown in Figures V-IX. Liquid hydrocarbon yields improve with temperature up to about 280°C when carbon formation starts to occur. Liquid hydrocarbon yields also increase with pressure up to an imposed constraint of 125 psig. Further increases in yield appear possible at high pressures. The effect of catalyst loading (residence time) appears to flatten out at the higher values investigated (corresponding to 15-20 seconds residence time). The feed composition study indicates an optimum pyrolysis gas composition of about 21% olefins, 21% carbon monoxide, and 28-42% hydrogen in a mixed pyrolysis gas stream.

The organic liquid product using the modified Fischer Tropsch catalyst was basically the same for most runs and paraffinic in nature (like diesel, kerosine or jet fuel). To produce a high octane gasoline, a post fixed bed reforming step was added using a commercial catalyst. Potential chemical reactions are listed in Table X. Results of an experiment on this system are shown in Figures X and XI. Here, to achieve commercial octane ratings, approximately a 20% loss in liquid yields occurs. Studies are in progress (prehydrogenation, isomerization) to minimize this loss. The high pressure in the reformer is achieved by the relatively economical task of pumping a liquid feed.

An analysis of liquid hydrocarbon phase from the Fischer Tropsch reactor and reformer is given in Table XI. As indicated, the Fischer Tropsch product is

Table VI. PYROLYSIS GAS
(mole %)

H_2	14-33
C_2H_4	5-10
CO	40-55
CH_4	13-17
C_2H_6	1-2
CO_2	3-8

Heating Value = 500—600 Btu/SCF

Table VII. LIQUID FUELS SYNTHESIS STUDIES

1. CATALYST SCREENING — COMPLETED
2. FACTOR STUDIES (T, P, θ, FEED COMP.) — COMPLETED
3. OPTIMIZATION — CONTINUING
4. FRACTION STUDIES — CONTINUING
5. ALTERNATIVE CATALYSTS — STARTING
6. MATHEMATICAL MODELS — CONTINUING
9. ENVIRONMENTAL ASSESSMENT — CONTINUING
8. NORMAL PROPANOL UTILIZATION — CONTINUING

Table VIII. MODIFIED FISCHER TROPSCH REACTIONS

Ethylene Incorporation

$$C_2H_4 + (n - 2) CO + (3n - 4) H_2 \rightarrow C_nH_{2n+2} + (n - 2) H_2O$$

Hydropolymerization

$$CO + 1/2(n - 1) C_2H_4 + 3H_2 \rightarrow C_nH_{2n+2} + H_2O$$

Oxo Reaction

$$C_2H_4 + CO + H_2 \rightarrow C_2H_5CHO$$
$$C_2H_5CHO + H_2 \rightarrow C_3H_7OH$$

Table IX. CATALYST SCREENING
(Fischer Tropsch Reactor)

Ethylene Polymerization
$ZnCl_2$, $AlCl_3$, H_3PO_4

Commercial Catalysts
Harshaw Ni-0101, Ni-1404, Co-0124, Co-1506,
Girdler G-67, G-64

Fischer Tropsch Catalysts
Co-K/Kieselguhr
Co-Th-Mg-Cu-Na/Kieselguhr
Co-Th-Mg-Al-Na
Co-Th-Mg-Na/Al_2O_3
Co-Th-Mg-Na/Super Filtrol
Co-Th-Mg-Na/Kieselguhr
Co-Th-Mg-Cu-Na/Silica gel

Blended Catalysts
$CoCO_3$ + alumina
Fe_3O_4 + alumina
Co_3O_4 + alumina

Table X. REFORMING REACTIONS

Isomeriationn
n-paraffins $\rightleftharpoons$ i-paraffins
alkyl-cyclopentanes $\rightleftharpoons$ alkyl-cyclohexanes

Dehydrocyclization

$$\left.\begin{array}{c}\text{n-paraffins}\\[1em]\text{i-paraffins}\end{array}\right\} \rightleftharpoons \left\{\begin{array}{c}\text{alky1-cyclohexanes}\\[1em]\text{alkyl-cyclopentanes}\end{array}\right. + H_2$$

Dehydrogenation
alkyl-cyclohexanes $\rightleftharpoons$ alkyl-benzenes + H_2

Hydrocracking

$$\left.\begin{array}{l}\text{n-paraffins}\\\text{i-paraffins}\\\text{cycloparaffins}\end{array}\right\} + H_2 \rightarrow \text{cracked products}$$

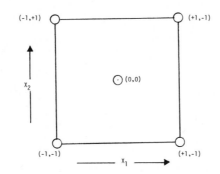

X_1 = reactor temperature, °C
X_2 = feed rate, lbs/hr

Uncoded:	-1	0	+1
X_1	580	688	788
X_2	3.1	4.8	.7.8

Figure 4.   Pyrolysis factorial experiment

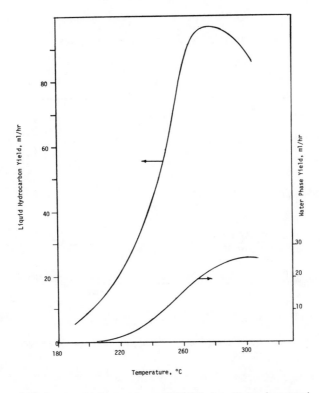

Temperature, °C

Figure 5.   Temperature vs. yield (Fischer–Tropsch reactor)

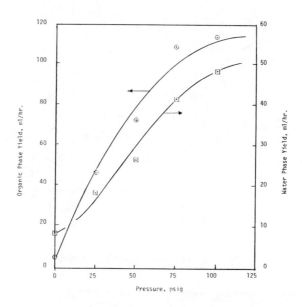

Figure 6.   Pressure vs. yield (Fischer–Tropsch reactor)

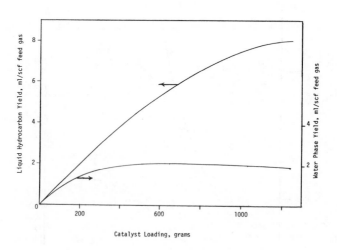

Figure 7.   Catalyst loading vs. yield (Fischer–Tropsch reactor)

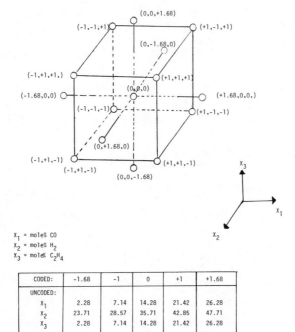

X_1 = mole% CO
X_2 = mole% H_2
X_3 = mole% C_2H_4

CODED:	-1.68	-1	0	+1	+1.68
UNCODED:					
X_1	2.28	7.14	14.28	21.42	26.28
X_2	23.71	28.57	35.71	42.85	47.71
X_3	2.28	7.14	14.28	21.42	26.28

Figure 8.   Fischer–Tropsch factorial and central composite extension points

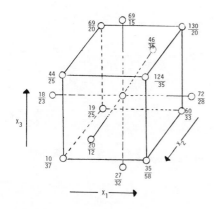

At base point, $\frac{38}{28}$, $\frac{44}{24}$, $\frac{39}{27}$, $\frac{41}{26}$

Figure 9.   Experimental results (Fischer–Tropsch)

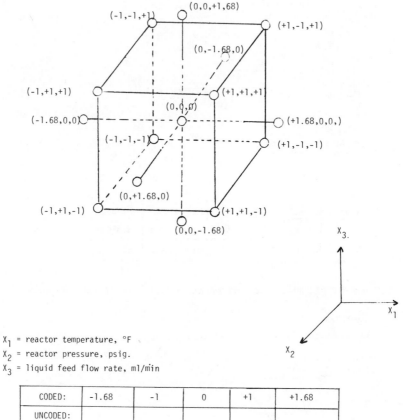

X_1 = reactor temperature, °F
X_2 = reactor pressure, psig.
X_3 = liquid feed flow rate, ml/min

CODED:	-1.68	-1	0	+1	+1.68
UNCODED:					
X_1	808	825	850	875	892
X_2	416	450	500	550	584
X_3	0.16	0.5	1.0	1.5	1.84

Figure 10.   Reformer factorial and central composite extension points

very paraffinic in nature. For the reforming step, the octane rating can be manipulated easily (but with a drop in yield with increase in octane number). Typical analysis of the Fischer Tropsch reactor and reformer off-gas are shown in Tables XII and XIII. Typically, the olefins and hydrogen are depleted with some excess of carbon monoxide. Appreciable hydrocracking apparently occurs in the reformer resulting in an extremely high quality gas with heavy concentration of the low molecular weight normal paraffins.

An analysis of the alcohol-water phase from the Fischer Tropsch reactor is shown in Table XIV. The normal propanol fraction could be marketed separately (1979 price approximately $2.22/gal) or blended with the gasoline product to form a "gasohol" fuel. Normal propanol has an octane rating of about 92 and is more miscible with gasoline than methanol or ethanol. A sufficient quantity of normal propanol is available to produce roughly a 10 vol % mixture of alcohol in gasoline. The process then would have the unique distinction of producing an alcohol-gasoline blend where both components are obtained from renewable resources. This could have large benefits with regard to tax advantages and allocations.

SCALE UP AND COMMERCIALIZATION

The next logical step for the process would be a scale up to about 10 tons/day. A minimum commercial scale would be approximately 300 tons/day. Process flow diagrams and economic projections have been prepared by Argonne National Laboratory and Mittelhauser Corporation and are contained in an interim report (7). The proper role of the research scale unit and pilot plant are listed in Table XV. As indicated, the research unit is best suited for detailed factor and optimization studies. The pilot plant is best suited for endurance testing on equipment and catalysts at fixed conditions specified from research scale studies. An additional role of the pilot plant would be to upgrade material and energy balances and commercial scale economic projections, incorporate remaining commerical scale processing features (recycle streams, etc.) and produce a sufficient quantity of product for testing.

A basic problem existing for all biomass conversion schemes is a playoff between feedstock density and economy of scale. In general, feedstock availability within a reasonable transportation distance will dictate that a conversion facility be limited to no more than about 3000 tons/day of dry feedstock. It is not economically realistic to expect to move solid waste material more than about 100 miles. This will produce no more than 300,000 gals/day of hydrocarbon liquid fuel. For a plant of this size, the product line will have to be limited. While this certainly is appropriate in many cases, an

Table XI. EXPERIMENTAL LIQUID FUELS ANALYSIS
(Undistilled)

Sample:	80 Octane	120 Octane	Unreformed
Composition (wt %)			
n-Paraffins $C_3 - C_5$	12.5	8.4	0.7
$C_6 +$	30.0	17.9	54.6
i-Paraffins $C_4 - C_5$	7.6	4.9	2.1
$C_6 +$	15.7	26.6	11.0
Cycloparaffins	0.1	0.85	0.5
Olefins	0.8	3.0	17.0
Aromatics	22.7	31.4	5.7
Unknowns	10.5	6.9	8.4
Specific Gravity	0.7296	0.7931	0.7486
ASTM Boil. Pt. (°F)			
10%	86	93	199
90%	422	404	640
Heating Value (Btu/lb)	22409	21046	22113
Octane No.	80	120	— —
Cetane No.	— —	— —	70

Table XII. FISCHER TROPSCH OFF GAS
(mole %)

H_2	<1
CO	39
CO_2	7
CH_4	46
C_2H_4	<1
C_2H_6	7

Heating Value = 700 Btu/SCF

Table XIII. REFORMER OFF GAS
(mole %)

H_2	trace
CH_4	21
C_2H_6	20
C_3H_8	23
C_4H_{10}	36

Heating Value = 2300 Btu/SCF

Table XIV. FISCHER TROPSCH WATER PHASE
(wt%)

H_2O	83.76
EtOH	0.54
i-PrOH	0.36
n-PrOH	14.74
i-BuOH + n-BuOH	0.16
Unidentified	0.43

Table XV. RESEARCH VS. PILOT PLANT FUNCTIONS

Research Scale (25 lb/hr)	Pilot Plant (10 ton/day)
1. EQUIPMENT DESIGN AND DEVELOPMENT	1. EQUIPMENT IMPROVEMENTS
2. FACTOR STUDIES	2. ENDURANCE TESTING
3. OPTIMIZATION	3. RECYCLE STREAMS
4. PRODUCT CHARACTERIZATION	4. PRODUCT APPLICATIONS TESTING
5. MATERIAL AND ENERGY BALANCES	5. REFINED MATERIAL AND ENERGY BALANCES
6. PRELIMINARY ECONOMICS	6. COMMERCIAL SCALE ECONOMICS

Table XVI. OPERATING CONDITIONS

	Pyrolysis	Modified Fischer Tropsch	Reformer
Temperature, °C	600-800	250-300	490
Pressure, psig	0-5	125	400
Residence Time, sec.	<2	18	11

Table XVII. PERFORMANCE

	Pyrolysis	Fischer Tropsch	Reforming
Present:	CO = 40-50 mole %	40 gal/ton	32 gal/ton
	H_2 = 14-33	cellulose	cellulose
	C_2H_4 = 5-10		
Optimal:	CO = 21 mole %	100 gal/ton	80 gal/ton
	H_2 = 28	cellulose	cellulose
	C_2H_4 = 21		

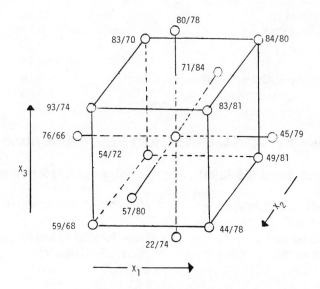

At base point: 68/78, 68/79, 70/78

Figure 11.   Experimental results (reformer)

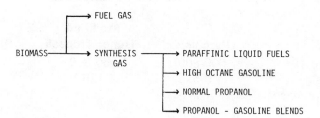

Figure 12.   Product options

alternative for the process described in this paper would be to convert the solids to Fischer Tropsch liquids in plants located at the source of feedstock and then transport the liquids to existing oil refineries for processing. The obvious advantages are volume reduction, lower cost transportation, and utilization of existing facilities for final processing, additive incorporation, and blending.

SUMMARY AND CONCLUSIONS

Liquid fuels equivalent to existing commerical products (kerosine, diesel, jet fuel, high octane gasoline) can be produced using biomass type feedstocks. A summary of potential products and operating conditions is shown in Figure XII and Table XVI. The present status of the project is indicated in Table XVII. Liquid hydrocarbon yields of 50-100 gal/ton of feedstock (dry, ash free) are to be expected. The process is characterized by:

(1) multi feedstock capability,
(2) mild operating condtions,
(3) minimal spearation steps,
(4) capability of producing several types of quality products,
(5) potentially minimal environmental problems.

Tasks underway at the research scale include alternate feedstock studies, gasification system optimization, waste stream characterization, and liquid fuels tailoring. A pilot scale facility (10 tons/day) is contemplated with eventual commercialization dependent on results at this stage.

ACKNOWLEDGEMENTS

The work in progress is supported by the U.S. Department of Energy (Contract No. EY-76-S-02-2982) and the Arizona Solar Energy Research Commission (Project No. 462-78). A number of graduate students have participated on the project: Wayne Fleming, Mark Wallington, Danny Lu, Robert Gough, Wayne Chung, Lee Scott, Louis Hsu, Edwin, Sabin, Michael Wang, Michael Hunter. The project technical manager is Mr. Edward P. Lynch of Argonne National Laboratory.

REFERENCES

1.   "CHEMRAWN I Faces Up to Raw Materials Future," *Chem. Eng. News* **1978** *56* (29) 28-31.

2.   "Tomorrow's Feedstocks: From Where Will They Come?" *Chem. Eng. Prog.,* **1977** *73* (12), 23-25.

3.   Albany Oregon, *Liquefaction Project,* Final Technical Progress Report, Bechtel Corp., USDOE Contract EG-77-C03-1338; April, 1978.

4.   Preston, G.T. "Symposium Papers", Clean Fuels from Biomass, Sewage, Urban Refuse and Agricultural Wastes, Sponsored by the Institute of Gas Technology, Chicago, Ill., 1976; 89-114.

5.   Meisel, S. L., et al. *Chemtech* **1976** *6* (2), 86-89

6.   Diebold, J. P.; and Smith, G. D., "Abstracts of papers", ASME 1979 Internatuional Gas Turbine and Solar energy Conference, San Diego, California, March, 1979.

7.   Kuester, J. L., Washington, D.C., March 1979, DOE Report COO-2982-38.

RECEIVED JULY 1, 1980.

9

Key Factors in the Hydrolysis of Cellulose

JEROME F. SAEMAN

Forest Products Laboratory, Forest Service, USDA, P.O. Box 5130,
Madison, WI 53705

If a successful cellulose hydrolysis process were available, the food problems of the world would diminish and wood sugar could form the basis of new enterprises providing an important source of chemicals and fuels. This has been known for more than a century. Some hundred to perhaps hundreds of millions of dollars have been spent on research dealing with cellulose hydrolysis. Billions have been invested in commercial plants for the hydrolysis of wood. Success, however, has been limited and future prospects are uncertain. A review of early wood hydrolysis processes was published by Sherrard and Kressman (1). A more recent review was published by Wenzl (2). The coverage here is selective rather than comprehensive. It identifies some key issues that determine the future usefulness of wood hydrolysis processes.

Cellulose is a polymeric carbohydrate $(C_6 H_{10} O_5)_x$ having the same elemental composition as starch and also yielding only glucose on complete hydrolysis. Cellulose consists of long chains of beta glucosidic residues linked through the 1,4 positions. Starch consists of alpha glucosidic residues linked through the same positions. These linkages are as similar as right and left hands, but differences in overall configuration cause cellulose to have a high crystallinity and hence a low accessibility to enzymes or acid catalysts. Starch by contrast is very readily hydrolyzed, suiting its role as a form of stored food.

Check back sections first.
Summaries.

This chapter not subject to U.S. copyright.
Published 1981 American Chemical Society

Cellulose fibers provide for the structural needs of plants. Hemicellulose, an easily hydrolyzed amorphous hetero-polymer yielding several different sugars on hydrolysis, is intimately associated with cellulose. Lignin, an aromatic three dimensional polymer, intersperses the fiberous constituents. Many woods consist of a little less than half cellulose and roughly a quarter each of hemicellulose and lignin. Cellulose is of the order of a hundred times as difficult to hydrolyze as starch. Lignin has little effect if any on the rate of acid hydrolysis, but it greatly inhibits enzymatic hydrolysis. This protects the plant against its ready use as a food by other organisms.

DILUTE ACID HYDROLYSIS OF CELLULOSE

Most simple oligosaccharides are quantitatively hydrolyzed by boiling dilute acid. Cellulose by contrast is hydrolyzed very slowly, and even on extended hydrolysis at higher temperatures, the maximum yield of recoverable sugar is very low. This was explained by Luers (3) who showed the cellulose hydrolysis in dilute acid involves consecutive first-order reactions of somewhat similar rates for the production and the decomposition of sugar. This observation formed the basis for the development of the Scholler percolation process. This process, which involves removal of sugar from the digester as it is formed, resulted in twice the yield obtainable by batch hydrolysis.

Luers erred, however, in concluding that changing the acid concentration and temperature changed the rate but not the course of the reaction nor the maximum yield attainable. Saeman (4) showed that the hydrolysis reaction is accelerated more than the decomposition reaction by both increased temperature and acid concentration. Hence, the sugar yield increases with both acid concentration and temperature. This observation is applicable to both the percolation process and to the much simpler batch process.

SINGLE-STAGE DILUTE ACID HYDROLYSIS

Figure I shows the maximum yield of "B" and the residual "A" in consecutive first-order reactions: $A \xrightarrow{k_1} B \xrightarrow{k_2} C$ for various ratios of $k_1 : k_2 = kr$. Parameters for cellulose hydrolysis were determined by Saeman (4) and redetermined by Kirby (5). It might be noted that despite analytical and experimental problems, the general model has been repeatedly confirmed.

Using Kirby's data and familiar first order kinetics, the relationship shown in Figure II was derived. This was tested at temperatures up to 280°C (corresponding to 916 lbs/sq in. equilibrium steam pressure) in 0.25 inch-diameter copper tubing reactors. In these batch experiments, 0.6 g of

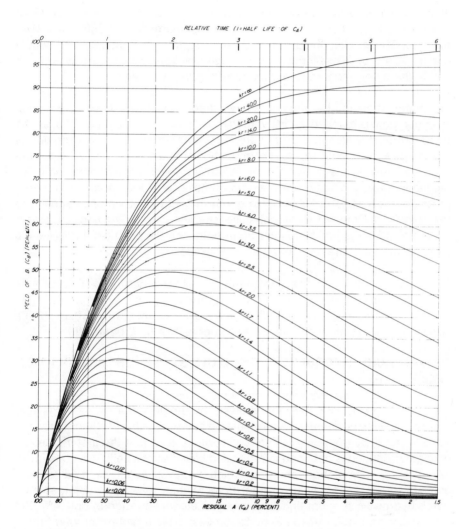

Figure 1.   The yield of product B *as a function of residual* A *in the consecutive*
$$k_1 \quad k_2$$
first-order reaction A → B → C *for various ratios of reaction rates* k_1 *and* k_2

prehydrolyzed wood was reacted with 1.5 ml of dilute sulfuric acid in an atmosphere of carbon dioxide. The reactors were heated with suitable precautions in a molten salt bath and quenched in water. As shown in Figure III, predicted general improvement in yield was maintained up to 260°C. At this point, the yield was 54%. The time to maximum yield at 260°C was 0.45 minutes. The model presumes isothermal reaction conditions, but with a reaction time of 30 seconds, the average reaction temperature was much below the batch temperature, thus resulting in lower yields.

An examination of the kinetics shows that if the lower limit of practical retention time is reached, it is advantageous to lower the acidity and raise the temperature. It is also evident the pretreatments which accelerate the rate of hydrolysis and increase the heat of activation are advantageous. Microtechniques are now available which can establish yields on milligram samples heated in glass capillaries with reaction times of a few seconds and internal pressure in excess of a 1000 lbs/sq in but they have not yet been applied to this problem. Pumped hydrocellulose slurries can be heated by steam injection and quenched by release of pressure, but the practicability of such a processing technique has not been established.

Grethlein recently confirmed Saeman's model by reporting a yield of over 50% sugar from cellulose using 1% sulfuric acid and a continuous-flow reactor with a residence time of 0.22 minutes at a temperature of 237 °C (6). Research seeking higher yields by this approach should be fruitful.

DILUTE ACID HYDROLYSIS BY MULTISTAGE OR PERCOLATION PROCESSING

Early attempts at dilute acid hydrolysis by single-stage processing were limited to yields of about 20% fermentable sugar because of the unfavorable k_1 to k_2 ratio. By reference to Figure I, it will be seen that the first increment of sugar produced is obtained in high yield based on the cellulose consumed. The Scholler percolation process, which removes sugar as it is formed, exploits this fact. The yield of sugar is more than twice that obtainable by batch hydrolysis. It is obtained, however, at a concentration of only 4%. The Madison modification of the Scholler process uses a higher temperature, a shorter reaction time, and continuous rather than periodic elution of sugar. Commercial plants of the Scholler type were built in Germany and Switzerland. One plant based on the Madison process was built in the United States. The greatest application of the process is in Russia where some 40 plants were built. The Russian development seemed to stem from a decision to favor initially the use of alcohol from non-food sources as a chemical raw material. Presently, the hydrolysis plants in Russia are used also for the production of food yeast.

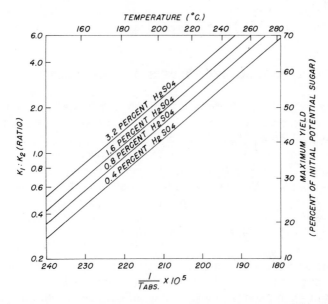

Figure 2.   The k_1 to k_2 ratio (log scale) as a function of reciprocal temperature. Scales for corresponding temperatures and yields from Figure 1 are superimposed.

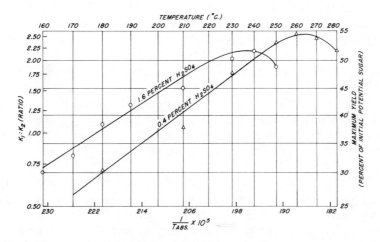

Figure 3.   Maximum yield of sugar from cellulose at two acid concentrations as a function of temperature. Experimental data plotted on grid used for Figure 2.

Wenzl (2) provides a description of the percolation process. An economic analysis by Katzen shows that in its present form, it has little chance of economic success (7,8). There is no obvious way to overcome the inherently high costs involved.

CONCENTRATED ACID HYDROLYSIS OF CELLULOSE

The low rate of cellulose hydrolysis in dilute acid is due to the inaccessibility of the cellulose crystallite. It was observed that concentrated acids result in disruptive swelling and solubilization of cellulose with negligible destruction of the carbohydrate. This in effect raises the k_1 to k_2 ratio. The water-soluble product of such primary hydrolysis is readily hydrolyzed to sugar in high yield. The so-called Bergius method of wood hydrolysis extracts cellulose from prehydrolyzed wood by means of fuming hydrochloric acid. The acid is distilled away from the solubilized oligosaccharides which are then completely hydrolyzed by heating in dilute acid (2).

This strong acid process is straightforward but for a key factor — the cost of handling and recovering the acid. Gregor and Jefferies point out that advances in membrane technology may offer a solution to this problem (9).

HYDROLYTIC BEHAVIOR OF NATIVE AND MODIFIED CELLULOSES

Figure IV shows the course of hydrolysis of an array of native and modified celluloses (10). For convenience, the reaction was carried out in constant boiling hydrochloric acid. There is a large difference in reaction rate for cotton and pulp, and the mercerized or regenerated celluloses made from them. Higher yields of sugar can be obtained by selecting celluloses with a high hydrolysis rate, or by modifying cellulose in such a way as to accelerate the rate of hydrolysis. With improved knowledge of the mechanisms involved, it might be possible to reduce the rate of decomposition of glucose. Conditions or treatments which affect the heat of activation of the hydrolysis and the decomposition reactions can have particularly powerful effects.

In an effort to achieve a useful modification, cotton linters and a softwood high-alpha pulp were irradiated with high-energy cathode rays. In Figure V, it will be seen that random breakage of bonds reduced the degree of polymerization, destroyed some carbohydrates, and reduced the crystallinity of cellulose to achieve higher net yields of sugar (11). The benefits obtained, however, were not warranted by the costs.

Extreme grinding of cellulose in a ball mill has been shown to accelerate the enzymatic hydrolysis of lignocellulose and celluloses by dilute acid or by enzymes (12), but the cost is excessive.

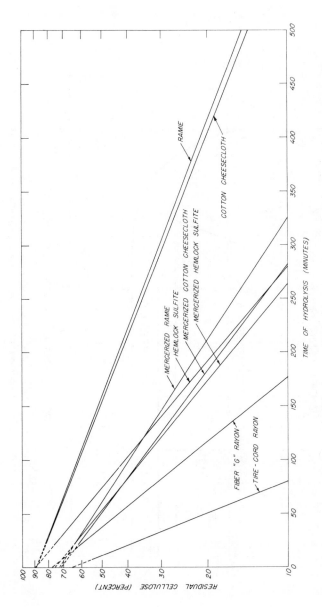

Figure 4. Relation of time of hydrolysis to residual cellulose (rate of hydrolysis is proportional to slope) (10)

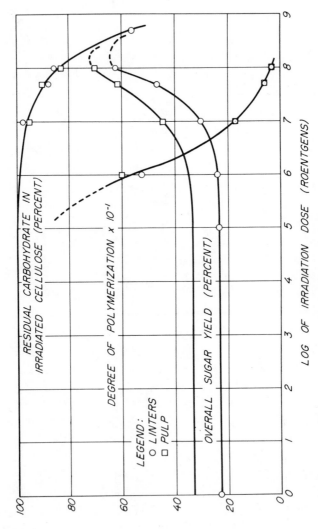

Figure 5. Overall sugar yield obtained by hydrolysis, degree of polymerization, and content of undecomposed carbohydrate as functions of irradiation dose (11)

Spano reported that cellulose pretreated by roll-milling with an energy input of 0.25 kilowatt per hour per pound was enzymatically hydrolyzed to the extent of 45% in 24 hours. The results of this were said to be encouraging (13).

ENZYMATIC HYDROLYSIS OF CELLULOSE

There is much excellent basic work underway on the enzymatic hydrolysis of cellulose. The key factor is the pretreatment required by lignocellulose before it is acted upon at an acceptable rate. In partially delignified pulps, the rate of hydrolysis rises with decreasing lignin content, but pulping the substrate is impractical.

Pretreatments mentioned previously include irradiation, ball-milling, and roll-milling. Some success in upgrading ruminant fodder has been achieved with steaming, treatment with dilute alkali, and digestion with aqueous sulfur dioxide solutions.

Intensive work directed to the development of inexpensive solubilization methods for increasing the yield of sugar from lignocellulose by acid or enzymatic hydrolysis is in progress at Purdue University. A recent report showed that cadoxen is effective but cadmium presents unacceptable hazards (14). An approach in which prehydrolyzed lignocellulose is treated with 70% sulfuric acid followed by methanol has also been reported (14). The flow chart calls for recycling 2.5 parts of sulfuric acid and 5.5 parts of methanol for each part of sugar produced. Another approach (15) calls for the use of a solvent consisting of sodium tartrate, ferric chloride, sodium sulfite, and sodium hydroxide. Such treatments increase the accessibility of enzymes to cellulose and increase the rate of acid hydrolysis, the k_1:k_2 ratio, and the yield of sugar. The cost of recovering the reagents, however, is a key factor. While there is widespread enthusiasm for the enzymatic hydrolysis of lignocellulose and many pretreatments facilitate the reaction, published process data do not permit economic assessment.

CONCLUSIONS

The outlook for fuels and chemicals from biomass is clouded because it depends on the outlook for energy. The most probable outlook for energy in the next decade is that it will be expensive but available, and consumption will continue to increase. While this situation may not constitute a crisis, it is "crisis prone." This latter fact justifies a careful consideration of biomass as a source of fuels and chemicals.

The price of biomass is critical in determining its potential. Because of lack of demand, current spot prices are often low. The price of a large quantity of wood (a thousand or more tons/day) for the life of a plant (decades) is not low. Katzen in 1975 (7) assumed that a large quantity of wood will have a minimum value of $24/ton (dry basis) set by its fuel equivalent, and the additional cost of assuring long-term supply would raise this to $36/ton. The cost of wood, together with labor and capital costs, will of course rise as the cost of petroleum rises.

The best use of biomass for all purposes requires realistic, discriminating, and selective setting of priorities.

Wood contributes most to our energy budget when it is used for the production of structural and fiber products, thus providing alternatives to energy-intensive materials. A ton of wood in the materials system contributes indirectly many times as much to our national energy budget as a ton of wood in the fuel system. The relative economic importance of forest products will increase as the energy shortage worsens, and as forest products industries accelerate their substitution of low-grade wood for fossil fuel.

After satisfying the needs for an energy self-sufficient wood industry, there could still be available some hundred or hundreds of millions of tons of biomass in the form of wood, together with a similar quantity of agricultural residues. The large-scale chemical conversion of such raw material will probably involve hydrolysis, but no available technology is considered economically viable. The key to future progress lies in basic studies, and an evolutionary approach to possible large-scale integrated chemical utilization. Cellulose hydrolysis by means of enzymes, strong acid, or dilute acid is preferably preceded by a prehydrolysis to separate easily hydrolyzed hemicellulose constituents. As an evolutionary step, prehydrolysis itself might be a viable route for a limited quantity of special products.

Prehydrolysis

Prehydrolysis is a simple and well-known step requiring little further basic studies, but there are solvable technical problems in obtaining the prehydrolyzates in favorable concentration.

Prior to the establishment of a fully integrated hydrolysis operation, there might be opportunities to produce and put to use hardwood or crop-residue prehydrolyzates. There are now in prospect boiler plants which will burn over 1,000 tons of wood/day. Preceding combustion, wood can be prehydrolyzed to yield a stream rich in pentoses for subsequent conversion to xylose, xylose derivatives, furfural, yeast, or other feed and fermentation products (16). The residual wood, about three-fourths of the incoming weight, can then be burned for process steam and additional steam as a coproduct. Such an operation would be an evolutionary step toward an integrated hydrolysis plant.

Pretreatment for Enzymatic Hydrolysis

While there has been much progress in the study of cellulases, the applications of such technology have been limited by a lack of economical pretreatment of the lignocellulose. Without such pretreatment, hydrolysis is slow and incomplete. The value of enhanced enzymatic attack on lignocellulose is not limited to the production of sugar and chemicals. The same procedures would be applicable to the increased digestibility of coarse fodder by ruminants.

While government-supported research now emphasizes the production of liquid fuels from biomass, commercialization might be reached sooner by cooperation with those interested in cellulose digestion by ruminants. Experience in the practical upgrading of coarse fodder would be directly applicable to the hydrolysis of biomass by cellulases.

The Strong Acid Hydrolysis of Cellulose

The high yield and hence higher purity of sugar obtained by strong acid hydrolysis of cellulose makes it an attractive process, but the lack of a recovery system for strong acid complicates the outlook. The improvement of membrane technology will probably proceed because of potential applications to many problems. Application to cellulose hydrolysis adds justification for intensified work in the field.

The Dilute Acid Hydrolysis of Cellulose

The rapid, high-temperature hydrolysis of cellulose seems to get less attention than it deserves. The sequence involved is simple; the incoming wet material need never be dried. The consecutive first-order reactions involved can be studied with adequate precision in very simple equipment. While the outlook for the process, based on meager presently available data, is marginal, studies can be conducted to increase the ratio of the rate of sugar production to destruction and hence the yield, to decrease the temperature and pressures involved, and to increase the recovery of coproducts.

REFERENCES

1. Sherrard, E. C.; Kressman, F. W. *Ind. Eng. Chem.* **1945,** *37,* 5.

2. Wenzl, H.F.J. "Chemical Technology of Wood"; Academic Press: New York, 1970.

3. Luers, H.Z. *Angew. Chem.* **1930,** *43,* 455; **1932,** *45,* 369.

4. Saeman, J.F. *Ind. Eng. Chem.* **1945,** *37,* 43.

5. Kirby, A. M. M.S. Thesis, University of Wisconsin, 1949.

6. Grethlein, H. E. "Proceedings", the Second Annual Fuels From Biomass Symposium sponsored by Rensselaer Polytechnic Institute, June 1978; Rensselaer Polytechnic Institute: Troy, N.Y., 1978; Paper No. 26.

7. Katzen, R., Associates. NTIS Accession No. PB 262 489. Natl. Tech. Inf. Serv., Springfield, Va. 1975.

8. Saeman, J. F. "Symposium Papers", Clean Fuels from Biomass and Wastes Symposium sponsored by the Institute of Gas Technology, Orlando, Florida, 1977. Institute of Gas Technology: Chicago, 1977.

9. Gregor, H. P., and Jefferies, T. W. *Annals of New York Academy of Sciences,* **1979,** pp 273-87.

10. Millett, M. A.; Moore, W. E.; Saeman, J. F. *Ind. Eng. Chem.,* **1954,** *46* (7), 1493.

11. Saeman, J. F.; Millett, M. A.; Lawton, E. L. *Ind. Eng. Chem.* **1952,** *44* (12), 2848-52.

12. Millett, M. A.; Effland, M. J.; Caulfield, D. F. *Adv. Chem. Ser.,* in press.

13. Bungay, H. R. and Walsh, T. J., Eds. *Fuels from Biomass Fermentation Newsletter.* Rensselaer Polytechnic Institute: Troy, N.Y., April and July 1978.

14. Ladisch, M. R.; Ladisch, C. M.; Tsao, G. T. *Science* **1978,** *201,* 743.

15. Tsao, G. T. "Proceedings", the Second Annual Fuels from Biomass Symposium sponsored by Rensselaer Polytechnic Institute, Troy, N.Y., June 1978; Rensselaer Polytechnic Institute: Troy, N.Y., 1978; Paper No. 30.

16. Harris, J. F. "Applied Polymer Symposium"; John Wiley & Sons, Inc.: New York, 1975; Vol. 28, pp 131-44.

RECEIVED JUNE 18, 1980.

10

Perspectives on the Economic Analysis of Ethanol Production from Biomass

HARRY J. PREBLUDA and ROGER WILLIAMS, JR.[1]

Roger Williams Technical And Economic Services, Incorporated,
P.O. Box 426, Princeton, NJ 08540

Our objective is to clarify the many misapprehensions pertaining to the practical use of ethanol from biomass to extend motor fuel. While one might agree that biomass can be a renewable resource for non-polluting safe fuels, press reports have questioned the practicality of making large quantities of ethanol as a fuel source not only from rice, cereal grains or tuberous roots, but also from agricultural by-products, such as molasses, timber wastes, cheese whey, pineapple, waste paper or even garbage. From the many articles, there has appeared a potpourri of facts and fallacies (1,2). At all levels of our Federal and State Governments as well as within the automotive and petroleum industries, people have taken sides on the alcohol question.

Opposing views are often voiced within the same organization. A leading university economics professor disagreed with the engineering department and questioned the feasibility of alcohol for transportation fuel (3). However, we want to take a neutral position and point out some of the pitfalls in the thinking on this subject. Large-scale usage may some day correct the present day economic inequities. This will come from new breakthroughs in fermentation and engineering technology to increase yields and reduce costs.

[1] Deceased

0097-6156/81/0144-0199$05.00/0
© 1981 American Chemical Society

Motor Fuel Alcohol

In 1894, Professor Hartman at the Laboratory of the German Distillery, Deutschen Landwirtshafts-Gesellschaft, Leipzig, Germany, was among the first to use alcohol as a fuel in competition to petroleum. In later years, scientists in other countries found that alcohol/fuel blends have shortcomings. Storage tank moisture can contaminate the mixture so that there is alcohol and gasoline separation. Under certain conditions, impurities in denaturing agents accelerate corrosion. Methanol is a denaturant for ethanol in many countries and usually can be used without interfering with the effectiveness of fuel systems. In some parts of the world, impurities in local-area gasoline can react with small amounts of water in the alcohol to accelerate galvanic action on fuel systems with dissimilar metals.

Henry Ford was once questioned as to what would happen to his automobile "buggy" business if petroleum supplies should dwindle. He said, "We can get fuel from fruit, from that of sumac by the roadside or from apples, weeds, sawdust — almost anything. There is fuel in every bit of vegetable matter that can be fermented. There is enough alcohol in a year's yield of potatoes to drive the machinery necessary to cultivate the field for a hundred years...And it remains for someone to find how this fuel can produced commercially — better fuel at a cheaper price than that we now know."

Ford hosted the Dearborn Conferences of Agriculture, Industry and Science in the 1930's. This was the beginning of the Farm Chemurgic Council which pioneered the use of renewable resources as industrial raw materials. Fermentation alcohol for motor fuel was a major topic at the Dearborn Conferences (4). A plant at Atchison, Kansas soon followed in October 1936 with the first attempt to market an alcohol/gasoline blend in the United States. During 1938-1939, twenty million gallons of alcohol/gasoline blends were sold through independent dealers and farm bureaus in ten western and midwestern states (5). Nebraska alone had as many as 250 dealers. At about this time, American auto makers were shipping vehicles and tractors to the Philippines with special engines designed to use alcohol from sugar cane. This equipment became popular in the Philippines when gasoline prices were too high.

Auto engines burning straight alcohols are prone to poor cold weather starting. Much background has been built up for over 50 years on both methanol and ethanol by racing car enthusiasts using these blended fuels in their special engines (6,7). Only recently has there been serious thought to changing the design of the automobile engine for handling either methanol or ethanol. Prestart heating of some kind may have to be used to overcome the sluggish performance of special engines for these fuels in cold climates.

Gasohol in Brazil

Currently, Brazil is leading the World in trying to decrease dependence on imported fossil fuels. A bold government program there centers around the idea of constructing 170 fermentation plants and distilleries. After working on alcohol for over 50 years, Brazilian research had reached a standstill until the rise of oil prices in 1973. Although the economics on this project are far from favorable as yet, Brazil is going forward to nationalize the "gasohol" movement. The politicians in Brazil feel their farsightedness should pay off when oil prices go up. In the meantime, there are many new jobs for their unemployed. It should be kept in mind that Brazilian alcohol cost would be much higher if our U.S. wage scale were used. Plans are also progressing in Brazil for intensive cultivation of cassava (manioc). More than a dozen new alcohol plants will be built requiring cassava as a raw material. This tuberous root can thrive on poor soil conditions in Brazil with low rainfall and is unlike cane since it can be harvested year round. Stillage by-product from manioc could be used to make methane or a high protein animal feed. Incidentally, cassava, being a perennial, can grow for several years while the roots accumulate starch. It differs from sugar cane in that cassava processing requires some hydrolysis before fermentation. Also, sugar cane decomposes when left in the fields too long. Because of starches and fibers, cassava is more stable to weathering. It has not had much opportunity as yet for genetic improvement.

Brazilian officials have apparently overlooked the possibility of using ethanol or methanol for conversion to hydrocarbons such as gasoline using the Mobil process(8). This process can convert ethanol to gasoline directly, thus allowing the use of hydrocarbon fuel from renewable resources in existing cars without engine modification. This scheme would not require a separate ethyl alcohol/gasoline blend distribution system.

Biomass Not Complete Answer

The opinions of farmers, legislators and the public on the use of biomass from alcohol have been debated and the oil companies have had a difficult course to steer(9). It is not generally realized that oil companies get into practically all facets of energy. In addition to coal and petroleum developments, they are in other activities such as solar, wind, atomic, tidal and renewable energy sources. They also want to know how alcohol can be best used as a source of energy. Some American petroleum companies have taken a long-range view and made breakthroughs for high alcohol yields from cellulosic wastes using special fermentation technology.(10)

The U.S. Department of Energy has been evaluating the pros and cons of alcohol fuels (11). DOE's main conclusion has been that both ethanol and methanol can contribute to the energy resources of this country by extending liquid fuel supplies. DOE appears to be commited to developing alcohol fuel. If nationwide market penetration of alcohol fuels takes place, present major fuel suppliers will undoubtedly have to participate. Over the long term it is expected that methanol will offer lower cost possibilities than ethanol. Although the U.S. Department of Agriculture vigorously supports the development of gasohol in the U.S., it urges Congress to be wary of proposals that would commit huge amounts of U.S. feed grains to gasohol (12).

Mention should be made of our concern about distribution costs. The DOE position papers gloss over distribution costs of alcohol/gasoline fuels. Also overlooked is the possible economic gain by going from coal to synthesis gas to methanol to gasoline using the Mobil process for the latter route (8). This could avoid the problem of new storage facilities and get around the need for another fuel distribution system.

Over 90% of the present U.S.A. non-beverage ethyl alcohol production comes from petroleum or natural gas-derived ethylene synthesis. Less than 10% of the remaining alcohol market comes from fermentation of grains, fruit, and sulfite liquors. Using a round number figure of 100 billion gallons of gasoline per year, the industrial ethanol production in the U.S.A. amounts to about one-third of one percent of motor fuel used by vehicles on the highway. It has been estimated that if *all* the available farmland were used for growing agricultural crops in excess of those needed for food production, the ethyl alcohol produced from these renewable crops and residues would meet only 8% of our nation's liquid fuels energy needs for transportation. Unless ethyl alcohol from biomass is subsidized for political reasons or for national security purposes, it will not be the fuel of choice to be used in large quantities (13,14). Of course, there will be breakthroughs in improved crop yields, processing time and other energy savings. Limited use will take place in local geographic areas where the fermentation of off-grade grains to ethyl alcohol can be supported by subsidy, tax credits or loan guarantee. The U.S. has plenty of corn inventory presently because of the excellent 1978 carryover and improved outlook for the 1979 crop. At first glance, it appears that alcohol from fermented grain could leave us less dependent on petroleum supplies. Yet there are cautious meteorologists who expect the drought cycle to hit our country in the next few years. We recall vividly the days of the dust storms and lack of rain in the corn belt.

In a special U.S.D.A. report, it has been questioned whether we could afford to divert sizable quantities of grain for motor fuel purposes (13). It has been estimated that startup costs for plants and distilleries would be between $15-$17 billion for making 100 billion gallons of gasohol blend per year. This does not include a direct added subsidy of $10.4 billion a year to make the product competitive with gasoline prices.

A new method (15) for extracting wheat gluten offers some possibilities for alcohol fermentation of wheat starch by-products in geographic areas where this grain would be advantageous. The new technology called the "Raisio Alfa-Laval" process is described as a "unique closed system for wheat fractionation substantially increasing the yield of high quality starch and vital gluten while permitting pollution-free processing." With high beef prices ahead, the world demand for gluten in human food is expected to increase.

Gasohol and the Beef Industry

The beef producers feel that the gasohol development in the U.S. could hurt its industry (16). A nationwide gasohol program would bring higher prices and higher production costs to be passed along to consumers. There would be an additional cost to producers in the form of higher taxes to pay for the large government subsidies needed for the gasohol program. The beef producers are also worried about regional livestock production shifts. Stockmen would try to relocate near distilleries to be close to a low-cost source of by-product feed. The greater feed use of distillers grains could slow down the livestock cycle. U.S.D.A. researchers think that the high fiber value of distillers feed might require a longer digestive phase and thus push consumer beef prices higher. Overall livestock production would be expected to decrease. Also the expected 10 billion gal/yr subsidized alcohol market would sharply increase feed prices and food grains. Incidentally, where would the subsidies come from? Who would pay for them? The 35 million tons of dried distillers feed grains from the gasohol program each year would depress soybean meal prices in such a way that there would be radical changes in that industry. The soybean crushing industry for animal feed would be supplanted primarily by one producing food oils and special products intended for human use in the export market. Another worry of beef producers — what would happen in the event of short U.S. food crops? Also, is there enough capacity for both food and gasohol production? What is the real answer to these questions?

Land Program Possibility

Jawetz (17) presented an interesting idea as a renewable resource raw material potential for ethanol. It offers the farmer an opportunity to grow specific crops using the millions of acres in the Federal Government "set aside" program in exchange for guaranteed minimum prices of farm crops. Instead of leaving the land fallow and idle, Jawetz has suggested growing a crop that could be used by a distiller to make alcohol. The subsidy would revert to the distiller. Ethanol based on $2.00/bushel of corn would then cost only $0.45/gal at the distillery after allowing the distiller a subsidy credit of approximately $0.54/gal of ethanol from the "set aside" program and diverted land. However, this program might raise some questions by conservationists. They could claim that the constant planting of crops on land would have a tendency to run down the soil unless there was heavy fertilization.

While pondering answers to these questions, let us examine the possibilities of using other raw materials such as sugar cane or cane molasses for ethanol production. We know that of the many types of plant material that can be fermented, the greatest energy yield is obtained when sugar cane is fermented. The sugar cane plant is considered to be one of the most effective to fix solar energy. If ethanol is to be made from a cane source, it must be produced in large quantities from available land to make it competitive and not have constraints with other competing crops. In the future, the economics can be improved by fermenting the hydrolyzed bagasse fiber if it is not used directly as fuel. It has also been suggested that total fermentable material in cane juice be made into alcohol without crystallizing sucrose.

Whole Cane Process

Rolz of Guatemala (18) reported on the "Ex-Ferm" process whereby ethanol is fermented directly from small pieces of whole chopped cane. The accumulation of ethyl alcohol during the fermentation leaches out other non-sucrose components and breaks down the solid fibre matrix of the cane. The simultaneous extraction and fermentation provides many advantages for the cane grower to become an alcohol producer. Rolz projects a manufacturing cost of $0.643/gal of alcohol vesus $0.795/gal for the conventional standard technology exclusive of charges for return on original investment.

Sugar Market Implications

Vacillation and mistrust in Washington led to the plummeting of sugar prices prior to August 1978. The combination of record world stocks of 30 million tons of sugar and low prices at the time stimulated the thinking of government officials in cane-producing, energy-beleaguered countries to be more favorably disposed to make alcohol for fuel or gasohol programs. Another factor has entered the picture for countries basic in cane sugar and molasses production. The large-scale development in the U.S. of high fructose corn syrup (HFCS) using glucose isomerase to prepare sweeteners from corn starch is having a dramatic effect on cane sugar markets. HFCS will probably replace most of the cane sugar we now import. However, it will not replace the beet and cane sugar we now grow in the U.S. Indirectly, this will release cane acreage especially in the developing nations which import petroleum (gasoline). These nations are now thinking of alcohol production from their cane sugar or cane molasses. Currently, sugar production is shrinking in the U.S. The 1979 crop is expected to be down by as much as 500,000 short tons and Congress is expected to raise support prices.

Attention to By-Products

Greater attention is presently being given to some of the by-products of sugar processing and fermentation. With rising fertilizer and labor costs, the sugar cane growers in the State of Sao Paulo, Brazil have been returning filter-press cake to their growing fields. Several mills have decreased usage of organic fertilizers when using filter-press cake. Brazilian environmental protection laws have recently prohibited alcohol producers from dumping distillery slops or wastes into the rivers which they have been doing for some time. Applying these wastes to the growing fields and irrigation systems is proving beneficial. In Europe, both beet and cane fermentation residues are evaporated to approximately 65% dry matter and are referred to as "vinasses" (19). Western European production of these materials in 1978 was close to 680,000 tons. Alcohol fermentation vinasses have been used in Europe for animal feed because of appetite-stimulating properties. There has also been a large-scale post-harvest use of vinasses on fields.

Prior to World War II cane residues from U.S. alcohol fermentation were incinerated to make potash for fertilizer use. Both during and after World War II, millions of pounds of dry and also condensed molasses fermentation products were profitably sold for speciality industrial use outside of the feed industry. These markets have been neglected and could readily be established again on an economic basis to reduce overall ethanol production costs from cane or beet molasses.

In looking at the fuel farming picture, one wonders just how much time any of these processes could buy in relation to the life of our fossil fuel supply. Biomass conversion to fuel does not appear to be a real immediate answer to our long-term energy problem because of the costs of the raw material; nevertheless, some breakthroughs appear on the horizon. The need for a year-round supply of raw material necessitates using cellulose from agricultural, forest or municipal solid wastes as the feedstock source. The high cost of ethanol from grain or sugar cane is due primarily to raw material which represents about two-thirds of alcohol production costs. In the U.S., there are tremendous tonnages of cellulosic materials derived from various urban and industrial organic wastes which can be obtained at low cost. To use this biomass for fermentation, it is most important to have it collected and available at a central location. Biomass enthusiasts often overlook the high cost of bringing these types of raw materials to the production plant.

Process Possibilities

Significant breakthroughs have recently taken place which will improve the economic picture for making alcohol from solid waste. Rutgers University research (20) on thermotolerant mutant strains of organisms producing cellulase of a higher order than reported heretofore should be mentioned. Important work on vacuum technology to overcome factors of alcohol inhibition limiting fermentation efficiency has been reported by Cysewski and Wilke (21) as well as Ramalingham and Finn (22).

Cysewski and Wilke suggested some process design fermentation schemes for continuous cell recycle and also vacuum fermentation processes for making 75,000 gallons of 95% ethanol/day. Using a reasonable yeast by-product credit of $0.10/lb, their net estimated production cost appeared to be $0.823/gal for the continuous cell recycle system as compared to $0.806/gal for the vacuum fermentation. This compares favorably with the current price of synthetic 95% ethanol selling around $1.40/gal.

Spano (23) at the U.S. Army Laboratories has reassessed the economics of cellulose process technology for production of ethanol using urban waste as a cellulosic substrate. The results of this study are encouraging. With impoved cellulose productivity and by-product credits of $0.54/gal of ethanol, Spano has been able to get the lowest estimated cost down to $0.89/gal of 95% ethanol using a unit cost of $0.11/gal of alcohol for the cellulosic material.

Hoge (24) has suggested a novel ethanol process using steam-sterilized impure cellulosic fractions of solid municipal wastes treated with a mixture of enzyme and yeasts. The combined enzymatic digestion of cellulose to sugars and the fermentation of sugars to alcohol takes place in a common reaction vessel. The system has several reactors operating at atmospheric pressure. Each reactor is intermittently connected to a shared recirculation system having a flash chamber from which alcohol is distilled from the reaction mass. The reaction temperature of the fermentation is also controlled by the vacuum. The unique design of this process extends the life of the enzyme and reduces cost. Repeated reuse of the enzyme also makes the process attractive. Enzyme cost has been holding back commercial-scale cellulose digestion. Ethanol cost in the Hoge process is estimated around $0.40/gal, exclusive of selling expenses, administrative costs and profits based upon an annual production of 32.7 million gallons of ethanol. Recent unpublished data from Cornell University showed that vacuum removal of alcohol from the Hoge process improved yields per unit of cellulase enzyme used.

Another noteworthy development is going on at Purdue University. A combination of solvent pretreatment and enzyme hydrolysis of cellulose is used to obtain a very high yield of sugar for the production of alcohol (25). This opens great opportunities for making use of the one billion tons of cellulosic residues available each year from cornstalks, wheat straw, sugar mill bagasse, sawmill rejects, packaging residues, logging residues, animal feedlot wastes, city trash, etc. There is no question that the diminishing supplies of our non-renewable materials will accentuate the demand for resources that are renewable. We should be practical in our thinking. Let us remember that if *all* the corn grown in the World and *all* the wheat and *all* the other world crops grown by farmers were to be converted into alcohol, it would be six to seven percent of the energy equivalent of the world crude oil production (Table I). The calculations in Table I are based on the assumption that upon fermentation, approximately half the energy of grain is converted to ethanol. This is exclusive of natural gas or coal energy sources.

The idea of renewable sounds great until you put it into perspective and look at the economics (Table II). Most ecologists just don't. The economic disparity of fermentation alcohol as fuel at the service station pump compared to gasoline at today's prices of around $1.00/gal is self-evident without even going into Btu or performance ratings.

Yet in the midst of this pessimism, there is some hope to at least make use of industrial and urban waste material to extend our fuel supplies (26).

We feel that our greatest opportunities may yet come from breakthroughs involving the cross disciplines between the chemist, engineer, agricultural engineer, microbiologist, geneticist, biochemist, agronomist and last but not least, the *economist.*

As Winston Churchill once said, "We have not reached the beginning of the end, but perhaps we have reached the end of the beginning."

Table I. ANNUAL WORLD GRAIN SUPPLY ALCOHOL EQUIVALENCE AND CRUDE OIL USE

CRUDE OIL   1975 19.5×10^9 barrels
5.8×10^6 Btu/barrel

      Total 113.1×10^{15} Btu's
113.1 Quads

CORN   1975   324,721,000 metric tons (358×10^6 short tons)
380 Kcal/100 grams   6859 Btu/lb

WHEAT   1975   355,985,000 metric tons (392×10^6 short tons)
330 Kcal/100 grams   5957 Btu/lb

RICE   1975   348,374,000 metric tons (384×10^6 short tons)
370 Kcal/100 grams   6679 Btu/lb

SOYBEAN 1975   68,900,000 metric tons (76×10^6 short tons)
360 Kcal/100 grams   6498 Btu/lb

Btu = .25198 Kilogram — calorie   3.97 Btu/.22 lbs.
Pounds = 454 grams   18.05 Btu/lb

CORN 716×10^9 lbs. $\times 6.859 \times 10^3 = 4.9 \times 10^{15}$ Btu's = 4.9 quads

WHEAT 784×10^9 lbs. $\times 5.957 \times 10^3 = 4.7 \times 10^{15}$ Btu's = 4.7 quads

RICE 768×10^9 lbs. $\times 6.679 \times 10^3 = 5.1 \times 10^{15}$ Btu's = 5.1 quads

SOYBEAN 152×10^9 lbs. $\times 6.498 \times 10^3 = .9 \times 10^{15}$ Btu's = .9 quads
                                     15.6 quads

$15.6 \times 0.50 = 7.8$ quads = alcohol equivalent of WORLD GRAINS
$\dfrac{7.8}{113.1} = 6.9\%$ of CRUDE OIL USE

Table II. FERMENTATION ETHANOL AS FUEL

¢/GAL	CORN	WOOD WASTE	SUGAR CANE	CASSAVA
Net Raw Material Cost	44	68	56	68
Processing Cost	30	40	20	33
Capital Charges	40	82	30	36
SUB-TOTAL	114	190	106	137
Distribution	5	5	5	5
Dealer Margin	8	8	8	8
PUMP PRICE (excluding taxes)	127	203	119	150

REFERENCES

1. Anderson J. "Gasohol Gasohowl", *N.Y. Daily News*, July 5, 1978.

2. Klass, D. L. "Symposium Papers", Energy from Biomass and Wastes Symposium sponsored by the Institute of Gas Technology, Washington, D.C., Aug 1978; Institute of Gas Technology: Chicago, 1978.

3. Anderson, E. V. *Chem. Eng.* **1978,** *56* (31), 8-12, 15.

4. Reese, K. M. *Chem. Eng. News* **1979,** *57* (35), 56.

5. Hind, J. D. *Chem. Eng. News* **1978,** 56 (35), 43.

6. Williams, Roger, Jr. "Methanol — The Energy Chemical", presented to the World Trade Institute and Chemurgic Council on Renewable Resources, New York, May 1976.

7. Williams, Roger, Jr. "Methanol — Markets vs. Price". Testimony before State Affairs Committee, Alaska State Senate, Juneau, Alaska, 1976.

8. Smay, V. E. *Popular Science* **1978,** *June,* pp 90-91; Weisz, P. B.; Marshall, J. F. "High Grade Fuels From Biomass — Analysis of Potentials and Constraints", International Symposium On Energy and Technology, International Association of Science and Technology for Development (I.A.S.T.E.D.), Montreux, Switzerland, June 19-21, 1979; Weisz, P. B.; Marshall, J. F. "Fuels From Biomass: A Critical Analysis of Technology & Economics"; Marcel Dekker: New York City, in press.

9. Pratt, H. T. *Chem. Eng. News* **1978,** *56* (41), 2,55.

10. Gauss, William F.; Suzuki, S.; Takagi, M. U.S. Patent 3990944, 1976; Huff, George F.; Yata, N. U.S. Patent 3990945, 1976; *Chemical Marketing Reporter* **1979,** *216* (8), 7.

11. United States Department of Energy, "Position Paper on Alcohol Fuels", March 1978; United States Department of Energy, "The Report of the Alcohol Fuels Policy Review", June 1979.

12. *Milling and Baking News* **1979,** *58* (28), 59.

13. United States Department of Agriculture, Economics, Statistics & Cooperatives Service, "Gasohol From Grain . . . The Economic Issues". Prepared for the Task Force on Physical Resources, Committee of The Budget, U.S. House of Representatives, Washington, D.C., January 19, 1978.

14. Klosterman, H.J.; Banasik, O. J.; Buchanan, M. L.; Taylor, F. R.; Harrold, R. L. *Farm Res.* **1978,** *35* (2), 3-8.

15. Kerkkonen, H. K.; Laine, K. M. J.; Alanen, M. R.; Renner, H. V. U.S. Patent 3951938, 1976.

16. Richter, J. *Beef* **1978,** *72.*

17. Jawetz, P. "Symposium Papers", Energy From Biomass and Wastes Symposium sponsored by the Institute of Gas Technology, Washington, D.C., Aug 1978; Institute of Gas Technology: Chicago, 1978; 14th Inter-Society Energy Conversion Engineering Conference, Boston, Mass. Aug 1979; American Chemical Society, Division of Petroleum Chemistry, Washington, D.C., September, 1979.

18. Rolz, C. Meeting of the American Chemical Society, Miami, September 1978. American Chemical Society: Washington, D.C., 1978.

19. Lewicki, W.*Process Biochemistry* **1978,** *78* (13), 12.

20. Montenecourt, B. S.; Eveleigh, D. E. "Hypercellulolytic Mutants and Their Role in Saccharification"; Rensselaer Polytechnic Institute; Troy, N.Y., June 20-21, 1978.

21. Cysewski, G. R.; Wilke, C. R. *Biotechnol. Bioeng.* **1977,** *19,* 1125; **1978,** *20,* 1421.

22. Ramalingham, A.; Finn, R. K. *Biotechnol. Bioeng.* **1977,** *19,* 583.

23. Spano, L. "Revised Economic Analysis of Cellulose Process Technology", *Fuels From Biomass Fermentation Newsletter;* Rensselaer Polytechnic Institute, Troy, N. Y., July 1978, Item 4, p 9.

24. Hoge, W. H. U.S. Patent 4009075, 1977; Finn, R. K. to Hoge, W. H., private report on laboratory results, 1979; Hoge, W. H. to Prebluda, H. S., private communication, 1979.

25. Ladisch, M. R.; Ladisch, C. M.; Tsao, G. T. *Science* **1978**, *201*, 743.

26. Williams, Roger, Jr.; Rains, W. A.; Prebluda, H. J. "Is Solid Waste a Viable New Jersey Raw Material?" New Jersey Academy of Science: Lawrenceville, N. J., April 7, 1979.

RECEIVED JUNE 18, 1980.

Chemicals from Biomass by Improved Enzyme Technology

GEORGE H. EMERT[1]

Gulf Oil Chemicals Company, P.O. Box 2900, Shawnee Mission, KS 66201

RAPHAEL KATZEN

Raphael Katzen Associates, 1050 Delta Avenue, Cincinnati, OH 44208

PART I - CELLULOSE TO ETHANOL PROCESS
Introduction

Gulf's biochemical research program began in 1971 with a search for alternate feedstocks for petrochemicals manufacture. Practically all of the organic chemicals industry relies on chemicals derived from fossil fuels — petroleum, natural gas or coal. We had become concerned even before the Arab embargo that increasing prices of petroleum and petroleum feedstocks would eventually make many of the materials now supplied by the chemicals industry too expensive for general use. It was becoming imperative to have alternate feedstock sources available.

Subsequent research led to the construction of a cellulose-to-ethanol pilot plant at Pittsburgh, Kansas. This pilot plant has a capacity of one ton of feedstock per day and has been in operation since January, 1976. Information gained from work done in our Merriam laboratories and the operation of the pilot plant has led us to recognize that renewable resources in the form of carbohydrates are an excellent source of a variety of chemicals now derived from fossil fuels.

Our plans for this technology in the very near future include the process design, procurement, fabrication, construction and operation of a 50-ton/day cellulosic waste conversion facility, followed by a commercial scale facility

[1] Current address: 415 Administration, University of Arkansas, Fayetteville, AR 72701

0097-6156/81/0144-0213$05.00/0
© 1981 American Chemical Society

utilizing 2,000 tons/day of cellulosic feedstocks to be operational by 1983. This bioconversion technology addresses itself to two important national problems. One is the removal or decrease in volume of solid waste. Secondly, it provides ethyl alcohol and other chemicals which can be utilized to supplant fossil fuel sources as a feedstock.

Direct Cellulose to Ethanol Process

The cellulose-to-ethanol process has five basic steps as shown in Figure I. They are: feedstock handling and pretreatment, enzyme production, yeast production, simultaneous saccharification/fermentation (SSF) and ethanol recovery. Cellulose is the most abundant organic material on the earth. It is annually renewable, and not directly useful as a foodstuff. It is a polymer of glucose linked β-1,4 as compared with the α-1,4 linked polymer starch which by contrast is easily digestible by man. There are three basic classes of potential cellulose feedstocks. These are agricultural by-products, industrial and municipal wastes, and special crops. The availability of these materials in the U.S. is shown in Table I. For economic reasons, we are concentrating our efforts on those materials that are collected for some other reason.

With respect to conversion to ethanol or other chemicals, there are four major factors to consider regarding the susceptibility of native cellulose to biodegradation. These are: its insolubility in water, particle size, extent of lignification, and crystallinity. Lignin, a polyphenolic, cement-like material, is found closely associated with naturally occurring cellulose, and not only inhibits access to cellulose linkages but in some cases is toxic to microorganisms. Lignin presence in kraft or sulfite pulped cellulosics is not a problem using our system. The crystal structure of cellulose has finite physical measurements which disallow the cellulolytic enzymes access to the β-1,4 linkage.

A wide variety of potential feedstocks has been tested at the pilot plant level. These include cotton gin trash, clarifier sludges, digester fines, digester rejects, straw, bagasse, and municipal solid waste. Several microbial, mechanical, and chemical pretreatments have been tested. Microbial pretreatments include: ligninase-producing organisms and xylanase-producing organisms. Mechanical pretreatments tested include hammer mills, rod mills, roller mills, ball mills, mullors, and attritors. Chemical pretreatments tested include acid, base, cadmium oxide, dimethyl suffoxide, ethylenediamine, tartrate, etc. Obviously, there are specific particle sizes, degrees of moisturization, and lignin contents which are optimum for conversion of the cellulose to chemicals.

Characterizations of potential feedstocks have been accomplished by measuring percent cellulose, lignin, ash, and acid detergent-soluble materials as shown in Table II. As an example, the compositions of pulp and paper wastes show approximately a 55% average cellulose content. Most of these materials have been partially delignified and thus show a moderate lignin content. High ash content would be a detriment for processing purposes since it is weight which has to be moved through the process along with substrate, thus using excess energy. Pretreatment of feedstock for subsequent processing may include mixing in water to obtain a uniform mixture and moisturization followed by either pasteurization or sterilization as necessary.

The second step in the direct ethanol process is that of enzyme production. The Gulf process utilizes a mutant strain of *Trichoderma reesei*, grown continuously to produce a complete cellulase system. The residence time is 48 hours. Enzyme production begins on a spore plate with subsequent scale-up to the enzyme production vessel size to be used. Our pilot plant facility has 300-gal enzyme reactors.

The cellulase system consists of β-1,4-endoglucanase, a cellobiohydrolase and a β-glucosidase. These activities are measured collectively for their ability to enzymatically catalyze the degradation of crystalline cellulose to glucose. The triple enzyme system does not have to be isolated or purified, but is used as a whole broth.

The third step in the direct ethanol process is to produce the yeast necessary for converting the glucose to ethanol. Varieties of organisms have been screened for compatibility with the *Trichoderma reesei* cellulase system. The optimum temperature for the cellulase system is 45° to 50°C. Most yeasts have a temperature optima less than that, e.g., 30° to 35°C. We have tested a variety of strains, including *Saccharomyces cerevisiae, S. carlsbergensis,* and *Candida brassicae.* A temperature of 40°C has been identified as optimum for the combined cellulase and yeast systems.

The fourth step in the direct ethanol process is considered to be key to the economic viability of the bioconversion of cellulose to chemicals. The pretreated cellulose slurry is simultaneously converted to glucose and the glucose to ethyl alcohol in the same vessel in a continuous or semi-continuous mode. The enzyme sample is the whole culture from the enzyme production vessel. The feedstock is a slurry of 7.5% to 15% cellulose. The yeast is either added as a cake or recycled as a cream.

When a comparison is made of simultaneous saccharification/fermentation, and saccharification alone with subsequent conversion to ethyl alcohol of the glucose, one finds a considerable enhancement of ethanol production as shown in Figure II. This enhancement amounts to a 25% to 40% increase in yield and is due to the removal of products formed during saccharification which inhibit the cellulase system. Glucose and cellobiose are feedback inhibitors of enzymes in the cellulase system. In SSF, when glucose is removed as fast as it is produced, there is no product inhibition; thus the enzymes function more efficiently. Figure III shows the projected ethanol production when paper and pulp mill waste is reacted with a standard amount of enzyme as protein which coincides with a 10% v/v inoculum of the enzyme production material. It is a simple matter to raise the conversion to 90% with slightly more protein, for example using a 20% v/v inoculum.

The last step in the direct ethanol process is recovery of ethanol. To do this, a slurry stripper was designed in which we could directly pump the mash of SSF into a stripping column and strip the alcohol using an upward flow of steam. Approximately 25% by weight ethanol concentration is obtained in this first step. This material can then be upgraded through a normal rectification system to either industrial grade, motor grade, or pharmaceutical grade ethyl alcohol.

There are four basic characteristics of fermentation processes which come to mind when comparisons are made with conventional petrochemical processing techniques. These are that fermentation occurs at essentially ambient temperatures and pressures, incurs low corrosion of equipment, inherently involves low reaction rates, and results in low product concentrations. The first two characteristics are advantageous and support using fermentation technology industrially. The second — low rates and low concentrations — are disadvantageous. It is the ability to manage or overcome these which determines whether or not a particular fermentation technology is appropriate for commercial use. Shown in Table III are the combined results indicating progress made in our research laboratories and pilot plant since 1975. The level of accomplishment in enzyme production and SSF ethanol yield shown for 1978 was utilized for the economic evaluation and energy calculations which are included in this paper.

PART II - TECHNICAL AND ECONOMIC EVALUATION

The basic flowsheet for the proposed commercial process is shown in Figure IV.

Table I. LIGNO-CELLULOSE AVAILABILITY
tons/year

Collected

Municipal Solid Waste	140,000,000
Pulp and Paper Mill Waste	3,000,000
Selected Agricultural Waste	1,000,000

Uncollected

Forest Residues	145,000,000
Wood Processing	20,000,000
Agricultural Residues	300,000,000

Table II. COMPOSITION OF CELLULOSICS

Wastes	Cellulose	% Dry Solids Lignin	Ash	Ads
Primary clarifier sludge	48	8	18	26
Secondary clarifier sludge	21	4	55	20
Deinking sludge	29	4	38	29
Super fines	62	6	8	24
Digester rejects	65	13	1	21
Digester fines	64	9	5	22
Municipal solid waste RDF	61	9	8	22
Bagasse	43	14	8	35

Table III. PROGRESS SUMMARY

	1975	1976	1977	1978	Design
Enzyme Production					
Residence Time	14 days	10 days	4 days	2 days	2 days
Enzyme Activity	— —	11%	51%	90%	90%
(Conversion of Cellulose)					
Saccharification-Fermentation					
Residence Time	— —	6 days	2 days	1 day	1 day
Cellulose Concentration	— —	6%	10%	8%	8%
Ethanol Production	— —	1%	1.8%	3.6%	3.6%
(Concentration)					

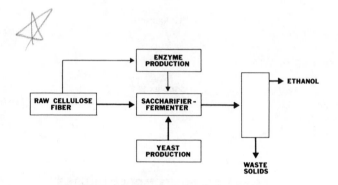

Figure 1.   Direct ethanol process

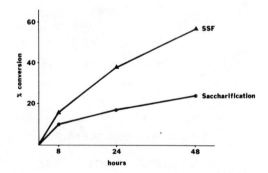

Figure 2.   Comparison of simultaneous saccharification/fermentation (SSF) and saccharification

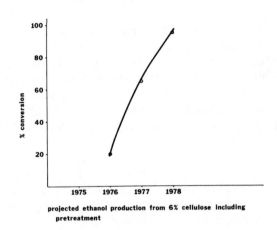

projected ethanol production from 6% cellulose including
pretreatment

Figure 3.   Merriam: ethanol production from pulp mill wastes

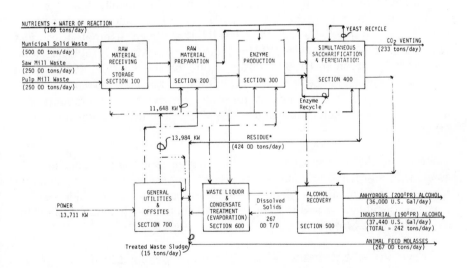

Figure 4.   1000 TPD feedstock

The nominal quantities shown are based on assumed availability of 1,000 tons/day (dry basis) of cellulosic wastes, containing approximately 55% cellulose. From the data studied, projected production of 75,000 gal/day of ethanol (190° pr basis) is estimated. Also, by-product/animal feed in the amount of 267 tons/day would be produced. Inherent in this process is the separation of insoluble solids, the organic content of which, primarily lignin and unconverted cellulose, would become the basic fuel for the plant, providing essentially all of the thermal energy and a major part of the motive (turbine-drive) energy. A mixed feedstock of municipal solid waste, pulp mill waste and sawdust was used as the basis for these balances. This, of course, may be varied regionally, and may include agricultural wastes and residues.

The essentially self-contained process requires only small amounts of fossil fuel, and purchased electric power, which may be generated from fossil fuel. Energy balances based on net fuel value of the raw materials and products (including fuel used to produce electric power) indicate an overall energy efficiency of 51% as indicated in Figure V.

Ongoing investigations have indicated a substantial economic advantage in converting 2,000 tons/day of cellulosic waste to 150,000 gallons of alcohol. This amount of waste material appears to be economically available at a number of locations in the U.S. Such a commercial facility, planned for construction during the period 1980-82, has been estimated to cost approximately $112 million (1981 costs). As indicated in Table IV, total production cost on a 100% investor equity capital basis is $0.70/gal of alcohol. The projected selling price during the first production year, 1983, is $1.44/gal ethanol, after credit is taken for animal feed by-products and allowances are made for taxes and a 15% after-tax return on investment for a 10-year plant life.

Since the facility is disposing of waste materials, and may be considered a solution to a growing national solids waste disposal problem — particularly the need for elimination of landfills — the project may be considered as a waste disposal facility and thereby might be financed to a substantial extent by municipal tax-free bonds.

As shown in Table V, on the basis of 80% municipal bond financing over 20 years for the fixed investment, and equity capital financing of the remaining fixed investment along with borrowed working capital from conventional sources, after taking credit for the animal feed by-product, the total operating cost is $0.59/gallon of alcohol and the projected selling price for a 15% after-tax return on investor's equity capital is $0.95/gallon in the year 1983. The 1983 selling price is atypical because the federal and state taxes are reduced due to previous losses in interest payment prior to startup of the plant.

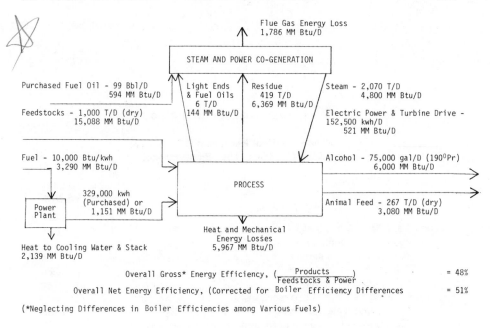

Flue Gas Energy Loss
1,786 MM Btu/D

STEAM AND POWER CO-GENERATION

Purchased Fuel Oil - 99 Bbl/D
594 MM Btu/D

Feedstocks - 1,000 T/D (dry)
15,088 MM Btu/D

Light Ends & Fuel Oils
6 T/D
144 MM Btu/D

Residue
419 T/D
6,369 MM Btu/D

Steam - 2,070 T/D
4,800 MM Btu/D

Electric Power & Turbine Drive -
152,500 kwh/D
521 MM Btu/D

Fuel - 10,000 Btu/kwh
3,290 MM Btu/D

329,000 kwh
(Purchased) or
1,151 MM Btu/D

Power Plant

Heat to Cooling Water & Stack
2,139 MM Btu/D

PROCESS

Alcohol - 75,000 gal/D (190^{0}Pr)
6,000 MM Btu/D

Animal Feed - 267 T/D (dry)
3,080 MM Btu/D

Heat and Mechanical
Energy Losses
5,967 MM Btu/D

Overall Gross* Energy Efficiency, ($\frac{Products}{Feedstocks\ \&\ Power}$) = 48%

Overall Net Energy Efficiency, (Corrected for Boiler Efficiency Differences = 51%

(*Neglecting Differences in Boiler Efficiencies among Various Fuels)

Figure 5. Cellulose alcohol process: energy balance, daily basis

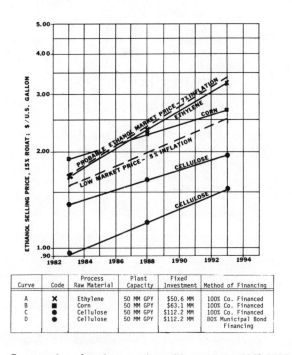

Curve	Code	Process Raw Material	Plant Capacity	Fixed Investment	Method of Financing
A	X	Ethylene	50 MM GPY	$50.6 MM	100% Co. Financed
B	■	Corn	50 MM GPY	$63.1 MM	100% Co. Financed
C	●	Cellulose	50 MM GPY	$112.2 MM	100% Co. Financed
D	●	Cellulose	50 MM GPY	$112.2 MM	80% Municipal Bond Financing

Figure 6. Comparative ethanol economics, selling price for 15% ROIAT

Table IV. CELLULOSE ALCOHOL 2000 T/D

Commercial Plant Investment (1981 Costs)
50 MM USGPY (190° Proof) Ethanol

	1981 $ MM
Section 100 - Raw Materials Receiving & Storage	8.95
200 - Raw Material Preparation	11.65
300 - Enzyme Production	8.34
400 - SSF	16.80
500 - Alcohol Recovery	10.42
600 - Waste Liquor (feed molasses) & Condensate Treatment	13.98
700 - General Facilities and Offsites	
Including Dryer	
Boiler	
Power Generation	
Power Distribution	
Scrubber	
Cooling Tower	
Wells	
Water Treatment	
Offsites	
Office and Labs	
Maintenance & Stores	
Alcohol Storage and Shipping	
Fire Protection	
Yard Piping	
Miscellaneous	
Total Section 700	
	27.46
Total Installations	$ 97.60
+15% Contingency	14.60
Total Investment	$112.20

Table V. CELLULOSE ALCOHOL

OPERATING COST AND SELLING PRICE ESTIMATES (1983) BASE CASE — WITH ENZYME RECYCLE

50 MM U.S. GALLONS/YR PRODUCTION

| | 10-YEAR AMORTIZATION — WITH FEED BY-PRODUCT 100% COMPANY FINANCING | | 20-YEAR AMORTIZATION — WITH FEED BY-PRODUCT 80% MUNICIPAL BOND FINANCING | |
	Annual ($MM)	$/Gallon	Annual ($MM)	$/Gallon
Fixed Charges				
Depreciation (D) Straight Line	11.22	0.224	5.61	0.112
License Fees	0.02	0.000	0.01	0.000
Maintenance 4% of TFI	4.49	0.090	4.49	0.090
Taxes & Insurance 2% of TFI	2.24	0.045	2.24	0.045
Subtotal	17.97	0.359	12.35	0.247
Raw Materials, Chemicals				
MSW — 1,000 OD T/D x 330 D at $14.00/OD T	4.62	0.093	4.62	0.093
SMW — 500 ED T/D x 330 D at $21.00/OD T	3.47	0.069	3.47	0.069
PMW — 500 OD T/D x 330 D at $14.00/OD T	2.31	0.046	2.31	0.046
Nutrients, Chemicals, etc	9.68	0.193	9.67	0.193
Subtotal	20.08	0.401	20.07	0.401
Utilities				
Electric Power 219.4 MM kwh at 0.42 $/kwh	9.23	0.185	9.23	0.185
Fuel 64,000 bbl/yr at $21.00/bbl	1.35	0.027	1.35	0.027
Subtotal	10.58	0.212	10.58	0.212
Labor				
8 — Supervisory staff ($43,750/yr)	0.35	0.007	0.35	0.007
11 — Lab and Office Staff ($27,275)	0.30	0.006	0.30	0.006
58 — Operators, Laborers ($31,900)	1.85	0.037	1.85	0.037
Subtotal	2.50	0.050	2.50	0.050
Total Production Cost	51.13	1.022	45.50	0.910
By Product Credit ($120/ton)	(21.02)	(0.420)	(21.02)	(0.420)
Sales, Freight, G&AO ($0.10/gall)	5.0	0.100	5.00	0.100
Total Operating Costs	35.11	0.702	29.48	0.590
Interest on Working Capital (10%)	0.00	0.000	1.00	0.020
Interest on Loan (65%)	0.00	0.000	5.50	0.110
Previous Taxable Losses Paid Back	0.00	0.000	5.50	0.110
Federal and State Taxes (50%)	18.30	0.367	0.00	0.000
Net Profit After Taxes	18.30	0.367	6.20	0.124
Total Income and Selling Price	71.71	1.436	47.68	0.954

Table VI. SYNTHETIC AND GRAIN ALCOHOL INVESTMENT COSTS AND SELLING PRICES

Basis: 50 MM USGPY (190° Proof) Ethanol
Operating Cost Estimate for 1983

Investment	Synthetic MM$	Synthetic $/Gal	Grain MM$	Grain $/Gal
Total Fixed Investment (TFI)	50.6		63.1	
(For Construction 1980-1982)				
Working Capital	7.2		12.7	
	57.8		75.8	
Fixed Charges				
Depreciation (D), 10% of TFI	5.6	0.112	6.3	0.126
License Fees, (10 yr payout)	0.2	0.004	0.05	0.001
Maintenance, 4% of TFI	2.0	0.040	2.5	0.050
Taxes & Insurance, 2% of TFI	1.0	0.020	1.3	0.026
	8.8	0.176	10.15	0.203
Raw Materials				
Ethylene ($0.18/lb)	37.5	0.750	— —	— —
Corn ($3.00/bu)	— —	— —	60.0	1.200
Other	1.2	0.024	0.05	0.001
Utilities	11.9	0.238	13.9	0.278
Labor	0.9	0.018	2.3	0.046
Total Production Cost (TPC)	51.5	1.030	76.25	1.525
By-product Credit ($120.00/T)	— —	— —	(− 20.0)	(− 0.400)
Sales, Freight, G&AO ($0.10/gal)	5.0	0.100	5.0	0.100
Total Operating Costs	65.3	1.306	71.4	1.428
Additional Charges				
Federal & State Taxes	8.7	0.174	11.4	0.228
Net Profit	8.7	0.174	11.4	0.228
Alcohol Selling Price	82.7	1.654	94.2	1.884
(15% ROI After Taxes)				

Figure 6 shows the projected selling price for a 15% return on investment after taxes for the 50,000,000 gal/yr Gulf cellulose alcohol plant and for fermentation corn and synthetic ethylene-alcohol plants of the same capacity (Table VI). The ethylene costs are escalated at 9%, per industry projections, cellulosics at 7%, and according to USDA projections, corn at 5%. Feedstock costs used as a basis for these graphs are (starting 1983 as in Tables V and VI): MSW $14.00/oven-dried ton (ODT), SMW $21.00/ODT, Pulp mill wastes $14.00/ODT, Ethylene $0.18/pound and corn $3.00/bushel. Thus, the total feedstock cost per gallon of ethanol produced is $0.104 in the case of cellulose, $0.75 for ethylene, and $1.20 for corn. By-product credits used escalated from prices listed in 1983 at a 7% rate.

The two curves for cellulose alcohol represent the economics for 100% investor equity capital and for municipal bond financing. Also shown on the figure is the projected selling price for ethanol at 7% and 5% inflation rates. The cellulose alcohol curve with municipal bond financing shows the effect of interest losses prior to the first operating year. The third-year selling price is more representative of actual economics.

The technical feasibility has been demonstrated by the research work carried out at the Gulf Oil Chemicals Company. The commercial process economics, which are based on producing 50 million gal/yr of alcohol, show a more favorable selling price than grain fermentation and synthetic alcohols with 100% investor equity capital financing. With municipal bond financing, cellulosic waste alcohol yields much greater profitability or much lower selling prices to obtain a 15% return on investor equity.

REFERENCES

1.   Mooney, J. R., *Orange Disc,* **1977** 2.

2.   Emert, G. H.; Gum, E. G., Jr.; Lang, J. A.; Liu, T. H.,; Brown, R. D., Jr. *Adv. Chem. Ser.* **1974,** *136,* 79.

3.   Dyess, S. E.; Emert, G. H. "Encyclopedia of Chemical Processing and Design"; Marcel Dekker: New York, 1978; Vol. 7, pp 499-58.

4.   Takagi, M.; Abe, S.; Emert, G. H.; Yata, N. "International Symposium on Bioconversion of Cellulosic Substrates"; Ghose, T. K., Ed.; Institute of Chemical Engineering: New Delhi, India, 1977; p 551.

5.   Blotkamp, P. J.; Takagi, M.; Pemberton, M. J.; Emert, G. H. "Proceedings", the 84th American Institute of Chemical Engineers National Meeting; American Institute of Chemical Engineers: New York, 1978.

RECEIVED JUNE 18, 1980.

GASEOUS FUELS

Methane Production by Anaerobic Digestion of Bermuda Grass

DONALD L. KLASS and SAMBHUNATH GHOSH

Institute of Gas Technology, 3424 South State Street, Chicago, IL 60616

It is now clear that every technically and economically feasible source of additional methane must be tapped to meet the growing demand for natural gas. One potentially large-scale source of methane is land- and water-based biomass which can be converted to substitute natural gas (SNG) by a variety of techniques. Because biomass is a renewable nonfossil carbon source that derives its energy from photosynthetic fixation of ambient carbon dioxide, the concept could lead to the development of perpetually available SNG supplies (1).

Perennial grasses have been suggested as one category of land-based biomass suitable for conversion to methane (2). Most perennial grasses can be grown vegetatively, and they reestablish themselves rapidly after harvesting. Also, more than one harvest can usually be obtained per year. The warm-season grasses are preferred over the cool-season grasses because their growth rate increases rather than declines as the temperature rises to its maximum in the summer months (3). In certain areas, rainfall is adequate to permit harvesting every 3 to 4 weeks from late February into November, and yields between 18 to 24 metric ton/ac-yr (8 to 10 short ton/ac-yr) of dry grass equivalent are believed to be attainable in managed grasslands (3).

0097-6151/81/0144-0229$0.5.25/0
© 1981 American Chemical Society

Our initial experimental work to study the conversion of grass to methane and the feasibility of developing small-scale installations for on-site use by the individual homeowner was done with common lawn grass, which consisted predominately of Kentucky bluegrass grown in Northern Illinois (4). Experiments carried out in laboratory digesters showed that the grass can be converted directly to high-methane gas under conventional anaerobic digestion conditions. Methane yields of 2.5 and 3.1 SCF/lb volatile solids (VS) added were observed at mesophilic temperatures and semicontinuous digestion conditions. This corresponded to energy recovery efficiencies as methane of about 28% and 34%. Alkali treatment of Kentucky bluegrass before digestion gave a methane yield and energy recovery efficiency of 4.6 SCF/lb VS added and 51%.

This paper summarizes the preliminary experimental work carried out with the warm-season grass Coastal Bermuda grass (*Cynodon dactylon*) to study its conversion to methane by anaerobic digestion. Bermuda grass is widely distributed throughout the tropical and subtropical countries of the world, and in the United States, is best adapted to the states south of a line connecting Virginia and Kansas (5). Coastal Bermuda grass, which tolerates more frost, makes more growth in the fall, and remains green much later than common Bermuda grass, grows tall enough to be cut for hay on almost any soil (5). From an overall standpoint of distribution and growth characteristics, Coastal Bermuda grass is a good candidate for biomass energy applications.

MATERIALS AND METHODS

Digesters

The digestion runs were carried out in the semicontinuous mode in which sequential wasting of a portion of the digester contents and feeding of grass slurry were performed on a daily basis under anaerobic conditions. Custom-made, 7-ℓ, cylindrical, flat-bottomed, Lucite digesters (7.5 in. ID) having working volumes of 5ℓ were employed. Mixing was provided by two 3-in. diameter, stainless steel, propeller-type impellers mounted on a single, top-driven, stainless steel shaft positioned in the center of the vessel. The impellers were mounted 3 in. and 6 in. from the bottom of the digester and were driven by an externally mounted 1/8-hp AC motor at 130 rpm. Four internal Lucite baffles 6 in. × 3 in. × 1 in. spaced 90 ° apart were attached to the digester wall at the 6 in. × 1/4 in. surface. The long dimension of each baffle was perpendicular to the flat bottom and totally immersed in the culture. The top of the digester was equipped with a thermometer well, a shaft seal on the stirring shaft, and feed, gas removal, and gas sampling ports. The bottom of the digester contained an effluent withdrawal port.

Temperature was controlled by mounting the digesters in a constant temperature cabinet in which preheated air was continuously circulated. The gas collection and measuring system for each digester was mounted outside the constant temperature cabinet and was similar in design to that described previously (6).

Digester Feeds

Coastal Bermuda grass was obtained from the North Louisiana Hill Farm Experiment Station in Homer, Louisiana. The station reported that the soil from which the grass was taken is classified as a Shubuta fine sandy loam, a soil type common to the Coastal Plains region of North Central Louisiana. The area was fertilized with 300 lb per acre of N as NH_4NO_3 on April 22, 1977, and with 60 lb of K_2O and 90 lb of P_2O_5 /acre on April 28, 1977. The grass was harvested on May 23-24, 1977 with a sickle bar mower, left on the ground the first day, raked into windrows the next day, and baled on the third day. The yield for this cutting was approximately 1.5 ton/acre. About 1400 lb (20 bales) were shipped to IGT by truck freight and arrived on June 30, 1977. The bales were stored under ambient conditions in an enclosed trailer. As received, the grass contained stems and blades 3 in. or longer in length.

A 300-lb sample of the grass was ground twice in an Urschel Laboratory Grinder (Comitrol 3600) equipped with a 0.030-in. cutting head, dry-mixed in a ribbon blender, and stored in a covered 20-gal plastic drum at room temperature. A typical particle size analysis of the ground grass is shown in Table I, and the physical and chemical properties of the as-received and ground grass as well as the ground grass after room-temperature storage for 3 months are shown in Table II.

Feed slurries were prepared fresh daily by blending the required amounts of ground grass and demineralized water. The properties of feed slurries prepared about 4 months apart are presented in Table III. The pH of the digester contents was maintained in the 6.8 to 7.2 range as much as possible by adding a pre-determined amount of 1.0 N NaOH solution to the feed slurry before dilution to the required volume with water. When added nutrient solutions were used, the compositions of which are shown in Table IV, pre-selected amounts were also blended with the feed slurries before dilution to the final feed volume.

Analytical Techniques

Most analyses were performed in duplicate; several were performed in triplicate or higher multiples. The procedures were either ASTM, Standard

Table I. PARTICLE SIZE ANALYSIS OF GROUND GRASS

U.S. Sieve Size, mm	Grass Retained on Sieve, wt%
1.18	0
0.60	0
0.297	36.2
0.250	55.4
0.212	74.6
0.180	83.3
0.149	87.9
0.105	92.6
0.063	98.5

Table II. PHYSICAL AND CHEMICAL CHARACTERISTICS OF GRASS

	As Received	After Grinding	After Storage and Grinding
Ultimate Analysis, wt%			
C	— —	47.1	47.5
H	— —	6.04	6.12
N	— —	1.96	— —
S	— —	0.21	— —
P	— —	0.24	— —
Ca	— —	0.30	— —
Na	— —	0.08	— —
K	— —	1.6	— —
Mg	— —	0.14	— —
Mn	— —	0.01	— —
Fe	— —	< 0.005	— —
Sr	— —	< 0.001	— —
Zn	— —	< 0.005	— —
Proximate Analysis, wt%			
Moisture	9.26	5.15	6.26
Volatile Matter	95.3	95.0	95.1
Ash	4.73	5.05	4.90
Organic Components, wt%			
Crude Protein	— —	12.3	— —
Cellulose	— —	31.7	— —
Hemicellulose	— —	40.2	— —
Lignin	— —	4.1	— —
High Heating Value,			
Btu/dry lb	— —	8,185	8,162
Btu/lb (MAF)	— —	8,616	8,583
Btu/lb C	— —	17,378	17,183
Bulk Density, lb/ft^3	4.70	23.76	— —

Table III. CHARACTERISTICS OF FEED SLURRY*

	July 23, 1977	November 14, 1977
Density, g/ml at 25°C	0.944	— —
Total Solids, wt% of slurry	1.59	1.59
Volatile Matter, wt% of slurry	1.51	1.50
Total Alkalinity, mg/ℓ as $CaCO_3$	211	240
Bicarbonate Alkalinity, mg/ℓ as $CaCO_3$	137	143
pH	6.02	6.11
Conductivity, μmho/cm	1,030	1,060
Volatile Acids, mg/ℓ		
Acetic	104	115
Propionic	0	24
Butyric	0	0
Isobutyric	0	0
Total As Acetic	104	136
Chemical Oxygen Demand, mg/ℓ	35,000	32,630
Ammonia N, ppm as N	3.5	5.0

* Formulated for loading rate of 0.1 lb VS/ft^3 -day, 12-day detention time, 5 ℓ culture volume.

Table IV. COMPOSITION OF ADDED NUTRIENT SOLUTIONS

Component	Mixed Nutrient Formulation, g/ℓ	Ammonium Chloride Solution, g/ℓ
NH_4 Cl	3.0	120.0
$NaH_2 PO_4$	20.0	— —
KI	2.0	— —
$FeCl_3$	2.0	— —
$MgCl_2$	2.0	— —
$CoCl_2$	0.25	— —
$NaMoO_2$	0.10	— —
$CuCl_2$	0.10	— —
$MnCl_2$	0.10	— —
N concentration, mg/ml	7.85	31.42

Methods, or special techniques as reported previously (6). Cellulose, hemicellulose, and lignin in the grass and digested solids were determined by the methods of Goering and van Soist (7).

Data Reduction

Gas yield, methane yield, volatile solids reduction, and energy recovery efficiency were calculated by the methods described previously (6). All gas data reported are converted to 60°F and 30 in. of mercury on a dry basis.

Inoculum, Start-Up and Operation

Development of the mesophilic inoculum for the grass digester runs was started on June 22 and 29, 1977 by accumulating daily effluents from existing laboratory digestion runs operating on giant brown kelp and on mixed primary-activated sewage sludge. The kelp and sludge digesters were operated at 35°C and detention times of 18 and 5.6 days, and loading rates of 0.1 and 0.8 lb VS/ft^3-day, respectively. The effluents from these digesters were collected in other digesters until 8.75 ℓ of the biomass culture and 7.50 ℓ of the sludge culture had been accumulated. Each digester was then operated semicontinuously for 10 days to stabilize its performance. On July 15, 1977, 1.75 ℓ of biomass culture and 0.75 ℓ of sludge culture were anaerobically transferred to each of two Bermuda grass digesters which were thus started with 2.5 ℓ of mixed inoculum. These digesters were then operated in the semicontinuous mode with a daily feeding and wasting schedule designed to increase culture volumes by 10% per day to a volume of 5.0 ℓ while maintaining the detention time and loading constant at about 15 days and 0.1 lb VS/ft^3-day. The digesters were fed with kelp, sludge, and grass. The ratio of kelp VS to sludge VS was maintained at 70:30. The percentages of the kelp and sludge were gradually decreased while the percentage of grass was gradually increased during this transition. The culture volume of 5.0 ℓ was attained on July 23, 1977 and the kelp and sludge were completely displaced by Bermuda grass by August 19, 1977. Digestion was then continued at the selected operating conditions with grass feed only.

Steady-state digestion was defined in this work as operation without significant changes in the gas production rate, gas composition, and effluent characteristics. Usually, operation for two or three detention times established steady-state performance.

With the exception of Run 1, which did not achieve steady state, selected steady-state results are shown in Table V. Run 1 is the experiment started as indicated above to establish a baseline without added nutrients. Runs 5 and

Table V. SUMMARY OF SELECTED STEADY-STATE DATA

	Run 1[d]	Run 2	Run 3	Run 4	Run 5	Run 6	Run 7	Run 8
Operating Conditions								
Digester Volume, ℓ	7	7	7	7	7	7	7	7
Working Volume, ℓ	5	5	5	5	5	5	5	5
Agitation Schedule[a]	C	C	C	C	C	C	C	C
Agitation Type[a]	M	M	M	M	M	M	M	M
Feeding Frequency[a]	D	D	D	D	D	D	D	D
Nutrients Added[a]	O	MN	MN	MN	N	N	N	N
Temperature, °C	35	35	35	35	35	35	35	55
pH[b,c]	6.8	6.9	6.9	6.9	6.8	6.8	6.8	6.4
Loading Rate, lb VS/ft^3-day	0.10	0.10	0.10	0.10	0.10	0.10	0.10	0.10
Detention Time, day	12	12	12	12	12	12	12	12
Total Solids in Feed Slurry, wt%	1.59	1.59	1.59	1.59	1.59	1.59	1.59	1.59
Volatile Solids in Feed Slurry, wt%	1.51	1.51	1.51	1.51	1.51	1.51	1.51	1.51
C/N Ratio in Feed Slurry	24.0	16.3	12.3	8.3	12.3	8.3	6.3	12.3
Caustic Requirements, m·q/ℓ Feed	72	66	66	103	81	81	36	42
Gas Production[c]								
Gas Production Rate, vol/vol-day	0.313	0.459	0.407	0.398	0.350	0.464	0.587	0.527
Gas Yield, SCF/lb VS added	3.13	4.59	4.07	3.98	3.50	4.64	5.87	5.27
Methane Yield, SCF/lb VS added	1.92	2.68	2.45	2.45	2.20	2.90	3.51	2.73
Methane Concentration, mol%	61.4	58.3	60.3	61.7	63.7	62.4	59.8	51.8
Coefficient of Variation, Methane Yield	10	7	6	10	9	13	9	10
Efficiencies								
Volatile Solids Reduction, %	20.0	29.3	26.0	25.4	22.1	29.7	37.5	33.7
Energy Recovered as Methane, %	22.6	31.5	28.8	28.9	25.8	34.1	41.2	32.1
Effluent Volatile Acids, mg/ℓ as HOAc[c]	1,989	2,056	1,300	2,123	2,540	1,159	354	2,273

[a] "C" denotes continuous agitation. "M" denotes continuous mechanical mixing. "D" denotes daily feeding and wasting cycles. "O" denotes no nutrients added to feed slurry. "MN" denotes mixed nutrient solution added to feed slurry. "N" denotes ammonium chloride solution added to feed slurry.
[b] pH maintained in indicated range by periodic NaOH additions.
[c] Mean values.
[d] Did not achieve steady state.

6, 7 were sequentially derived from Run 1. Runs 2, 3, and 4 were sequentially derived from a replicate of Run 1. Typical performance of one of the runs (Run 7) over an extended time is shown in Figure 1.

Thermophilic Run 8 was initiated by anaerobically accumulating the effluent from Run 7 until 5.0 l had been collected. The culture was maintained in the batch mode at 55°C for a few days, and then semicontinuous operation with grass was started at a detention time of 50 days and a loading rate of 0.02 lb VS/ft^3-day. The ammonium chloride nutrient solution (Table IV) was added to the feed slurry at a rate of 1.0 ml of solution per 0.02 lb/ft^3-day loading. The loading rate was gradually increased at a constant detention time; 0.1 lb VS/ft^3-day loading rate was attained in 10 days. The detention time was then gradually reduced to 12.0 days over a 10-day period. Operation of Run 8 was then continued at the target conditions.

Dewatering Tests

Gravity sedimentation tests were conducted by a modified AEEP method (8) in which a 400-ml sample of effluent was examined in a 1-l graduated cylinder giving a fluid depth of 140 mm. The height of the interface between the thickened sludge and clarified supernatant is plotted versus time. Vacuum filtration tests were conducted by a modified AEEP method (9) in which a 0.05 ft^2 circular Lucite leaf covered with an Eimco No. NY-415 monofilament filter cloth was used in a 1-l beaker containing 417 ml of effluent sample.

DISCUSSION

Feed Properties

All of the digestion runs were carried out with small particle-size grass to facilitate maximum gas yields and production rates in the laboratory work. Some moisture loss was observed on grinding as expected, and no significant changes were detected on room-temperature storage of the ground grass (Table III). Several characteristics of the particular lot of grass and resulting feed slurries examined in this work indicated that anaerobic digestion under conventional conditions might not provide good methane fermentation. The mass ratios of C/N (24), C/P (196), C/Ca (157), and C/Mg (336) in the dry grass solids and the COD/N ratio (105-112) in the feed slurry appear too high when compared with the corresponding ratios supplied by suitable substrates such as sewage sludge and giant brown kelp. The quantities of ammonia N (3.5 mg/l), Ca (48 mg/l), Na (13 mg/l), and Mg (23 mg/l) in the feed slurries formulated to meet the desired loading rate and detention time conditions are also less than the stimulatory concentrations recommended for anaerobic treatment of waste (10). The concentrations of N, P, Na, Ca, and Mg might thus have to be adjusted to promote adequate methane production. Also, the relatively low pH (6.0) and bicarbonate alkalinity (137 mg/ℓ as $CaCO_3$) of the feed slurry indicate poor buffering capacity and potential problems with pH control during digestion.

Mesophilic Digestion

Representative data from Run 1, which did not achieve steady state, as previously mentioned, are shown in Table V. This represents the baseline run without added nutrients. It was found that to maintain pH in the 6.8-7.2 range, 72 meq NaOH/ℓ of feed slurry was required; this raised the sodium ion concentration in the digester to about 1670 mg/ℓ. Overall, the results of Run I were poor. The methane yield, volatile solids reduction, and energy recovery efficiency as methane in the product gas were low, and the volatile acids concentration in the digester effluent was high.

The performance of Run 1 and the compositional data indicating possible nutritional deficiencies led to the evaluation of Runs 2, 3, and 4 at the same operating conditions as Run 1 except that the mixed nutrient solution in Table IV was added to the feed slurry to raise the concentrations of the nutrients. Sufficient nutrient formulation was added to reduce the C/N ratios of Runs 2, 3, and 4 to 16.3, 12.3, and 8.3, respectively. Substantial improvements were observed in the performance of these runs, but the volatile acids concentrations in the digester effluents were still high. Also, there did not seem to be a correlation between gas production and the concentration of added mixed nutrient solution.

Significant improvement in digester performance was also observed when pure NH_4Cl nutrient solution (Table IV) was added to the feed slurry as shown by the results in Table V for Runs 5, 6, and 7. In these experiments, there was a good correlation of methane yield, volatile solids reduction, and energy recovery efficiency with the concentration of added nitrogen. The higher the added nitrogen concentration up to the highest concentration evaluated (C/N ratio 6.3; calculated ammonia N, 1,190 mg/ℓ), the higher the gas yield. Figure 2, which includes the data from Runs 1, 5, 6, and 7 as well as data from three other runs not shown in Table V, illustrates this correlation. No inhibition by added nitrogen was observed although it might be expected at concentrations above 1,500 mg/ℓ (10).

Run 7 exhibited the best methane yield of 3.51 SCF/lb VS added and the highest volatile solids reduction and energy recovery efficiencies of Runs 1 to 7. Also, the volatile acids concentration in the digester effluent is in the range

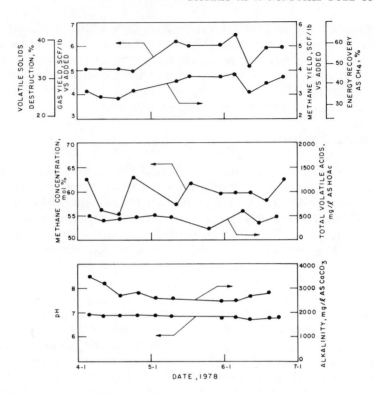

Figure 1.   Typical performance of Run 7

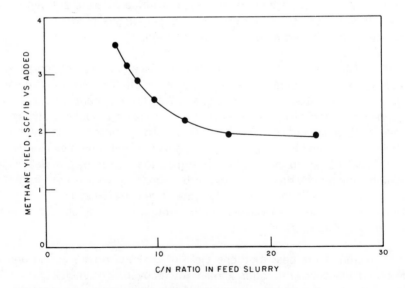

Figure 2.   Effect of ammonium chloride added to feed slurry on methane yield:
35°C, 0.1 lb VS/ft³-day, 12-day detention time

expected for balanced digestion. The plot in Figure 2 supports the conclusion that the particular lot of Coastal Bermuda grass evaluated in this work was nitrogen-limited under anaerobic digestion conditions, and that continued addition of ammonium chloride up to the highest concentration studied (C/N ratio of 6.3) appeared to have a stimulatory effect on methane production. Since it is known that fertilization methods and dosage rates affect the nitrogen content of Coastal Bermuda grass (11), a tradeoff analysis would have to be performed to establish the incremental benefits of increased fertilization vs. nutrient addition if an integrated production-harvesting-gasification system were designed to manufacture methane.

A reasonably good linear correlation was found between energy recovery efficiency as methane in the product gas and volatile solids reduction as shown in Figure 3. This type of correlation has also been found to exist for giant brown kelp after correction was made for any hydrogen in the product gas (6).

Thermophilic Digestion

One run, Run 8, was carried out to study the effect of digestion at thermophilic temperatures. The results shown in Table V were obtained with supplemental nitrogen additions at the same concentration as that used in Run 5. Run 8 exhibited about 50% higher gas production and volatile solids reduction than Run 5, but due to the lower methane content in the product gas, the energy recovery efficiency of Run 8 was only about 24% higher than that of Run 5. It is apparent also that the high volatile acids concentration in the digester effluent of Run 8 does not indicate balanced digestion.

Carbon and Energy Balances

It was difficult to calculate carbon and energy balances for the digester runs performed with Coastal Bermuda grass because of the addition of relatively large quantities of alkali for pH control and of nutrients. These additives contributed to ash weights. Two techniques were used to circumvent these problems. One assumed that added NaOH was converted to $NaHCO_3$ on ashing at 550°C and remained in the ash, and that added NH_4Cl was completely volatilized on ashing. These assumptions are probably not strictly true. The other technique relied upon experimental measurement of ash and volatile solids in the dry digested solids.

The details of one set of calculations by both techniques are illustrated in Chart 1 for Run 1 which was performed with only added alkali. The best balance, 101% carbon and 100% energy accounted for, was obtained by the

Chart 1. Carbon and Energy Balance For Run No. 1

Accounted For: Feed Carbon 101%*, 94%†
Feed Energy 100%*, 93%†

100 lb Grass → (35°C, 0.10 lb VS/ft^3-day, 12-day DT; 40.0 meq NaOH/l. feed, 20.0% VS Reduction) → **Digested Solids** + **Gas**

100 lb Grass	Digested Solids: Method 1*	Digested Solids: Method 2†	Gas
5.15 lb H_2O	72.09 lb VS	72.09 lb VS	3.13 SCF gas/lb VS added
90.11 lb VS	24.57 lb Ash	16.06 lb Ash	x90.11 lb VS added
4.74 lb Ash	96.66 lb	88.15 lb	=282.0 SCF gas
100.00 lb			282.0 + 379=0.744 mole gas
Carbon	**Carbon**	**Carbon**	**Carbon**
94.85 x 0.471 = 44.7 lb	96.66 x 0.376 = 36.3 lb	88.15 x 0.376 = 33.1 lb	0.744 x 12.0 = 8.9 lb
Energy	**Energy**	**Energy**	**Energy**
94.85 x 8,185 = 776,300 Btu	96.66 x 6.237 = 602,900 Btu	88.15 x 6.237 = 549,800 Btu	3.13 x 0.614 x 1,012 x 90.11 = 175,300 Btu

*Ash calculated by assuming 40.0 meq NaOH added/l. feed slurry for pH control converted to $NaHCO_3$ on ashing at 550°C and 0.417 l. feed equal 417 g feed. Thus:

$$4.74 \text{ lb original ash} + \left[\frac{40.0 \frac{\text{meq NaOH}}{\text{l. feed}} \times 0.417 \text{ l.} \frac{\text{feed}}{\text{day}} \times 0.084 \frac{\text{g NaHCO}_3 \text{ formed}}{\text{meq NaOH}}}{417 \frac{\text{g feed}}{\text{day}} \times \frac{1.59 \text{ wt \% TS in feed}}{100} \times \frac{5.05 \text{ wt \% ash in TS}}{100}} \right] 4.74 = 24.57 \text{ lb ash}$$

†Ash and VS experimentally determined to be 18.22 wt % and 81.78% on dry digested solids.

technique that assumed the added alkali was converted to bicarbonate in the ash. The technique that relied on experimental ash and volatile solids determinations gave results that accounted for less than the feed carbon and energy. But this did not occur in the other calculations, the results of which are summarized in Table VI. Higher and lower deviations from 100% occurred with each technique.

Properties of Effluent and Digested Solids

As already pointed out, the volatile acids concentrations in the effluents from most of the runs were high. The detailed breakdown of the individual acids and other properties are summarized for the feed slurry and effluents from Runs 1, 7, and 8 in Table VII. The effects of the added NaOH and NH_4Cl on the alkalinity, ammonia nitrogen, and specific conductivity are in the expected directions. The conversion of non-ammonia nitrogen to ammonia nitrogen in Run 1, which was conducted without added NH_4Cl, during the digestion process is evident; the ammonia nitrogen increased from 3.5 mg/ℓ in the fresh feed slurry to 76 mg/ℓ in the effluent.

Gravity sedimentation and vacuum filtration tests were conducted on the unconditioned effluent from Run 7. The effluent exhibited rapid settling velocities (Figure 4) and the filtration characteristics were excellent (Table X).

The properties of the digested solids from Runs 1, 7, and 8 are compared with those of the dry feed solids in Table IX. For the total digested solids, carbon content and heating values decreased and the ash content increased as expected during digestion. The heating values of the digested solids per mass unit of carbon, however, did not exhibit a general decrease on digestion. On this basis, the heating value of the digested solids from Run 1 was slightly less than the corresponding value of the feed solids, while the heating values of the solids from Runs 7 and 8 were higher. These trends are probably the result of different rates of biodegradability of the organic components in the grass, each of which would be expected to have different heating values.

To attempt to acquire information on the degradabilities of the major classes of organic components in the grass, the conversion data summarized in Table X were derived from the compositional data in Table IX by assuming that a decrease in the concentration of any organic component was caused by gasification. This assumption is not strictly true, but it permits first-approximation calculations of relative biodegradabilities. Hemicellulose, which was present in the highest concentration, was converted to gas at the highest yield with one exception, while the crude protein fraction and the lignin fraction gave the lowest gas yields. The exception appears to be

Table VI. SUMMARY OF CARBON AND ENERGY BALANCES

	Accounted For	
	Feed Carbon, %	**Feed Energy, %**
Run 1	94.0^a, 101^b	93.4^a, 100^b
Run 7	85.7^a, 115^b	96.7^a, 116^b
Run 8	107^a, 116^b	106^a, 116^b

[a] Calculated from experimental determinations for moisture, volatile solids, ash, carbon, and heating values of feed and digested solids, and yield and composition of product gas. Volatile solids in digested solids calculated from percent volatile solids reduction.

[b] Calculated from parameters in footnote "a" except that ash in digested solids estimated by assuming original ash in feed is in digested solids, that NaOH used for pH control is converted to $NaHCO_3$ on ashing at 500° C and remains in ash, that that NH_4 Cl, if added, is volatilized on ashing.

Table VII. COMPARISON OF FEED AND DIGESTER EFFLUENT SLURRIES

Parameter*	Feed	Run 1	Run 7	Run 8
pH	6.0	6.8	6.8	6.4
Total Alkalinity, mg/ℓ as $CaCO_3$	221	3,182	2,571	2,906
Bicarbonage Alkalinity, mg/ℓ as $CaCO_3$	137	1,525	2,276	1,013
Ammonia Nitrogen, mg/ℓ	3.5	76	738	500
Specific Conductivity, μmho/cm	1,030	5,000	12,540	6,250
Volatile Acids (Filtrate), mg/ℓ				
Acetic	104	1,461	238	1,821
Propionic	0	541	126	323
Butyric	0	43	5	71
Isobutyric	0	42	9	125
Valeric	0	25	2	72
Isovaleric	0	54	3	18
Caproic	0	0	0	14
Total As Acetic	104	1,989	354	2,273

* Mean Values

Table VIII. VACUUM FILTRATION CHARACTERISTICS OF UNCONDITIONED DIGESTER EFFLUENT, RUN 7
TEST TEMPERATURE 26°C

Effluent		Cake		Yield*	
TS, wt%	VS, wt% of TS	TS, wt%	VS,wt% of TS	Dry Cake, lb/ft^2 -hr	Filtrate, lb/lb dry cake
1.66	78.8	16.8	96.9	12.3	115
1.66	78.8	16.3	97.1	13.1	155

* 30 sec cycle time, 6 sec form time, 12 sec drying time, 12 sec removal time, 20 in. Hg.

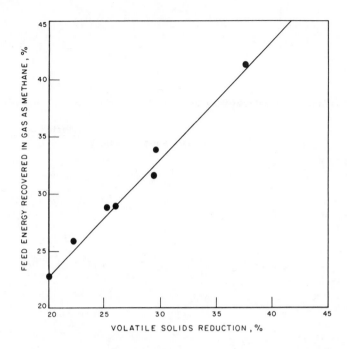

Figure 3.   Feed energy equivalent recovered in gas as methane vs. volatile solids reduction at 35°C

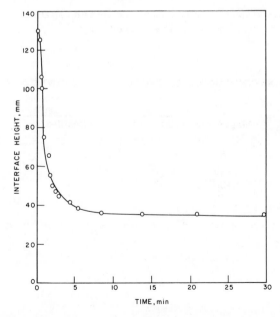

Figure 4.   Interface height vs. time for gravity settling of unconditioned effluent from Run 7

Table IX. COMPARISON OF DRY FEED AND DIGESTED SOLIDS

		Digested Solids*		
	Dry Feed	Run 1	Run 7	Run 8
Ultimate Analysis, wt%				
C	47.1	37.6	30.2	43.9
H	6.04	4.78	5.43	5.50
N	1.96	2.34	6.65	2.34
Proximate Analysis, wt%				
Moisture	5.15	— —	— —	— —
Volatile Matter	95.0	81.8	78.8	80.2
Ash	5.05	18.2	21.2	19.8
Organic Components, wt%				
Crude Protein (Kjeldahl Nx6.25)	12.3	14.6	— —	— —
Hemicllulose	40.2	17.7	15.9	37.9
Cellulose	31.7	17.5	13.6	19.5
Lignin	4.1	7.4	4.5	13.6
Heating Value				
Btu/dry lb	8,185	6,237	6,021	7,721
Btu/lb (MAF)	8,616	7,625	7,641	9,627
Btu/lb C	17,378	16,588	19,957	17,588

* Prepared by evaporation of total effluent to dryness on steam bath, pulverization, and drying in an evacuated desiccator to a constant weight.

Table X. COMPARISON OF ORGANIC COMPONENT CONVERSION

Component	Mass Ratio In Feed Solids	Run 1		Run 7		Run 8	
		Gasified, wt%	Ratio*	Gasified, wt%	Ratio*	Gasified, wt%	Ratio*
Hemicellulose	1.0	55.1	1.0	67.3	1.0	16.2	1.0
Cellulose	0.79	43.8	0.63	64.5	0.76	45.3	2.2
Crude Protein	0.30	0	0	— —	— —	— —	— —
Lignin	0.10	0	0	9.3	0.01	0	0

* Expressed as mass of indicated component gasified per unit mass of hemicellulose gasified.

thermophilic Run 8, in which less hemicellulose was converted than cellulose. The higher level of unconverted hemicellulose than cellulose in Run 8 may be an artifact caused by conversion of cellulose at thermophilic conditions to derivatives which are detected in the hemicellulose fraction. The experimental data also indicate that a small amount of the lignin fraction was converted in Run 7. Overall, the polysaccharide fraction is more biodegradable than the protein and lignin fractions. The low biodegradability of lignin under anaerobic digestion conditions is expected, while the higher biodegradability of the hemicellulose might be predicted because it has a higher reactivity to acid and alkali than cellulose. The lower biodegradability of protein with respect to the monosaccharides and polysaccharides in giant brown kelp has been reported (12).

Thermodynamic Estimates

The enthalpy of digestion, product gas composition, and methane yield for the lot of grass examined in this work were estimated as shown in Chart 2. The process is projected to be slightly exothermic, -173 Btu/lb grass reacted, which agrees well with the slight exothermicity reported for kelp (6). Assuming that 20% of the carbohydrate fraction and 7% of the protein present in the grass would be converted to new bacterial cells by anaerobic fermentation and hence not be available for methane production by a single pass through the fermentor, the maximum theoretical yield of methane is given by (12):

$$(1 \text{ lb VS added} - 0.176 \text{ lb VS to cells}) \left(\frac{7.92}{0.95} \frac{\text{SCF CH}_4}{\text{lb VS}} \right) = 6.87 \frac{\text{SCF CH}_4}{\text{lb VS-pass}}$$

The highest experimental yield obtained in this work is 3.5 SCF/lb VS added (Run 7) or 51% of the theoretical maximum value. Thus, considerable yield improvements are still possible.

Comparison With Other Substrates

The gas yields and volatile solids reduction and energy recovery efficiencies from Coastal Bermuda grass are compared in Table XI with those obtained from other substrates under similar mesophilic digestion conditions. The nitrogen-supplemented Bermuda grass was converted to methane at about the same efficiencies as giant brown kelp and primary sewage sludge. The non-supplemented Bermuda grass, however, afforded considerably lower methane yields and efficiencies than these substrates. Untreated Kentucky bluegrass was somewhat better in performance than Coastal Bermuda grass.

Chart 2. Summary of Thermodynamic Calculations

I. <u>Grass Composition</u>

C, 47.10 wt% (dry)
H, 6.04
N, 1.96
S, 0.21
O, 39.64*
Ash, 5.05

II. <u>Empirical Formula</u> $C_{3.92} H_{5.99} O_{2.48} N_{0.14} S_{0.007} Ash_{5.0}$, Mol. Wt. 100

III. <u>Heat of Digestion, Gas Composition, Methane Yield</u>

    a. Assuming N and S can be neglected and CH_4 and CO_2 are the only products:

$$C_{3.92} H_{5.99} O_{2.48} + 1.182\ H_2O = 2.089\ CH_4 + 1.831\ CO_2$$

    Gas composition: 53.3 mol% CH_4, 46.7 mol% CO_2

    Methane Yield: 7.92 SCF CH_4/lb grass reacted (max.)

    b. For 1.00 lb dry grass reacted:

    Input = 8,185 Btu

    Output= 8,012 Btu

    Heat of Reaction: 8,012 - 8,185 = -173 Btu

---

*By difference.

Table XI. COMPARISON OF COMPOSITIONS AND SELECTED
STEADY-STATE ANAEROBIC DIGESTION RESULTS

Compositional Parameter	Coastal Bermuda Grass		Kentucky Bluegrass[c]	Giant Brown Kelp[f]	Primary Sludge[g]
C, wt%	47.1		45.8	27.8	43.75
H, wt%	6.04		5.9	3.73	6.24
N, wt%	1.96		4.8	1.63	3.16
Moisture, wt%	5.15		7.8	88.8	94.12
Volatile matter, wt%	95.0		86.5	57.92	73.47
Ash, wt%	5.05		13.5	42.08	26.53
High heating value, Btu/dry lb	8,185		8,052	4,620	8,537
High heating value, Btu/lb (MAF)	8,616		9,309	7,977	11,620
High Heating value, Btu/lb C	17,378		17,581	16,619	19,513
Digestion Conditions					
Mechanical Agitation[a]	C	C	C	C	C
Feeding Frequency[a]	D	D	D	D	D
Nutrients Added[b]	O[d]	N[e]	O	O	O
Temperature, °C	35	35	35	35	35
pH	6.8	6.8	7.1	7.0	7.0
Loading Rate, lb VS/ft^3-day	0.10	0.10	0.13	0.10	0.10
Detention Time, day	12	12	12	12	12
C/N Ratio in Feed Slurry	24.0	6.3	9.54	17.1	13.8
Gas Production					
Gas Production Rate, vol/vol-day	0.313	0.587	0.55	0.662	0.78
Gas Yield, SCF/lb VS added	3.13	5.87	4.20	6.62	7.8
Methane Yield, SCF/lb VS added	1.92	3.51	2.54	3.87	5.3
Methane Concentration, mol%	61.4	55.9 ·	60.4	58.4	68.5
Efficiencies					
Volatile Solids Reduction, %	20.0	37.5	25.1	43.7	41.5
Energy Recovered as Methane, %	22.6	41.2	27.6	49.1	46.2

[a] "C" denotes continous agitation, "D" denotes daily feeding and wasting cycle.
[b] "O" denotes no nutrients added to feed slurry. "N" denotes ammonium chloride added to feed slurry.
[c] Run B of Ref. 4.
[d] Run 1 of Table 5.
[e] Run 7 of Table 5.
[f] Run 2 of Ref. 12.
[g] Experimental data obtained with primary thickened sewage sludge in laboratory digesters under standard high-rate conditions. Sludge obtained from Metropolitan Sanitary District of Greater Chicago.

SUMMARY AND CONCLUSIONS

Bermuda grass (*Cynodon dactylon*) is one of the high-yield warm-season grasses that has been suggested as a promising raw material for conversion to methane. Experimental work performed with laboratory digesters to study the anaerobic digestion of Coastal Bermuda grass harvested in Louisiana and having a C/N ratio of 24 is described. Methane yields of about 1.9 SCF/lb of volatile solids (VS) added were observed under conventional mesophilic high-rate conditions. When supplemental nitrogen additions were made, the methane yields increased. This observation along with the compositional data compiled on the grass used in this work indicated that the nitrogen content of the unsupplemented grass was insufficient to sustain high-rate digestion at the higher yield level. However, as the C/N ratio was reduced by addition of ammonium chloride, the methane yield continually increased up to 3.5 SCF/lb added at the lowest C/N ratio examined (6.3) even after relatively high concentrations of ammonium nitrogen were measured in the effluent. It appears that the added nutrient had a stimulatory effect on methane production above the point where nitrogen was not limiting. Thermophilic digestion with supplemental nitrogen additions afforded methane yields of about 2.7 SCF/lb VS added. Carbon and energy balances were calculated and the relative biodegradabilities of the organics were estimated.

It was concluded from this work that Coastal Bermuda grass can be converted to high-methane gas under conventional anaerobic digestion conditions The performance of the particular lot of grass studied was substantially improved by supplemental nitrogen additions.

ACKNOWLEDGEMENT

The authors wish to express their appreciation for the financial support of the work described in this paper by United Gas Pipe Line Co., and especially for the many valuable discussions and suggestions provided by Dr. Victor Edwards and Robert Christopher of United. The authors also appreciate the assistance supplied by Mike Henry, Al Iverson, Frank Sedzielarz, and Janet Vorres, who performed the experimental digestion studies, and by James Ingemanson and Robert Stotz and their staff who performed many of the chemical analyses. Special thanks is given to Mr. Dawson Johns of North Louisiana Hill Farm Experiment Station for supplying the Bermuda grass and information on its production.

REFERENCES

1. Klass, D. L. *Chemtech* **1974,** *4,* 161-68.

2. Klass, D. L. 169th National Meeting, American Chemical Society, April 1975; *Energy Sources* **1977,** *3* (2), 177-95.

3. InterTechnology Corp. October 1975, American Gas Association Project IU-114-1, Final Report.

4. Klass, D. L.; Ghosh, S.; Conrad, J.R. "Symposium Papers", Clean Fuels From Biomass, Symposium sponsored by the Institute of Gas Technology, Orlando, Fla., January 1976; Institute of Gas Technology: Chicago, Ill., 1976, pp. 229-52.

5. Burton, G. W. In "Forages The Science of Grassland Agriculture"; Highes, H. D.; Heath, M. E.; Metcalfe, D. S., Eds.; The Iowa State College Press: Ames, Iowa; Chapter 24.

6. Klass, D. L.; Ghosh, S. "Symposium Papers", Clean Fuels From Biomass and Wastes, Symposium sponsored by the Institute of Gas Technology, Orlando, Fla., January 1977; Institute of Gas Technology: Chicago, Ill. 1977; pp 323-51.

7. Agricultural Research Service, Washington, D.C., December 1970, United States Department of Agriculture, Agriculture Handbook No. 379.

8. Association of Environmental Engineering Professors, "Environmental Engineering Unit Operations and Unit Processes Laboratory Manual", O'Connor, J. T., Ed., III-1-1, July 1972.

9. Ibid., V-2.

10. McCarty, P. L. *Public Works* **1964** (November), 91-4.

11. Johns, D. M. "Fertilization of Coastal Bermuda Grass on a Coastal Plain Soil", North Louisiana Hill Farm Experiment Station, Homer, Louisiana.

12. Klass, D. L.; Ghosh, S.; Chynoweth, D. P. 175th National Meeting, American Chemical Society, Anaheim, Calif., March 1978.

RECEIVED JUNE 18, 1980.

Advanced Digestion Process Development for Methane Production from Biomass–Waste Blends

SAMBHUNATH GHOSH and DONALD L. KLASS

Institute of Gas Technology, 3424 South State Street, Chicago, IL 60616

Biomass and organic wastes (discarded biomass or biomass-derived material) may supply up to 15% of the U.S. energy needs by the end of this century via gasification or other conversion schemes [1]. One conversion process that is expected to play a role in producing methane from biomass and wastes is anaerobic digestion. Commercial production of substitute natural gas (SNG) and medium-Btu fuel gas by anaerobic digestion of waste materials has already started in the U.S. and other countries [2]. Increased usage of the anaerobic digestion process for methane production is, however, hindered because of the low reaction rate and conversion efficiency of the conventional digestion process. New digestion techniques are needed to improve conversion rates and efficiencies so that the full potential of anaerobic digestion as a methane-producing process can be realized.

Starting with the crude septic tank, a number of improved process configurations, including "standard-rate" digestion, stage digestion, "high-rate" digestion, and the anaerobic contact process, have evolved during the nearly 100 years of application of the anaerobic digestion process to sewage sludge stabilization. However, the design requirements of even one of the best anaerobic sludge stabilization modes, high-rate digestion, result in large expensive plants that are difficult to justify for commercial SNG production. An additional problem with the conventional digestion process is

0097-6156/81/0144-0251$07.00/0
© 1981 American Chemical Society

that 40 to 70% of the feed organics remain unconverted and must be disposed of at substantial cost. Considerable research, such as that reported by Pohland and Ghosh (3), Gavett (4), Ort (5,6), Switzgable (7), Klass *et al.* (8), Ghosh *et al.* (9), Ghosh and Klass (10,11), Haug (12), and others (13) has therefore been directed to the development of better digestion methods.

Considerable work has been done at the Institute of Gas Technology since 1971 to develop advanced digestion methods and process configurations for the conversion of various organic feeds to high-Btu fuel gas (8-11). In this paper, we will describe a few selected advanced digestion concepts and the results of their application to biomethanation of a mixed biomass-waste blend. The term "advanced digestion" as used in this paper includes unconventional fermentation modes and process configurations for improved methane production rate and yield.

The research reported here consisted of a laboratory evaluation of several advanced digestion methods and a selected biomass-waste blend. The objective of this work was to search for an optimum bioconversion system configuration, and the ultimate goal is to apply this configuration to the production of SNG. Specifically, the advanced digestion techniques studied were digestion of pretreated feed, recycling of digester effluent and product gas, aerobic posttreatment and recycling of digester effluent, and two-phase digestion.

The experimental plan consisted of:

- Conventional high-rate digestion under baseline operating conditions of 35°C digestion temperature, 0.1 lb VS/ft^3 -day loading and a 12-day detention time.

- Mesophilic (35°C) and thermophilic (55°C) digestion of feed subjected to mild alkaline (sodium hydroxide) pretreatment with recycling of spent caustic for fresh feed treatment and neutralization of treated feed with digester gas to minimize acid neutralizer requirement.

- Mesophilic (35°C) digestion with product gas recycling.

- Mesophilic (35°C) digestion with recycling of aerobically posttreated digester effluent.

- Two-phase digestion.

MATERIALS AND METHODS

Digester Feeds

Four feedstocks, water hyacinth (*Eichhornia crassipes*) and Coastal Bermuda grass (8) *(Cynodon dactylon)* indigenous to the Gulf Coast, and sewage sludge and municipal solid waste (MSW) were considered for bioconversion to methane. Conventional mesophilic (35°C) high-rate digestion of a biomass-sludge blend prepared with these materials showed that the mixed feed was superior to the single feeds in terms of nutritional balance and methane yield and production rate (14-16). Utilization of a biomass-waste blend has several advantages. Use of a blend facilitates feedstock supply on a year-round basis, and a mixed feed may be superior to biomass alone in terms of process economics. In addition, use of a biomass-waste blend provides the opportunity for simultaneous energy recovery and waste stabilization in an optimized integrated system. The feed used for the work described in this paper is a mixture of sewage lagoon effluent-grown water hyacinth, Coastal Bermuda grass, the combustible fraction of municipal solid waste, and mixed activated-primary sludge blended in the mass ratio of 32.3:32.3:32.3:3.1 on a volatile solids (VS) basis. This particular ratio was selected based on the projected availability of these feed components for a commercial plant in the Gulf States area. The biomass and MSW components were finely ground before blending.

The mixed feed had a median particle size of 0.25 mm. It had moisture, VS, carbon, nitrogen, phosphorus, sulfur, hydrogen, calcium, sodium, potassium, magnesium, cellulose, hemicellulose, lignin, and crude protein contents and a high heating value of 88.97, 82.78 (of total solids), 43.14, 1.64, 0.43, 0.31, 5.60, 1.23, 0.78, 1.05, 0.22, 37.5, 31.8, 4.6, and 10.1 wt %, and 7,445 Btu/lb (dry), respectively. The mixed feed had the empirical formula $C_{3.595} H_{5.545} O_{1.979} N_{0.117} P_{0.014} S_{0.010} Ash_{17.22}$ at a molecular weight of 100, and a theoretical methane yield on anaerobic digestion of about 6.4 SCF/lb VS added (14).* Also, it was determined by long-term batch digestion tests that this mixed feed had an ultimate anaerobic biodegradability or volatile solids destruction efficiency of 66% (14).

Digester feed slurries were prepared by diluting a weighed mass of the feed according to the loading rate with distilled demineralized water to a volume determined by the hydraulic detention time of the run.

* Assumes 30% of volatile solids converted to cells on one pass through the digester.

Digesters

The digestion runs were conducted in custom-made, flat-bottomed, cylindrical Plexiglas digesters having four vertical baffles mounted 90° apart on the inside walls to prevent vortexing of the culture during mechanical agitation by two stainless steel impellers at 130 rpm. The digesters were heated by placing them in a constant-temperature chamber or by thermistor-controlled heating tapes. All digesters were geometrically similar in construction. The culture volumes of the various digesters are indicated in the tabulations of the experimental results. Except for the two-phase system, all digesters were manually fed once per day after withdrawing an equal volume of digester effluent. As described later, the two-phase system was equipped with an automated feeding system.

ADVANCED SYSTEM CONFIGURATION AND OPERATION

Digestion of Pretreated Feed

The mixed hyacinth-grass-MSW-sludge feed was pretreated with caustic soda solution at mild temperatures and pressures in an attempt to improve feed biodegradability and methane yield. Treatment temperatures (5°, 25°, 55°, 100°, 121°C) and pressures (atmospheric and 30 psig) and dilute caustic concentrations were selected for the pretreatment studies for reasons of lower reaction vessel costs, reduced energy and chemical inputs, and to keep the salt concentration in the digester low. Alkaline treatment under these conditions is expected to be a cost-effective method for increasing methane production from cellulosic feeds (14,17).

Digestion of the caustic-treated feed was conducted at selected mesophilic (35°C) and thermophilic (55°C) temperatures. Flow diagrams depicting the experimental sequence are presented in Figures I and II. The processing steps included mixing of the undiluted feed with caustic solution, pretreatment at the chosen temperature, pressure, and time; diluting the treated feed with distilled demineralized water and neutralization of the digester feed slurry with hydrochloric acid or digester gas; and digestion of the pretreated neutralized feed. In some runs, the digester feed slurry was neutralized with the digester gas to reduce neutralizer acid requirement and digester salinity, and to increase the bicarbonate alkalinity (buffer capacity). In still other runs, the caustic-treated feed was vacuum filtered, and a portion of the filtrate containing the spent caustic solution was recycled for fresh feed pretreatment. Filtrate recycling reduced the caustic requirement for feed pretreatment, the acid requirement to neutralize the pretreated feed, and the salt concentration in the feed slurry and digester.

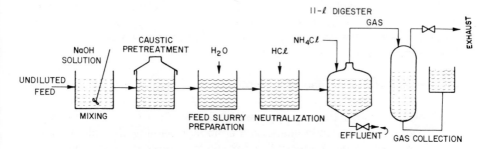

Figure 1.   Experimental sequence for mesophilic digestion of caustic-treated feed

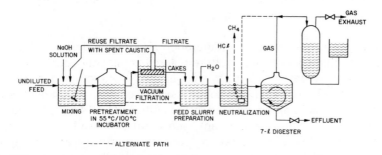

Figure 2.   Experimental sequence for thermophilic digestion of caustic-treated feed

Digestion With Product Gas Recycling

The system used to conduct the gas recycling studies is shown in Figure III. Digester gas was cleaned and dried by passing it through a glass-wool particulate trap, a water-cooled condensate trap, and a $CaSO_4$-filled gas-drying column. The clean, dry gas was recycled at a selected recycle ratio (defined as the ratio of recycled gas flow rate in dry standard cubic feet at 60°F and 30 in. Hg to the digester gas production rate in dry standard cubic feet) into the digesting culture through a glass diffuser. Experiments were conducted at gas recycle ratios ranging from about 2 to 125. The pressure difference across the gas pump corresponding to these recycle ratios ranged from about 1 to 9 psi.

Digestion With Recycling of Aerobically Posttreated Digester Effluent

The purpose of this study was to examine the effect of aerobic biological posttreatment of digester effluent on the biodegradability of recalcitrant feed components passing unconverted or partially converted through the digester. Presuming that aerobic treatment of the digester effluent improved biodegradability, recycling of the treated solids was expected to increase methane production rate and yield with simultaneous and enhanced reduction of the effluent solids and soluble organics load discharged from the overall anaerobic-aerobic system.

In the experimental process configuration (Figure IV), digester effluent was aerobically treated in a 14 *l* culture volume semicontinuously, or in a 2 *l* culture volume batch activated sludge unit. Aeration and mixing were accomplished by diffused aeration. Dissolved oxygen was monitored by a lead-silver galvanic probe inserted in the culture. The 14 *l* activated sludge unit was a two-compartment tank, the aeration chamber of which was separated from the adjacent settling zone by a vertical plate. This unit was used for semicontinuous operation at aerator detention times greater than 2 days. Settled sludge withdrawn from the bottom of the settler was recycled to the digester. The 14 *l* unit was operated at 35°C with an air flow rate of 1 *l*/min. Various runs were conducted at selected aerator detention times and sludge recycle ratios (defined as the volume of recycle sludge divided by the volume of the digester feed).

The 2 *l* batch unit was operated under conditions of limited aeration (0.12 *l*/min) at an ambient temperature of about 25°C. The fill-and-draw activated sludge unit was fed daily with 1,667 ml of fresh digester effluent after withdrawing an equal volume of mixed liquor aerated for 24 hours. Selected volumes of the aerator mixed liquor were recycled to the digester to

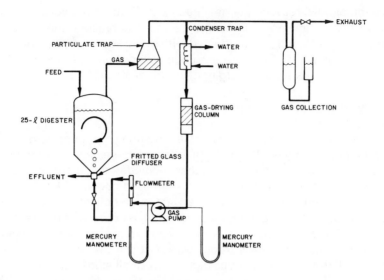

Figure 3. Gas recycling system for mesophilic digestion

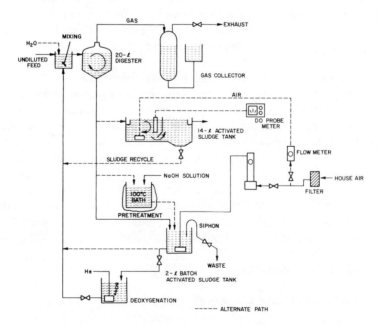

Figure 4. Recycling system for aerobically-posttreated digester effluent

obtain the desired recycle ratio (defined as the ratio of the volume of mixed liquor to the volume of fresh feed). In some experiments, the aerator feed (fresh digester effluent) was pretreated with dilute sodium hydroxide solution for 24 hours at 100°C. The recycle sludge was deoxygenated with a helium purge in some runs before charging it to the digester.

Two-Phase Digester

Two-phase digestion has considerable potential for increasing methane production rate and yield (3,10,18,19). It is a multi-stage, high-rate digestion process in which acidogenic and methanogenic fermentations are optimized in separate digesters. The two-phase system used in our work consisted of a completely mixed acid-phase digester and a completely mixed methane-phase digester or a methane-phase anaerobic filter packed with Raschig rings (Figure V). The acid digester was gravity fed from a sealed overhead feed reservoir having a helium or argon blanket above the feed slurry. Caustic-treated feed was delivered to this digester in small slugs up to 70 times per day by a timer-operated valve to obtain operating conditions closely approaching those of continuous feeding. Effluent from the acid digester overflowed directly to the completely mixed methane-phase digester. Alternatively, part of this effluent was vacuum filtered, and the filtrate was fed continuously to the anaerobic filter from a constant-head Mariotte bottle supplied with helium to fill the reservoir gas phase.

The packed-bed anaerobic filter had a gross volume of 18.5 ℓ and a void ratio of 0.63. Filter effluent was recirculated continuously to the inlet end at a recirculation ratio of 5.1 (defined as the ratio fo the recycle flow rate to the daily feed flow rate) to dilute the incoming feed and accelerate the transport of digestion products out of the culture. The filter had a hydraulic detention time of about 2.33 days (defined as the gross volume divided by the daily feed flow rates).

The two-phase system feed was supplemented with external nitrogen, phosphorus, and magnesium to ensure that feed hydrolysis, acidification, and gasification were not nutrient limited.

RESULTS AND DISCUSSION

Conventional High-Rate Digestion

Steady-state performance of conventional high-rate mesophilic digestion of the biomass-waste blend at the baseline loading and detention time is presented in Table I. The methane yields from replicate baseline runs ranged

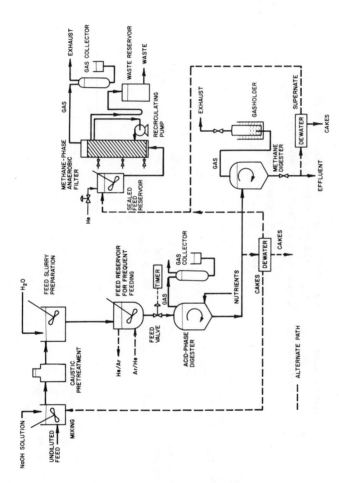

Figure 5.   Two-phase system for biomass–waste blend

Table I. CONVENTIONAL MESOPHILIC (35°C) DIGESTION OF HYACINTH-GRASS-MSW-SLUDGE BLEND UNDER BASELINE AND NON-BASELINE CONDITIONS

	Baseline Runs*			Run
	5M	6M	3MA/2	3MA/7
Operating Conditions				
Culture Volume, ℓ	5	5	20	20
Loading, lb VS/ft^3-day	0.1	0.1	0.1	0.18
Detention Time, days	12	12	12	9
Gas Production				
Rate, std vol/vol culture-day	0.550	0.550	0.570	0.868
Methane Content, mol %	62.0	63.9	60.7	58.7
Methane Yield, SCF/lb VS added	3.40	3.53	3.45	2.88
Effluent Quality				
Volatile Acids, mg/ℓ as acetic	42	35	27	6
Efficiency				
VS Reduction, %†	33.3	33.6	34.5	29.9
Energy Recovery in Collected Methane, %	38.3	39.7	38.8	32.4

* Runs 5M and 6M were replicates.
† Calculated by formula suggested by Klass and Ghosh (21).

between 3.4 and 3.5 SCF/lb VS added. These yields were between 53.1 and 54.7% of the theoretical methane yield. As expected for digesters of similar geometry and mixing configuration, culture volume had no discernible effect on methane production (compare Runs 5M, 6M, and 3MA). Nutritional studies, the details of which were presented in an earlier paper (14), showed that conventional digestion of the biomass-waste feed was not nutrient or growth-factor limited. The performance of subsequent advanced digestion runs was evaluated with reference to the baseline performance data reported in Table I.

Methane yield decreased to 2.4 SCF/lb VS added (43.8% of theoretical yield) when the loading rate was increased to 0.18 lb VS/ft^3-day and the detention time decreased to 9 days (Table I). These data indicate that conventional digestion process efficiency would decrease substantially as the loading rate is increased and the detention time is decreased beyond the baseline values of these parameters. Unconventional or advanced digestion methods are thus needed to overcome the limitations of the conventional high-rate process.

Digestion of Pretreated Feed

As pointed out previously, only about 66% of the hyacinth-grass-MSW-sludge feed VS was determined to be biodegradable under long-term batch digestion conditions. About 52% of the biodegradable organics and 32% of the cellulose component of the feed were gasified at the baseline operating conditions of 0.1 lb VS/ft^3-day loading and a 12-day detention time (14). These data suggest that 34% of the feed organics resisted anaerobic degradation, and that only about one-half or less of the biodegradables could be gasified by conventional high-rate digestion. One probable explanation for the low bioconversion efficiency was that hydrolysis and acidification limited the digestion of the highly fibrous lignocellulosic feed (14). If this is the case, then considerable improvement would be expected when the feed is pretreated chemically to hydrolyze the complex polymeric substances or to convert them to a form suitable for subsequent enzymatic hydrolysis.

Acid or alkaline treatments of particulate feeds have been shown to improve digester gas yields (17,22-25). Acid hydrolysis was not used in our work because severe reaction conditions are required, and there is considerable decomposition of the hydrolytic products under these conditions (26). Dilute alkaline pretreatment was evaluated because alkali was shown to be more effective in promoting hydrolysis of cellulosic biomass than acid (17,26). It is postulated, for example, that sodium hydroxide breaks down the cross-linked lignin macro-molecules surrounding the cellulose fibers into alkali-soluble lower-molecular-weight units. In this way, the cellulose fibers are exposed for enzymatic hydrolysis during anaerobic digestion. Also, it has been suggested that alkali treatment hydrolyzes ester bonds between the uronic acids of hemicellulose and lignin (27), thereby enhancing the biodegradability of hemicellulose.

As already mentioned, digestion runs with caustic treated feed were conducted under a variety of operating conditions. The results of a few selected runs with pretreated feed are presented in Table II. The data in this table show that the highest mesophilic (35°C) methane yield, 4.10 SCF/lb VS added, from digestion of the pretreated feed at a loading of 0.1 lb VS/ft^3-day and a 12-day detention time was obtained with wet feed pretreated with 3 wt % caustic solution. However, the mesophilic methane yield (3.92 SCF/lb VS added) from feed treated with 1 wt % NaOH solution was not significantly different. Also, neutralization of the pretreated feed or addition of external nitrogen to the digester did not affect the mesophilic methane yield. These observations indicate that an increase in mesophilic methane yield up to 20% may be expected at a loading of 0.1 lb VS/ft^3-day and a detention time of 12 days with alkaline pretreatment.

Thermophilic (55°C) digestion of the caustic-treated feed at 0.4 lb VS/ft^3-day loading and a 6-day detention time showed the same methane yield of about 4 SCF/lb VS added as observed during mesophilic digestion of the pretreated feed at a 0.1 lb VS/ft^3-day loading and a 12-day detention time (Run 5T/24, Table II). However, the thermophilic gas production rate was 4-5 times the mesophilic rate. Also, the caustic requirement for feed pretreatment was lower for the thermophilic run owing to the recycling of the spent caustic solution.

A thermophilic methane yield of about 3.7 SCF/lb VS added, which was still larger than the baseline yield, was observed at a loading rate of 0.43 lb VS/ft^3-day and a detention time of 5.5 days when the caustic-treated feed was neutralized to pH 10 with digester gas instead of to pH 9 with hydrochloric acid(Run 5T/26, Table III). This reduced yield resulted from increased loading and decreased detention time. As expected, a higher bicarbonate alkalinity could be maintained when the digester was charged with pretreated feed neutralized with digester gas. In addition, neutralization of the alkaline feed with product gas eliminated the need for neutralizing acid and provided a method of carbon dioxide removal from the digester gas. This technique should provide a reduction in the cost of feed treatment and digester gas cleanup.

Table II. MESOPHILIC (35°C) AND THERMOPHILIC (55°C) DIGESTION OF
HYACINTH-GRASS-MSW-SLUDGE BLEND PRETREATED WITH
CAUSTIC SODA SOLUTION

| | Mesophilic Digester Feed | | Thermophilic Digester Feed | |
	Run 6MA/13[a]	Run 6MA/22[b]	Run 5T/24[c]	Run 5T/26[d]
Neutralization Method	No neutralization of caustic-treated feed.	Whole feed to pH 8-8.4 with acid.	Mixed vacuum filter cakes plus filtrate to pH 9 with acid.	Mixed vacuum filter cakes plus filtrate to pH 10 with acid.
Operating Conditions				
Culture Volume, ℓ	10	10	5	5
Loading, lb VS/ft^3-day	0.1	0.1	0.4	0.43
Detention Time, days	12	12	6.0	5.5
Gas Production				
Rate, std vol/vol culture day	0.606	0.670	2.851	2.911
Methane Content, mol %	59.5	58.1	56.4	56.0
Methane Yield, SCF/lb VS added	4.10	3.92	4.00	3.68
Effluent Quality				
pH	6.84	6.71	7.35	7.43
Volatile Acids, mg/ℓ as acetic	20	21	99	140
Total Alkalinity, mg/ℓ as CaCO$_3$	2618	2433	6075	6298
Bicarbonate Alkalinity, mg/ℓ as CaCO$_3$	2592	2412	5993	6240
Efficiency				
VS Reduction, %	41.9	41.0	43.1	40.0
Energy Recovery in Collected Methane, %	46.1	44.1	45.0	41.4

[a] The fresh feed was treated with 140 ml of 3 wt % NaOH solution at 55°C for 24 hr. The treated feed was not neutralized, and no NH$_4$Cl was added.

[b] Feed pretreatment same as Run 6MA/13 except that 140 ml of 1 wt % NaOH solution was used at 25°C for 24 hr. The treated feed was neutralized with HCl to the indicated pH and supplemented with 10 ml of 120-g/l NH$_4$Cl solution.

[c] Fresh feed was treated with 160 ml of 3 wt NaOH solution at 100°C for 24 hr. The treated feed was vacuum filtered, and about 158 g of filter cake and 200 ml of filtrate were fed to the digester after dilution to the proper volume and neutralization with HCl to pH 9.

[d] Fresh feed was treated in the same way as in Run 5T/24, except that the alkaline feed slurry was neutralized by bubbling a portion of the digester gas (Ave 9.2 ℓ , range 7.19-11.55 ℓ) through it. The average pH of the digester feed after gas neutralization was 9.98 (range 9.31-11.55).

Table III. EFFECT OF PRODUCT GAS AND POSTTREATED EFFLUENT RECYCLING ON MESOPHILIC (35°C) DIGESTION OF HYACINTH-GRASS-MSW-SLUDGE BLEND

Recycled Material and Amount	Run 2MA/11 Digester Gas; 39 gas recycle ratio.[a]	Run 1MA/13 Aerated effluent; 56 vol % recycle.[b]	Run 1MA/18 Aerated digester effluent after deoxygenation; 257 vol % recycle.
Operating Conditions			
Culture volume, ℓ	20	20	20
Loading, lb VS/ft^3-day	0.1	0.1	0.1
Detention Time, days	12	12	12
Gas Production			
Rate, std vol/vol culture-day	0.596	0.630	0.687
Methane Content, mol %	60.5	61.5	56.7
Methane Yield, SCF/lb VS added	3.95	3.83	3.88
Effluent Quality			
pH	7.06	6.69	6.80
Volatile Acids, mg/ℓ as acetic	8	9	20
Total Alkalinity, mg/ℓ as CaCO$_3$	3997	2826	4667
Bicarbonate Alkalinity, mg/ℓ as CaCO$_3$	3997	2819	4645
Efficiency			
VS Reduction, %	39.7	37.9	41.6
Overall VS Reduction, %	39.7	—	55.0
Energy Recovery in Collected Methane, %	44.4	43.1	43.7

a   Recycle ratio is std ℓ recycled gas/std ℓ digester gas produced-day.
b   Daily digester effluent was aerated (air flow rate 0.12 ℓ/min) in a 2.0-ℓ culture volume, batch, activated sludge unit operated at ambient temperature. A volume of 1200 ml of aerator content was recycled with 467 ml of fresh feed daily.
c   Same operating conditions as in Run 1MA/13, except that the recycle aerator sludge was deoxygenated by bubbling helium through it.

The experimental data suggest the following sequence of operations for improved digestion of the hyacinth-grass-MSW-sludge feed:

1.   Caustic treatment of undiluted feed for 24 hr at 100°C with 3 wt % NaOH and recycling of spent caustic solution. Fresh caustic is added at the rate of about 3 meq per gram of VS. Spent caustic recycle flow rate is 75 vol % of the feed slurry flow rate.

2.   Dewatering of the pretreated feed. Dewatered cakes and the balance of the filtrate after recycling are fed to the digester.

3.   Neutralization of alkaline feed slurry with digester gases.

4.   Thermophilic digestion of the feed at a loading rate of 0.4 lb VS/ft^3-day and a detention time of 6 days.

Digestion With Product Gas and Posttreated Effluent Recycling

The second major advanced system was concerned with the recycling of product gas and liquid effluent to the digester. The effluent was posttreated before recycling. The product gas was dehydrated before recirculation to the digesting culture. The aqueous effluent was subjected to various forms of posttreatment including sonication, simple heat treatment for 30 min at 270°F and 25 psig, and activated sludge treatment to improve the biodegradability of the undigested solids. Recycling of sonicated and heat-treated sludge to the digester did not increase methane yield above the baseline yield. Gas recycling and activated sludge posttreatment, which effected a higher system VS reduction, will be discussed here.

Gas Recycling

Recycling of product gases through the digesting culture was expected to increase methane production because of additional methane fermentation from increased reduction of carbon dioxide and because the sweeping action of the recycled gas might be expected to accelerate removal of the gaseous products surrounding the microorganisms thereby minimizing end-product repression.

Digester gases were recycled at various recycle ratios from 2 to 125. The best methane yield of 3.95 SCF/lb VS added, which was about 15% higher than the baseline yield, was observed at a gas recycle ratio of 39 (Table III). The gas-phase dilution rate at this recycle ratio was about 0.2 hr^{-1}. The existence of an optimum gas recycle ratio is rationalized as follows: As the gas recycle ratio is increased, the opportunity for carbon dioxide reduction and sweeping of the gaseous digestion products out of the digester increases thereby stimulating methane production. However, as the gas recycling ratio is increased further, the culture is increasingly saturated with gaseous end products; this would tend to repress or inhibit additional methane production. The opposing effects of increasing gas recycle on methane production equal each other at the optimum gas recycle ratio which maximizes methane yield.

It should be noted that during gas recycling, the test digester was mechanically mixed as in the baseline control runs so that the effect of the recycled gas on test digester methane production could be evaluated. Since mechanical mixing alone was designed to provide complete mixing of the digester contents, the beneficial effect of gas recycling may not be attributed to effects such as substrate transport and substrate-microorganism contact. Mild mechanical mixing also proved to be beneficial during gas recycling because scum formation, which arises due to gas flotation of digester solids, was surprisingly not a problem even at very high gas recycle ratios. The reason for this was that mechanical agitation at a mixing Reynolds number of about 9,000 dispersed the surface solids and moved them down into the culture by the folding action of the two propeller mixers placed at heights of one and two impeller diameters above the digester bottom. The results of the gas recycling experiments showed that modest increases in methane yield can be obtained by recycling product gas at an optimum gas recycle ratio. Dual mechanical and gas mixing eliminated the scum problem associated with gas mixing alone.

Aerobic Sludge Posttreatment of Digester Effluent

The objective of aerobic posttreatment is to treat the "refractory" effluent organics to render them biodegradable, and to increase methane production by recycling the posttreated material for further digestion. Posttreatment may be preferred to pretreatment because it only treats the recalcitrant residue remaining after the biodegradable material is gasified and not the total solids in the feed. Thus, the organic loading rate on the posttreatment process is substantially lower than that for a similar pretreatment process. Also, while improving the biodegradability of the recalcitrant feed fraction, pretreatment may adversely affect the digestibility of feed components that are easily gasified in their original forms. Posttreatment obviates this problem.

Aerobic biochemical and chemical-biochemical posttreatments were investigated because several species of fungi and aerobic organisms are known to hydrolyze and degrade complex lignocellulosic substances to simpler substances (26). *Trichoderma viride, Cellulomonas,* and *Cytophaga hutchin-sonii, Cytophaga vulgaris* are examples of aerobic microorganisms that mediate these reactions (28,29). Because some aerobes cannot efficiently attack lignocellulosic cellulosic complexes (30,31), one run was also conducted in which the aerator feed (digester effluent) was predigested with hot caustic to make the fibers available for aerobic decomposition (28). Finally, activated sludge posttreatment of the digested residue was also conducted under conditions of "limited" aeration to arrest cellulose breakdown before the formation of monomeric products (hexose, pentose, etc.) which are readily oxidized aerobically and, thus, become unavailable for gasification), so that the products of partial aerobic digestion could be gasified anaerobically upon recycling to the digester. Evidence of partial cellulose degradation by limited aeration was presented by Kalnins (32), who showed that *Bacterium protoziodes* decomposed cellulose to dextrose with an unlimited supply of oxygen. However, when the supply of oxygen was limited, dextrose production was stopped, but the organism decomposed an increased quantity of cellulose to derive an amount of energy equivalent to that obtained during degradation of cellulose to dextrose. Thus, the advantage of limited aeration posttreatment is that it should limit degradation of the residual solids to forms that are lost by oxidation before recycling to the digester.

Posttreatment studies of digester effluent conducted in the 14-ℓ mesophilic (33°-34°C) aerobic sludge unit at detention times of 2.4 and 5.3 days, and an air flow rate of 1 ℓ/min produced settleable sludge which, when recycled to the digester at recycle ratios of 5.3 and 33.7%, effected digester methane yields based on fresh-feed volatile solids ranging between 2.81 and 3.51 SCF/lb VS added. These yields were lower than or equal to the baseline methane yield observed under the same digester operating conditions, which indicated that aerobic biological posttreatment at long detention times and a high air flow rate (producing residual aerator dissolved oxygen concentration of 0.7 to 2.8 mg/ℓ) did not enhance methane production. However, residue posttreatment in the 2-ℓ batch unit operated under limited aeration conditions (no residual dissolved oxygen in aerator) was superior in that recycling of this sludge at a 56% recycle ratio led to a modest increase in the digester methane yield (Run 1MA/13, Table III). The methane yield did not increase further when the recycle sludge was deoxygenated, the recycle ratio was increased from 56 to 257% (Run 1MA/18), or the feed to the aerator was pretreated with hot caustic. It is interesting to note that the volatile solids reduction increased when the aerobic posttreatment method was used as shown by the overall VS reduction of Run 1MA/18.

Two-Phase Digestion of Slurry Systems

Two-phase studies were undertaken to develop a multistage high-rate process superior to conventional digestion. Results of selected two-phase runs are presented in Tables IV and V. Two-phase digestion of the untreated feed at an overall detention time of 7.6 days and a loading rate of 0.26 lb VS/ft^3-day exhibited a total gas production rate of 1.35 std vol/vol of culture-day and an effluent volatile acid concentration of about 19 mg/ℓ (Run A30/M16, Table IV). Thus, the two-phase system had a gas production rate that was about 2.5 times that of conventional baseline digestion, and yet had about the same effluent volatile acid concentration as that observed during conventional digestion at a 12-day detention time. The methane yield for this run, however, was about 2 SCF/lb VS added, and the volatile acid (VA) yield from this system was estimated to be 0.31 (mass of VA as acetic divided by mass of VS added), which indicated that hydrolysis and acidification of the feed organics were inefficient. To improve this condition, Run A3C/M3C was conducted with caustic-treated feed. Gas production rate, methane yield, and acid yield from this run were 1.5 std vol/vol of culture-day, 3 SCF/lb VS added, and 0.49, respectively, all of which were substantially higher than those observed with the untreated feed. The volatile acid yield coefficient of 0.49 for the solid biomass-waste feed was lower than the acid yield of 0.73 reported by Ghosh and Pohland (33) for two-phase digestion of the simple soluble sugar, glucose, suggesting that it may still be possible to improve the VS-to-acid conversion efficiency beyond that realized by caustic treatment. Further acid yield or biodegradability increase is expected to be difficult to achieve and may require more severe feed pretreatment.

Run A7C/M7C in Table IV had the same operating conditions as the two other runs discussed above, but received only external nitrogen instead of external nitrogen, phosphorus and magnesium. Inspection of the data in Table IV shows that elimination of external phosphorus and magnesium from the feed did not affect two-phase process performance. This observation correlated with the results of the conventional digestion runs which showed that the biomass-waste blend used in this work was not nutritionally deficient. All two-phase process feeds, however, were fortified with external nitrogen to guard against any deficiency of this element due to loss that could occur during caustic treatment of the feed. Addition of external nitrogen would not be required in a commercial process utilizing a properly designed continuous pretreatment reactor.

To test kinetic potential of the two-phase system, additional runs were conducted at decreased detention times of 5 and 4 days, and increased loading rates (Table V). A temperature of 55°C was selected for the acid-

Table IV. EFFECT OF FEED PRETREATMENT ON TWO-PHASE MESOPHILIC (35°C)
DIGESTION OF HYACINTH-GRASS-MSW-SLUDGE BLEND

	Untreated Feed Run A30/M16	Caustic Treated Feed Run A3C/M3C	Run A7C/M7C
Operating Conditions*			
Acid-Phase Culture Volume, ℓ	16	16	16
Methane-Phase Culture Volume, ℓ	45	45	45
Acid-Phase Loading,			
lb VS/ft^3-day	1.0	1.0	1.0
Methane-Phase Loading,			
lb VS/ft^3-day	0.32	0.32	0.32
Acid-Phase Detention Time, days	2.0	2.0	2.0
Methane Phase Detention Time,			
days	5.6	5.6	5.6
Overall Loading, lb VS/ft^3-day	0.26	0.26	0.26
Overall Detention Time, days	7.6	7.6	7.6
External Nutrient Additions			
to Acid-Phase Feed	N,P,Mg	N,P,Mg	N
Gas Production			
Rate, std vol/vol culture-day			
Acid Phase	1.346	—	—
Methane Phase	0.924	—	—
Total	1.035	1.484	1.443
Methane Content, mol %			
Acid Phase	49.2	—	—
Methane Phase	59.0	—	—
Total	49.7	52.8	55.9
Gas Yield, SCF/lb VS added			
Acid Phase	1.34	—	—
Methane Phase	2.91	—	—
Total	3.98	5.63	5.50
Total Methane Yield,			
SCF/lb VS added	1.98	2.87	3.07
Effluent Quality			
pH			
Acid Phase	5.86	6.72	6.82
Methane Phase	6.44	7.03	7.15
Volatile Acids, mg/ℓ as acetic			
Acid Phase	1047	2440	1860
Methane Phase	19	287	148
Efficiency			
VS Reduction, %	24.2	34.2	33.4
Energy Recovery in Collected			
Methane, %	22.3	33.4	34.5

* No alkali was used for digester pH control. The acid-phase feeds for Runs A3C/M3C and A7C/M7C were pretreated for 24 hr with 2.56 ℓ of 1 wt % NaOH solution in a total volume of 4.8 ℓ under ambient conditions. The product gases from each phase were collected together for these two runs.

Table V. THERMOPHILIC (55°C) ACID-PHASE AND MESOPHILIC (35°C)
METHANE-PHASE DIGESTION OF PRETREATED HYACINTH-GRASS-MSW-SLUDGE BLEND

	Run A35/M21	Run A37/M23
Operating Conditions*		
Acid-Phase Culture Volume, ℓ	5	5
Methane-Phase Culture Volume, ℓ	20	20
Acid-Phase Loading, lb VS/ft^3-day	2.0	2.5
Methane-Phase Loading, lb VS/ft^3-day	0.44	0.56
Acid-Phase Detention Time, days	1.0	0.80
Methane-Phase Detention Time, days	4.0	3.2
Overall Loading, lb VS/ft^3-day	0.40	0.50
Overall Detention Time, days	5.0	4.0
External Nutrient Additions		
to Acid-Phase Feed	N	N
Gas Production		
Rate, std vol/vol culture-day		
Acid Phase	3.189	3.472
Methane Phase	1.553	1.740
Total	1.880	2.086
Methane Content, mol %		
Acid Phase	60.3	58.5
Methane Phase	59.6	59.1
Total	59.8	57.9
Gas Yield, SCF/lb VS added		
Acid Phase	1.53	1.40
Methane Phase	3.35	3.11
Total	4.70	4.25
Total Methane Yield, SCF/lb VS added	2.81	2.46
Effluent Quality		
pH		
Acid Phase	6.99	7.05
Methane Phase	7.05	7.05
Volatile Acids, mg/ℓ as acetic		
Acid Phase	1734	2255
Methane Phase	347	591
Efficiency		
VS Reduction, %	28.6	25.8
Energy Recovery in Collected Methane, %	31.6	27.7

* No alkali was used for digester pH control. The acid-phase digester feed was pretreated for 24 hr with NaOH solution at ambient conditions. The caustic concentration in the slurry was 142 meg/ℓ

phase digester because of the earlier observation that this thermophilic temperature improved gasification rates at higher loadings and shorter detention times without significantly affecting methane yield. As shown in Table V, an overall system gas production rate of 1.9 std vol/vol culture-day, a methane yield of 2.8 SCF/lb VS added, and an effluent volatile acid concentration of about 350 mg/ℓ were observed at a system loading and detention time of 0.4 lb VS/ft^3-day and 5 days, respectively. As expected, gas production rate and effluent volatile acids concentration increased to about 2.1 std vol/vol culture-day and 600 mg/ℓ , and methane yield decreased to about 2.5 SCF/lb VS added when the system detention time was decreased to 4 days and the system loading was increased to 0.5 lb VS/ft^3-day. Volatile acid yields at the 5- and 4-day detention times were, however, about the same, 0.46 and 0.50, respectively. Comparison of acid production yield coefficients for the runs in Tables IV and V indicated that volatile solids conversion to volatile acids is significantly increased by alkaline feed pretreatment. Furthermore, it was observed that, while the VS-to-acid conversion efficiency of the pretreated feed remained the same as the system detention time was decreased from 7.6 to 4 days, the gas production rate increased by 50% compared to a 17% decrease in methane yield. These observations indicate that it is desirable to operate the two-phase system at a detention time of 4 days or less, and to couple this system to a cell mass recycling device (e.g., anaerobic settler) to prevent methane yield reductions and volatile acids accumulation associated with short detention times. It should be noted, however, that settling of relatively concentrated digested biomass-waste slurry and anaerobic settler operation were problematic and appeared impractical in light of our experience with custom-designed laboratory settlers. An alternate approach, which has the same effect of maintaining higher cell density and increasing solids retention time (SRT) as in a settler, is a packed bed anaerobic filter. With this reactor, it is be possible to conduct digestion at short hydraulic retention time (HRT) and still obtain high methane yield and low effluent volatile acid concentration because of the high SRT.

Packed-Bed Anaerobic Digester

The packed-bed anaerobic digester, commonly referred to as an anaerobic "filter", was operated with filtrates from the vacuum filtration of the acid-digester effluent. Filtrate was used because unfiltered acid-digester effluents tended to clog the packed bed. The feed to the packed-bed digester had volatile acids concentrations between 1500 and 2000 mg/ℓ as acetic and a solids content of about 0.5 wt %. The digester was operated at an HRT of about 2.3 days and a loading rate of about 0.15 lb VS/ft^3-day based on the gross filter volume (1.5 days and 0.24 lb VS/ft^3-day when based on the void or culture volume). Effluent recirculation and culture temperature had significant effects on gas production (Table VI). Recirculation of the effluent from the top to the bottom of the bed increased gas production rate by 76%, while also increasing the methane yield by 45%. Methane content was above 70 mol % with or without recirculation. Increase in mean digester temperature from 33° to 38°C decreased gas production rate by about 20% and methane yield by about 25%. Depending on the operating condition, volatile acids concentrations in the filter effluent varied between 90 and 190 mg/ℓ (Table VI), which were still lower than those in the slurry methane digester operated at a higher HRT of 5.6 days (Table V).

It should be pointed out that the performance of the packed-bed methane digester is more dependent perhaps on the influent volatile acids concentration than the overall VS loading rate. Thus, filter performance with acid-digester filtrate might be expected to be better than that obtained with methane-digester effluents having comparable volatile acids content.

Hypothetical Multi-Stage Digestion Process

The ultimate objective of this research was to synthesize a hypothetical biomass-waste pilot process design based on the results of our investigation of promising advanced digestion modes. The work presented here provided information on the effects of alkaline pretreatment, alkali recycling, digestion temperature, and digester gas recycling; heat, sonication, and aerobic posttreatment and recycling of digested sludge under different aeration, sludge recycle, and aerator detention time conditions; slurry and packed-bed digestion; and two-phase digestion. The results of this work are suggestive of an advanced biomass-waste digestion system as depicted in Figure VI. Additional work to refine the pre- and posttreatment techniques and two-phase process optimization is better conducted in a pilot system similar to that of Figure VI.

The hypothetical system is necessarily a multi-stage system to accomodate pretreatment, multi-stage phasic digestion, and digested residue posttreatment. The hydraulic residence time in the total system is about 6 days allowing for 12 hr of dilute caustic pretreatment, 12-24 hr of thermophilic acid digestion, 2-5 days of slurry-phase mesophilic methane digestion, 2 days of mesophilic packed-bed methane digestion, and 12 hr of limited-aeration biological treatment of the digested sludge. Based on the data compiled in this work, methane yields up to 5.5 SCF/lb VS added are expected for this type of configuration.

Table VI. STEADY-STATE PERFORMANCE OF CONTINUOUSLY FED MESOPHILIC PACKED-BED METHANE DIGESTER

	Run 8MA/4	Run 8MA/5	Run 8MA/6
Operating Conditions*			
Temperature, °C	35	33	38
Loading, lb VS/ft^3-day	0.13	0.15	0.16
Detention Time, days	2.30	2.33	2.34
Effluent Recycle Ratio, %	0	510	510
Gas Production			
Rate, std vol/vol culture-day	0.495	0.872	0.698
Methane Content, mol %	73.5	72.3	68.7
Methane Yield, SCF/lb VS added	2.82	4.08	3.05
Effluent Quality			
pH	6.73	6.86	6.86
Volatile Acids, mg/ℓ as acetic	186	131	68
Efficiency			
VS Reduction, %	23.3	34.3	27.0
Energy Recovery in Collected Methane, %	31.7	45.9	34.3

* The packed-bed digester had a gross volume of 18.5 ℓ. The bed had a void ratio of 0.63. Filtrate from a slurry-phase acid digester effluent having a volatile acid content of 1550-2000 mg/ℓ as acetic acid and a solids content of about 0.5 wt % was used as feed. No pH control was used.

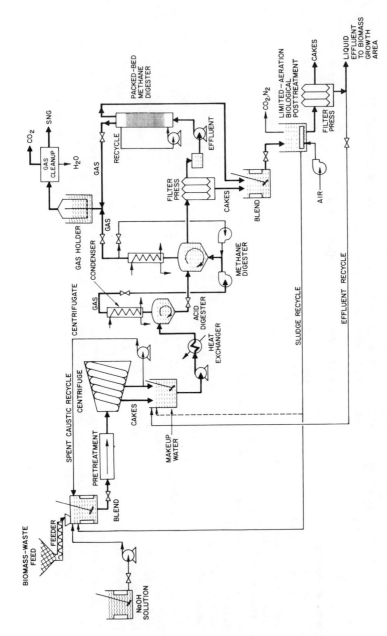

Figure 6.   Advanced biomass–waste digestion system

SUMMARY

A series of exploratory anaerobic digestion experiments was performed with a mixed biomass-waste feed to search for digestion configurations that provide improved performance over that of conventional high-rate digestion. The techniques studied were pretreatment of the feed with caustic soda, product gas recycling to the digester, recycling of aerobically treated digester effluent to the digester, two-phase digestion with complete mix acid- and methane-phase reactors, and packed-bed, methane-phase digestion of the effluent from an acid-phase reactor. Ambient-temperature pretreatment of the feed blend with dilute caustic and recycling of the product gas each afforded higher methane yields and volatile solids reduction efficiencies than high-rate digestion alone. It was found that spent caustic could be recycled for fresh feed pretreatment and that neutralization was not necessary before feeding to the digester. Two-phase digestion in the complete-mix reactors gave methane yields and reduction efficiencies about the same as those of high-rate digestion but at much higher loadings and reduced detention times thereby offering significant reductions in equipment size for the same throughputs. The use of a packed-bed anaerobic filter as a methane-phase reactor also showed considerable promise for operation at reduced detention times when the filter effluent was recycled to the filter inlet. Analysis of the data from the experiments conducted to study each advanced digestion technique indicates that an integrated series of unit processes consisting of dilute caustic pretreatment, thermophilic acid-phase digestion, mesophilic complete-mix and packed-bed methane-phase digestion, and limited-aeration aerobic treatment of the methane-phase effluents coupled with recycling should exhibit digestion efficiencies and methane yields near the upper practical limits.

ACKNOWLEDGEMENT

This research was supported by United Gas Pipe Line Company (UGPL), Houston, Texas. The project was done under the management of UGPL and is currently managed by the Gas Research Institute. The guidance and help of Mr. Robert Christopher and Dr. Victor Edwards, both of UGPL, were invaluable. The assistance of Dr. B. C. Wolverton of NASA, Mr. Dawson M. Johns of LSU, Mr. Robert Power of Waste Management, Inc., and the staff of the Metropolitan Sanitary District of Greater Chicago in providing the feed samples is appreciated. The authors also acknowledge the efforts of Janet Vorres, Michael Henry, Alvin Iverson, Mona Singh, Frank Sedzielarz, Phek Hwee Yen, and Ramanurti Ravichandran in collecting the experimental data.

REFERENCES

1. Bylinsky, G., *Fortune* **1979**, *100*, 78-79, 81 Sept. 24.

2. Klass, D.L. "Abstracts", Biomass for Energy, sponsored by the U.K. Sec. Int. Solar Energy Soc., London, England, July 3, 1979.

3. Pohland, F.G.; Ghosh, S. *Biotechnol. Bioeng. Symp. No. 2,* **1971,** 85-106.

4. Gavett, W. U.S. Patent 1 420 250, 1922.

5. Ort, J.E. U.S. Patent 3 981 800, 1976.

6. Ort, J.E. U.S. Patent 4 040 953, 1977.

7. Switzgable, H. U.S. Patent 4 053 395, 1977.

8. Klass, D.L.; Ghosh, S.; Conrad, J.R. U.S. Patent 3 994 780, 1976.

9. Ghosh, S.; Conrad, J.R.; Klass, D.L. Research Symposium of the 47th Annual Conference, Water Pollution Control Federation, Denver, Colo., Oct. 6-11, 1974.

10. Ghosh, S.; Klass, D.L. U.S. Patent 4 022 665, 1977.

11. Ghosh, S.; Klass, D.L. "Symposium Papers", Clean Fuels From Biomass, Sewage Urban Refuse, and Agricultural Wastes, Symposium sponsored by Institute of Gas Technology, Orlando, Fla., Jan. 27-30, 1976; Institute of Gas Technology: Chicago, 1976.

12. Haug, R.T. *J. Water Pollut. Control Fed.* **1977,** *49* (7), 1713.

13. Klass, D.L. "Proceedings", Bio-Energy World Congress and Exposition, Atlanta, Ga., April 21-24, 1980; Bio-Energy Council: Washington, D.C., 1980.

14. Ghosh, S.; Henry, M.P.; Klass, D.L. Second Symposium on Biotechnology in Energy Production and Conservation, Gatlinburg, Tenn., Oct. 3-5, 1979.

15. Klass, D.L.; Ghosh, S. Biomass as a Nonfossil Fuel Source, American Chemical Society and the Chemical Society of Japan Joint Chemical Congress, Preprints, Div. Pet. Chem., Am. Chem. Soc. **1979,** *24,* (2), 414-28.

16. Klass, D.L.; Ghosh, S. Fuels From Biomass, American Chemical Society National Meeting, Aug. 27-29, 1980, San Francisco, Calif., Preprints, Div. Fuel Chem., Am. Chem. Soc. **1980.**

17. Klass, D.L.; Ghosh, S.; Conrad, J.R. "Symposium Papers", Clean Fuels From Biomass, Sewage, Urban Refuse and Agricultural Wastes, Symposium sponsored by the Institute of Gas Technology, Orlando, Fla., Jan. 27-30, 1976; Institute of Gas Technology: Chicago, 1976.

18. Ghosh, S.; Klass, D.L. *Process Biochem.* **1978,** *13,* 15-24.

19. Heertjes, P.M.; Van der Meer, R.R. Purdue University Industrial Waste Conference, West Lafayette, Ind. May 8-10, 1979; Ann Arbor Science Publishers, Inc.: Ann Arbor, Mich., 1979.

20. Norrman, J.; Frostell B. *ibid.*

21. Klass, D.L.; Ghosh S. "Symposium Papers", Clean Fuels from Biomass and Waste, Symposium sponsored by the Institute of Gas Technology, Orlando, Fla., Jan. 25-28, 1977; Institute of Gas Technology: Chicago, 1977.

22. Buswell, A.M., *et al.* U.S. Patent 1 880 773, 1929.

23. Compere, A.L.; Griffith, W.L. Report under contract No. W-7405-eng 26, Oak Ridge National Laboratory, Oak Ridge, Tenn., June 1975.

24. McCarty, P.L. *et al.* "Proceedings", Microbial Energy Conversion, H.G. Schlegel; J. Barnea, Eds.; Erich Goltze KG: Gottingen, FRG, 1976.

25. Ghose, T.K.; Bisaria, U.S. *Biotechnol. Bioeng.* **1979,** *21,* 131-46.

26. Bellamy, W.D. *Biotechnol. Bioeng.* **1974,** *16,* 869-80.

27. Kirk, K. *Annu. Rev. Phytopathol.* **1971,** *9,* 185.

28. Bellamy, W.D. "Proceedings", International conference on Single Cell Protein; MIT Press: Cambridge, Mass., 1973.

278   BIOMASS AS A NONFOSSIL FUEL SOURCE

29. Imshenetskii, A.A.; Solntzeva, L. *Bull. Acad Sci. U.S.S.R.* **1936,** Ser. Biol. *6,* 1115.

30. Dunlap, E.E. 2nd International Conference on Single Cell Protein; M.I.T. Press: Cambridge, Mass., 1973.

31. Callihan, C.D.; Dunlap, E.E. Report No. SW-24c, U.S. Environmental Protection Agency, Washington, D.C., 1971.

32. Kalnins, A. Acta Univ. *Latvieness Lauksaimmiecibas Fakultat. Ser. I* **1930,** *11,* 221.

RECEIVED JUNE 24, 1980.

Methane Production from Landfills

An Introduction

EDWARD J. DALEY, IRA J. WRIGHT, and ROBERT E. SPITZKA

Brown and Caldwell Consulting Engineers, Resource Recovery and Energy
Conservation, 1501 North Broadway, Walnut Creek, CA 94596

The sanitary landfill process was developed to provide a means of disposal of wastes, particularly urban refuse and industrial solid wastes, in a manner that will not pollute the environment. In simple terms, this is accomplished by sealing the wastes to prevent interaction with the environment. Soil with low permeability provides the seal. The integrity of the seal is dependent upon the quality of the landfill operation, especially the cover soil compaction requirements. In a well-designed landfill, the movement of moisture and gas into or out of the interior of the landfill is significantly restricted to create a relatively closed environment for the wastes.

Within this closed environment can be found an extremely wide variety of materials. Though waste composition differs widely from place-to-place and from season-to-season, a typical landfill can be expected to have a composition similar to that shown in Table I (1). The waste contains considerable moisture (about 25 percent by weight) and a high concentration of organics. As a result of the collection, unloading, spreading, and compaction of wastes during landfilling, the fill material will be heterogeneous with a large surface area-to-volume ratio. It will be under compression due to the weight of the compacted trash and earth above it.

During the landfilling process, a certain amount of air will be trapped in the landfill interior along with the trash. The quantity of air cannot be determined, but can logically be expected to be significant. A small amount of free water

0097-6156/81/0144-0279$05.00/0
© 1981 American Chemical Society

Table I. TYPICAL LANDFILL WASTE COMPOSITION

Component	Bulk Composition Percent by Weight		Analysis Percent Dry Weight
Paper	42.0	Carbon	28.0
Wood	2.3	Hydrogen	3.5
Glass, brush, etc.	12.0	Oxygen	22.4
Leather	0.3	Nitrogen	0.33
Rubber	0.6	Sulfur	0.16
Plastic	0.8	Noncombustibles	24.9
Oil, paints	0.8	Moisture	20.7
Rags	0.6		
Dirt and fines	4.5		
Food wastes	12.0		
Noncombustibles	24.0		

is also likely. There are two primary sources for this water. It may result from water entrained in the waste that is released because of compaction, or it may result from rainfall occuring during the day prior to the placement of cover soil over the trash. In poorly designed landfills, it could also result from rainwater percolation through the cover soil or groundwater infiltration.

The temperature within the landfill is affected by the insulating quality of the soil (2). The landfill in many respects is an ideal environment for chemical and microbial activity. It provides a large quantity of organics with a large surface area-to-volume ratio, significant amounts of air and water, is under compression, and is very temperature-constant. As a result, the solid waste is subject to extensive chemical and microbial decomposition. Over a period of many years, these natural processes convert the organic fraction of the solid waste into a fairly inert waste material and generate liquid and gaseous byproducts, including methane. Landfill decomposition has been the subject of considerable scientific investigation over the last few years. Researchers are developing a better understanding of the mechanics of the physical processes involved and how they affect the generation of methane.

Chemical Factors

Although the chemical and microbial factors cannot, in reality, be understood separately, it is useful to discuss major chemical changes which take place within the landfill since they serve to accelerate the breakdown of the waste, thus facilitating microbial activity. Generally speaking, the major chemical reactions occur in the liquid phase. Contact with the solid waste results in a liquid heavily loaded with organics and inorganics. Typical processes occurring in a contaminated liquid environment, such as ion exchange and hydrolysis, increase the ability of the liquid to dissolve additional waste components. The oxidation-reduction potential of the liquid steadily increases, resulting in an environment that is highly conducive to chemical breakdown of complex molecules. This produces a fairly homogeneous substrate that is very susceptible to microbial action.

Microbial Factors

A recent report (3) included a review of biological decomposition within a landfill. This report serves as the basis for a brief summary of these microbial processes.

The solid waste placed in the landfill contains a wide variety of ubiquitous microorganisms. As the free oxygen in the landfill environment is consumed by the microorganisms, the overall process goes through aerobic, facultative, and finally anaerobic stages, as various groups of organisms dominate depending on their oxygen dependency. The separation of the phases is generally not distinct. This occurs because the phases are interrelated, and to an extent, interdependent. In addition, when considering the landfill as a micro-environment, it is clear that the state of activity can vary widely from location to location within a landfill due to the varying waste composition and age.

In the initial phase, aerobic microorganisms convert readily degradable organics in the waste to solid residuals, water, and a gas mixture that is primarily carbon dioxide. This process results in further modification of the landfill environment. Since the process is exothermic, the temperature within the landfill increases significantly. The carbon dioxide partially dissolves in the water and, along with organic acids released as waste material by the microorganisms, reduces the pH. (The chemical processes described in the previous section are thus fortified.) Very little methane is generated during this phase of the landfill decomposition.

The facultative microorganisms which become dominant as the oxygen is depleted by the aerobes are commonly called acid formers. As the name implies, organic acids are major products of their activity. Carbon dioxide remains the major major gaseous product, and less heat is generated. Due to the similarity of products generated in this phase and the aerobic phase, it is extremely difficult to distinguish between the two in practice. In many cases, this phase may actually never become dominant. However, its transitory function is very relevant to the overall decomposition process.

The anaerobic phase is of most interest because it results in the generation of gases which are typically rich in methane. This is the phase of the decomposition which has been studied most extensively. However, most of the work has not yet reached firm conclusions. The yet-to-be completely confirmed hypothesis is that three main groups of microorganisms are active in this phase. Microorganisms classed as fermentors act upon cellulose, lipids, and proteins, resulting in the production of a wide variety of organic acids as well as alcohols, hydrogen, carbon dioxide, ammonia, and sulfide. Another group, the acetogenics, (although they have not been confidently identified as such to date), apparently convert higher molecular weight organics to acetate and hydrogen. There is considerable speculation that hydrogen accumulation plays a major role in triggering the operation of the most important group of microorganisms, the methane formers. These microorganisms are capable of producing methane from carbon dioxide and hydrogen or formic, acetic, and possibly proprionic and butyric acids, using ammonia as a nitrogen source. They provide the last link in the decomposition process.

A very simplified scheme for conversion of cellulosics to methane in a landfill might be represented by:

Hydrolysis
$$(C_6H_{10}O_5)_x + H_2O \longrightarrow C_6H_{12}O_6$$

Acid Formation
$$C_6H_{12}O_6 \longrightarrow 3\ CH_3CO_2H$$

Methane Formation
$$CH_3CO_2H \longrightarrow CH_4 + CO_2$$

As with any over simplification of a complex process, its true character can be understood only by combining this representation with an appreciation of the details of the complete process. Our knowledge of landfill decomposition will be altered as new investigations shed more light on the subject. Basically, the process is similar to the anaerobic digestion of sewage and other wastes in water slurries but effectively corresponds to high-solids digestion.

GAS GENERATION AND RECOVERY

Generation Factors

Since landfill gas generation is a result of the complex natural processes summarized above, it is not surprising that the rate of methane generation varies widely between landfills. The maximum rate of generation requires a mixture of diverse nutrients and microorganisms as well as optimum environmental conditions such as near-neutral pH, adequate moisture, and moderate temperature (5,6,7). Generally, optimum conditions are seldom present in a landfill since one or more of the elements is absent. A common deficiency is nitrogen. In dry climates, moisture may be deficient. The methane content of the gas will be significantly retarded if pH levels drop too low or if oxygen is introduced into the landfill after the anaerobes have become established. Under optimum conditions, waste stabilization is obtained in 10 to 20 years and is characterized by the cessation of gas generation. In the absence of optimum conditions, it is very difficult to predict the time required for decomposition of the landfill to the point of stabilization. This condition may require up to 30 years. It is also difficult to determine when the landfill will reach the anerobic stage and begin to produce significant amounts of methane. Up to this point, the landfill gas is of little interest because of its low heating value. To date, a means of predicting the time interval for the generation of methane-rich landfill gas in significant quantities does not exist.

The uncertainty regarding methane generation rate is further compounded by the fact that there is no scientific agreement with regard to the total quantity of gas which will be produced in a landfill prior to waste stabilization. The reported information on landfill gas production is summarized in Table II to illustrate the wide variance in calculated, measured or estimated gas production. Additional study is needed to more accurately predict the quantity of gas which will be generated in a landfill.

In practice, the total gas generation potential will never be realized in a landfill. Generation losses will occur for a number of reasons. One source of carbon loss, particularly in a poorly designed landfill, results from liquid substrate (called leachate) percolating out of the landfill. Leachate from a landfill can have a biological oxygen demand (BOD) on the order of several thousand milligrams per liter, which represents the loss of a small amount of organics (13). An additional amount of substrate will pass out of the landfill with the gas, particularly during the operation of a gas recovery system. In one recovery system in California, the liquid condensate removed from recovered gas was found to be very high in organic acids (20). A certain amount of the substrate is also converted to cell matter by the microorganisms. This loss reportedly can be quite significant (21). Finally, a variable amount of gas will be unavailable for recovery as a result of gas migration, particularly for landfills located in permeable soils that permit lateral movement and venting to the atmosphere.

Because of these numerous uncertainties, attempts to predict the quantity of recoverable gas for a given landfill have to date been based on previous experience in existing landfill recovery systems. This empirical approach leaves much to be desired. A survey of existing systems indicated somewhat widespread use of an empirical factor of 200-600 cubic feet per minute of gas per million tons of refuse within a landfill during a peak generation period beginning one to two years after placement of the waste, a date which varies from point to point in a landfill, and extending for five to fifteen years. (22) However, there is a considerable amount of work being done in an attempt to develop an accurate numerical model for the prediction of the gas generation profile for a landfill. Those efforts with which the authors are familiar are being conducted by Pacific Gas and Electric Company in San Francisco, and Johns Hopkins University Applied Physics Laboratory.

Typical Gas Composition

Although the variation in gas composition reported in the literature and observed by the authors in past experience is large, there is in general a certain amount of similarity. In a typical landfill, the composition of the gas

Table II. TOTAL LANDFILL GAS (CARBON DIOXIDE AND METHANE) GENERATION FROM MUNICIPAL SOLID WASTE

Sources	Basis	Gas Production std cu m/kg
Anderson & Callinan (8)	Theoretical	0.41
Boyle (9)	Theoretical	0.45
Golueke (10)	Theoretical	0.30
Pacey (11)	Theoretical	0.12
Klein (12)	Lab Measurement	0.24
Hitte (13)	Lab Measurement	0.21
Pfeffer (14)	Lab Measurement	0.26
Schwegler (15)	Estimated	0.19
Alpern (16)	— —	0.53
City of Los Angeles (17)	Theoretical/ measurement in landfill	0.39
Bowerman, *et al.* (5)	Theoretical	0.40
Blanchet (18)	Theoretical/ measurement in landfill	0.13

can be predicted in approximate terms with a reasonable amount of certainty. During the first two phases of decomposition, the gas will be primarily carbon dioxide. In a classic example, a California landfill exhibited a carbon dioxide concentration of 90 percent during this period (23). As the methane formers become dominant, the gas typically becomes increasingly methane-rich, until a point of fairly constant methane generation is reached. From this point on, the gas can be expected to contain 40 to 50 percent carbon dioxide and 45 to 55 percent methane, although higher methane contents have been reported in very limited cases (1). In many instances, the combined concentration of carbon dioxide and methane will approach 99 percent, with other constituents present in very small amounts. There will also be a variety of trace gases which can at times be critical as a result of their corrosive, toxic or odor qualities. In other instances, hydrogen will be found in significant quantities, although normally only during short intervals prior to the anaerobic stage. This probably results from a temporary imbalance in the relative activity of the acetogenic and methane forming microorganisms. Nitrogen is quite often found in samples of gas from recovery systems and is often an indication of the presence of air in the landfill or air leaks in the recovery system. The nitrogen remains following the consumption of oxygen. Other compounds found in landfill gas in other than trace quantities are generally the result of anomalous waste composition. For example, the sulfur content of solid waste is normally too low to result in the formation of significant amounts of hydrogen sulfide (1), but unusual wastes that could be deposited within the landfill could contain more sulfur and cause high hydrogen sulfide concentrations in the gas. If air is not entrained in the recovery process and the waste does not contain abnormal quantities of nontypical compounds, the landfill gas can be expected to exhibit a higher heating value of 450 to 540 Btu/SCF. The gas can also be expected to be saturated with water vapor.

The composition of the gas will remain relatively constant through the period of peak generation. As the landfill nears the stabilization point, gas generation will begin to approach zero asymptotically. For this reason, the total period of gas generation is often described in terms of a half-life. During this period of decreasing generation, although the composition of the generated gas will probably remain constant, the low generation rate will result in low internal pressures and the seepage of air into the landfill. At this point, the gas withdrawn may contain significant concentrations of nitrogen and oxygen, because it will become difficult to collect the generated gas without collecting air as well.

Gas Recovery System Design

There is a dearth of information on the proper design of landfill gas recovery systems. Most existing reports emphasize the questions of gas generation and utilization. Several of the existing projects have been developed under proprietary conditions by private firms. The most illuminating report to date is a site study for the Mountain View landfill (24) in California, an EPA — Pacific Gas and Electric Company demonstration project. None of the other gas recovery system designs has been publicized to any extent and the U.S. Department of Energy has only recently become involved in funding studies related to landfill gas recovery system design.

Landfill gas recovery systems consist of a series of wells and lateral piping connected to a centralized vacuum pump. In simple systems, the pump is actually a centrifugal blower. In systems requiring compression of the gas to facilitate processing or transport, the compressor is normally used to provide the withdrawal vacuum as well. A system with decentralized vacuum pumps would be technically feasible, but has not yet been implemented. This approach may have merit if a low-Btu, small-scale gas turbine electric generator, which is being developed by Alpha National, proves feasible. The lateral piping could then be replaced with electrical transmission wires between generators located at each wellhead and the power user or a substation.

Horizontal collection pipes have been proposed for landfills presently being filled, but interference with ongoing equipment operation poses a major problem. In current practice, the wells consist of perforated pipes extending vertically into the waste to a point approximately 3/4 of the total depth of the landfill. The perforations are normally thin slots. The borehole is backfilled with rock or gravel to prevent waste from blocking the slots in the pipe casing. The perforations are located in the lower portion of the well to prevent the induction of air into the landfill from above the cover soil layer. A concrete seal is typically placed above the gravel. The upper portion of the borehole is backfilled with earth.

Plastic pipe is normally used for the well casing, although a few steel wells do exist. Steel wells are generally thought to be less desirable because of the corrosiveness of the wet gas and the tendency of the landfill to settle differentially as the waste decomposes. Polyvinyl chloride, polyethylene, and fiberglass are not subject to corrosion and are more flexible than steel. Plastic has been used for lateral piping in most designs due to its ability to accommodate the differential settlement of the landfill. Telescoping wells and expansion joints in the laterals have also been used. Laterals have been placed below the cover soil, in the cover soil layer, and above the cover, but are always blanketed with earth to prevent expansion and contraction that would result if the plastic pipe were exposed to ambient temperatures. The earth also tends to keep the pipe warm, thereby reducing the amount of water which condenses in the pipe. This water can be a problem in low flow rate systems, and considerable effort has been expended in the design of acceptable water traps. Valves are placed at key points in the system to control the amount of vacuum at each wellhead.

A difficult part of a landfill gas recovery system design is the determination of the well spacing. The area of influence of each well, or the waste volume around the well casing from which the gas is extracted, is determined by: (1) the permeability of the waste and intermediate soil layers within the landfill which vary widely between landfills, and (2) the level of applied vacuum. But the area of influence can fluctuate since the internal pressure of the landfill often tends to vary according to a diurnal cycle, and the necessary vacuum at the wellhead can change over time as the flow characteristics in the well, the rock backfill, and the waste immediately surrounding the well change. The area of influence is normally determined using test wells. The heterogeneity of the landfill makes the extrapolation of test well results to other wells quite approximate. Unanticipated pockets of water in the landfill tend to restrict the well influence and are one example of factors that can confound the designer. In a deep landfill, e.g., 100 to 200 feet deep, an area of influence 200 to 800 feet in diameter would be expected. For shallower landfills, the area of influence would be proportionately smaller, since it would be necessary to apply a smaller vacuum to avoid entraining air from above the landfill. Until an accurate means of determining the area of influence is developed, recovery system designs will continue to be quite conservative.

There is an overall aspect of recovery system design that is often overlooked. The generation of the gas is a natural process with inherent limitations. The landfill should not be considered as an unlimited source of gas. In the design and operation of a landfill gas recovery system, every attempt should be made to withdraw the gas at a rate no greater than the natural rate of gas generation within the landfill. Experience in Mountain View, California has shown that the generation of methane rich gas in a landfill can actually be completely thwarted by excessive extraction rates. This limitation, as well as the additional limitations and uncertainties discussed in previous sections, must be taken into account in the design of the landfill gas processing and utilization systems.

Table III. SUMMARY OF EXISTING LANDFILL GAS RECOVERY SYSTEMS

Project	Owner/Operator	Process Description	Project Status	Remarks
Ascon — Wilmington, CA	Watson Energy Systems	Condensate and particulate removal to produce 1.5×10^6 cu ft/day medium Btu gas for sale to Shell Oil refinery for process steam generation.	System has been operating since late 1978. Second unit under consideration.	Owners consider the project a financial success.
Azusa Western — Azusa, CA	Azusa Land Reclamation Company	Dehydration and acid removal for recovered gas and sale of 0.5×10^6 cu ft/day to a chemical company to generate process steam.	Gas has been recovered since early 1978. Current plans are to construct 10 MW power station and fire gas in internal combustion engines to generate electricity for City of Azusa.	Full potential of fill not tapped yet due to marketing problems.
Cinniminson — Newark, NJ	Public Service Electric and Gas Company	Sulphur and condensate removal to produce 250,000 cu ft/day medium Btu gas for sale to a nearby power steel plant for process heating.	System has been operating since fall 1979.	Presently expanding to 1 million cu ft/day.
City of Industry — City of Industry, CA	City of Industry	Dehydration and particulate removal to produce 0.5×10^6 cu ft/day of medium Btu gas for heating and cooling nearby convention center facilities.	In start-up phase.	Start-up has been delayed due to changes in the use of the gas.
Monterey Park — Monterey Park, CA	Getty Synthetic Fuels, Inc.	Dehydration and carbon dioxide removal to produce 4×10^6 cu ft/day of pipeline quality gas for sale to Southern California Gas Company for blending into pipeline.	Started up August 1979. Still in shakedown due to field capacity problems.	Second generation facility based on knowledge gained at Palos Verdes facility.
Mountain View — Mountain View, CA	Pacific Gas and Electric Company (prior phase included EPA, DOE now involved)	Dehydration and carbon dioxide removal to produce 0.5×10^6 cu ft/day of high Btu gas for blending into pipeline.	System operated as a demonstration facility mid-1978 to January 1980. Now in commercial mode.	Major emphasis of first phase was to demonstrate feasibility of gas recovery and utilization from relatively shallow landfills (40 ft deep).
Palos Verdes — Rolling Hills Estates, CA	Getty Synthetic Fuels, Inc.	Dehydration and carbon dioxide gas removal to produce 0.75×10^6 cu ft/day of pipeline quality gas for sale to Southern California Gas Company for blending into pipeline.	System has been operating reliably since 1977.	First commercial-scale landfill gas recovery and utilization facility in the United States.
Sheldon-Arleta — Los Angeles, CA	City of Los Angeles	Dehydration and particulate removal to produce 2.6×10^6 cu ft/day of medium Btu gas for steam generation at a nearby steam plant.	Commercial operation began in November 1979.	Extensive testing of firing landfill gas in internal combustion engines during early stage of project.

EXISTING SYSTEMS

Currently, there are presently eight landfill gas recovery systems operating in the United States. These eight systems are listed on Table III (22). Numerous projects are in the planning, design, or construction stages. Several of these may be operating by the end of 1980. Although there are still several aspects of the technology that could be improved, the recovery of landfill gas appears to be sufficiently well developed for widespread implementation. Since landfill gas recovery taps an energy source that is otherwise totally wasted, the present and future need for alternative energy sources makes it likely that this technology will be economically and institutionally attractive for many years to come.

REFERENCES

1. Farquhar, G.J.; Rovers, R.A. "Gas Production During Refuse Decomposition", *Water, Air and Soil Pollution* **1973,** *2,* 483-95.

2. Tchobanoglous, G. "Wastewater Engineering", 2nd Ed.; McGraw-Hill: New York, 1978.

3. Hekimian, K.K., *et al.* "Recovery, Processing, and Utilization of Gas From Sanitary Landfills", Washington, D.C., Feb 1979, EPA 600/2-79-001.

4. Mah, R.A.; Hungate, R.E.; Ohwaki, K. "Proceedings", Microbial Energy Conversion, H.G. Schlegel; J. Barnea, Eds.; Erich Goltze KG: Gottingen, FRG, 1976.

5. Bowerman, F.R.; Rohatgi, N.K.; Chen, K.Y.; Lockwood, R.A. "A Case Study of the Los Anglese County Landfill GAs Development Project", Washington, D.C., July 1977, EPA-600/3-77-047.

6. Ramaswamy, J.N. Ph.D. Dissertation, University of West Virgnina, Morgantown, West Virginia, 1970.

7. Songohuga, O.O. Ph.D. Dissertation, University of West Virgnina, Morgantown, West Virginia, 1969.

8. Anderson, D.R.; Callinan, J.P. "Proceedings", National Industrial Solid Wastes Management Conference, University of Houston, Houston, Texas, 1970; p 311.

9. Boyle, W.D. "Proceedings", Microbial Energy Conversion, H.G. Schlegel; J. Barnea, Eds.; Erich Goltze KG: Gottingen, FRG, 1976.

10. Golueke, C.J. "Comprehensive Studies of Solid Waste Management", Washington, D.C., 1971, Third Annual Report, EPA SW-10rg.

11. Pacey, J. "Proceedings", Congress on Waste Management Technology and Resource and Energy Recovery, 1976; EPA SW-8p, p 168.

12. Klein, S.A. *Compost Science* **1972,** *Jan/Feb,* p 6.

13. Hitte, S.J. *Compost Science* **1976,** *Jan/Feb,* p 26.

14. Pfeffer, J.T. "Reclamation of Energy From Organic Waste", Washington, D.C., 1974, EPA 679-74-016; NTIS Report PB-231 176.

15.   Schwegler, R.E. 11th Annual Seminar and Equipment Show, Govern-mental Refuse Collection and Disposal Assoc., Santa Cruz, California, Nov. 1973.

16.   Alpern, R. M.S. Thesis, California State University, Long Beach, Califor-nia, 1973.

17.   City of Los Angeles, Bureau of Sanitation, Research and Planning Divi-sion, "Estimation of the Quantity and Quality of Landfill Gas from the Sheldon-Arleta Sanitary Landfill"; Jan. 2, 1976.

18.   Blanchet, M.J. *et al.* "Treatment and Utilization of Landfill Gas", Moun-tain View Project Feasibility Study, EPA30-583.

19.   Cook, E.N.; Force, E.G. *J. Water Pollut. Control Fed.* **1974,** *46* (2), 381-85.

20.   Author's experience while extracting unreleased proprietary data from landfill owners.

21.   McCarty, P.L. *Public Works* **1964,** Sept., Oct., and Nov.

22.   Daley, E.J.; Nilson, J.A.; Spitzka, R.E.; Speed, N.A. Unpublished Draft Report, Nov. 1979; Brown and Caldwell.

23.   Beluche, R. Ph.D. Dissertation, University of Southern California, Los Angeles, California, 1978.

24.   Carlson, J.A. "Recovery of Landfill Gas at Mountain View, Engineering Site Study", Washington, D.C., May 1977, EPA530/SW-5872.

RECEIVED JUNE 20, 1980.

Product Distribution in the Rapid Pyrolysis of Biomass/Lignin for Production of Acetylene

MARTHA GRAEF, G. GRAHAM ALLEN, and BARBARA B. KRIEGER

Department of Chemical Engineering, University of Washington, BF–10, Seattle, WA 98195

Chemical feedstocks and fuels supplied by the petrochemical industry are in ever increasing demand. Due to the projected difficulty of traditional petroleum based resources meeting these demands, alternative materials and technologies are being examined. Although coal will be used as a raw material for these processes, it presents a number of environmental problems. In view of these problems, one might focus attention on the potential of pyrolyzing polymeric renewable resources such as agricultural and urban wastes (lignocellulosics) which have a higher hydrogen to carbon ratio than coal into useful monomeric chemicals and fuels. In particular, the relatively large amounts (1,2), of collected, low cost waste materials generated by the forest products industry represent a low ash, low sulfur, environmentally acceptable alternative feedstock to coal.

Dry hardwood and softwood lignocellulosics are chemically characterized as 50% C, 6% H, 44% O, less than 0.1% nitrogen, and 0.3% ash (3). Although the high oxygen and water content of woody waste materials has prevented wide usage of lignocellulosics as feedstocks in the past (4), the cellulose fraction (43%) and hemicelluloses (28% to 35%) containing the oxygen are utilized by the forest products industry, leaving the highly aromatic lignin fraction (22% to 29%) to be further processed (see Figure I).

0097-6156/81/0144-0293$05.00/0
© 1981 American Chemical Society

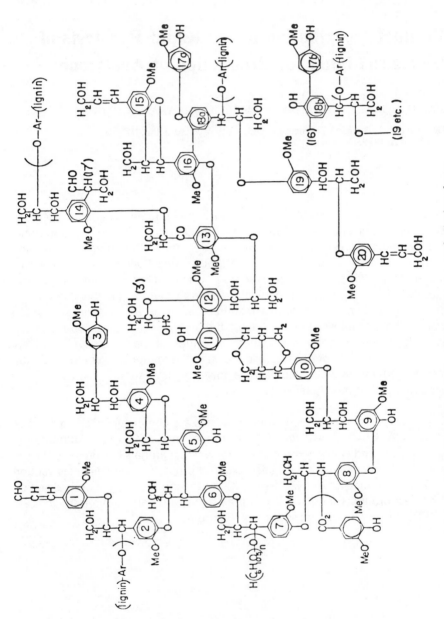

Figure 1.   Representative chemical structure of the lignin polymer

A deterent to deriving chemicals and fuels from polymers such as coal and lignin is precisely the wide variety of products, each of a relatively low yield, which are found in conventional pyrolysis processes. In attempts to control the product distribution, several workers (5-8) have studied the effects of heating rate, ultimate temperature, and quenching rate on pyrolysis products, especially the amount of char produced (8). It appears that rapid, severe degradation favors gas formation and can, in some cases, narrow the product distribution considerably (9). Extremely rapid, high temperature heating with rapid quenching can be provided by plasma heating (10-13). Recently, plasma heating has been successfully used in coal pyrolysis and a process unit is currently under study (14).

Pyrolysis studies have been carried out with lignocellulosics (8,9,15-24), but few have investigated the heating rate effect on product distribution and unfortunately, often only a single component or the large classes of compounds such as char, liquids (tars or condensible volatiles), and gases are reported. The detailed characterization of all products is seldom carried out (25). It is the intent of this article to characterize the products from microwave-induced plasma pyrolysis (rapid heating rate pyrolysis) of lignin as a prelude to determination of the kinetics and economics of chemical production from pinewood Kraft lignin, a waste material of the pulp and paper industry.

PREVIOUS STUDIES

Extensive surveys of the literature on microwave pyrolysis appear in Che (26), Bittman (27), and Fu and Blaustein (28-31), who studied the reactions of a complex molecule, coal, in a microwave-induced plasma. Most other studies, however, have been conducted with simple low molecular weight compounds such as CO and CO_2 (32) and simple hydrocarbons (27,33). These studies, conducted with relatively expensive chemicals of high purity, lend insight into the nature of the plasma reactions, but contribute little to the role heating rate plays in pyrolysis or to economic evaluation of plasma processing. In addition, most of these studies have used resonance cavity microwave applicators which are limited to small size by design and are exclusively research tools. As discussed in Bosisio (34), waveguide applicators and more recent designs (35) allow reactor configurations that are suited to possible industrial scale-up. Discussion of the effect microwave applicator design has on field intensity and ultimate product distribution is beyond the scope of this study. At least two patents exist concerning microwave degradation of waste (36,37) and hazardous materials (38) but specific reactions are not discussed. The degradation of lignin in a discharge is limited to work by Goheen (9) in an arc and Zaitsev (39) who used a tungsten electrode discharge system. The emphasis in both articles was to study only a single class of products such as the gases or char.

Simple thermal pyrolysis of lignocellulosics with or without additives has been reviewed by several authors (15-17). In contrast to microwave pyrolysis, reasonably extensive conventional pyrolysis product characterization has been conducted for certain types of biomass (15-25) and some of these results will be compared to those cited here. To these authors' knowledge, no single study on microwave pyrolysis (plasma or di-electric loss mode) has identified the components of all product fractions nor their relative amounts (40,41). The work reported here has been extended by others (43,44) to include pyrolysis studies of biomass fractions and other types of biomass with the emphasis on detailed product characterization, formation kinetics, and effect of transport rates.

EXPERIMENTAL

Microwave Application System and Reactor

The microwave circuit used in this study is shown in Figure II. The Gerling-Moore variable 0-2.5 kW power generator operates at 2450 MHz. Three crystal detectors in the circuit measure forward, reflected, and transmitted power and are mounted in the S-band wave-guide (3.8×7.6 cm). The latter two power measurements are recorded continuously. Constant forward power and magnetron protection are ensured by the presence of a 3-port circulator. The impedance of the circuit is manually matched with the E-H field tuner. The reactor is placed through the wave guide as shown.

The gas handling and reaction system is shown in Figure III. The reactor (I.D. 18 mm) is made of Vycor fused to standard tapered glass. Carrier gases (Argon, Airco, CP grade; Helium, Liquid Air, CP grade; and Hydrogen, Liquid Air, CP grade) were dried before flowing through the reactor. Indulin AT, a pinewood, kraft lignin (Westvaco Company) was extracted with ether in a Soxhlet extractor for 48 hours, dried under vacuum, and stored over $CaSO_4$. The dry lignin powder was pellitized (average density, 1.1 ± 0.05 g/cm^3), weighed, and placed on a hollow quartz pedestal in the reaction zone. Resulting microwave field intensities are about 50 to 100 W/cm^2. This value is calculated assuming the initial indicated absorbed power impinges on the pellet cross-section. The entire system was evacuated and checked for leaks prior to beginning the reaction. Liquid nitogen or ice baths were used to collect condensible products.

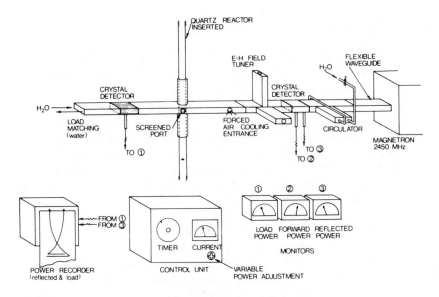

Figure 2.   Microwave circuit

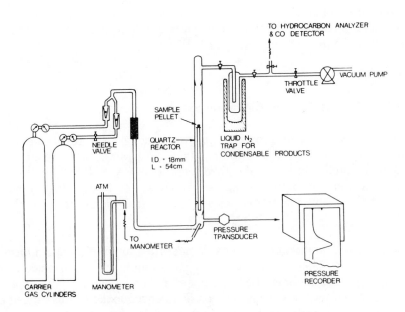

Figure 3.   Gas handling and reactor schematic

Analytical Scheme

The so-called permanent gas fraction was routinely analyzed on a 6-ft × 1/8-inch diameter Supelco Porapak Q column using a Perkin-Elmer Model 3920 gas chromatograph (G.C.). The flowrate of the G.C. carrier, helium, was 30 ml/min. The bridge current was set for 175 mA and the thermal conductivity detector temperature was maintained at 200°C. CO and CO_2 peaks were quantitatively analyzed at room temperature while the assymmetry of the acetylene peak necessitated elution at 100°C. The presence of hydrogen was determined on a molecular sieve 13X column at room temperature. Since acetylene was not separated from ethylene, confirmation of acetylene was made on a Supelco Porapak T column (10 ft × ¼ inch) at room temperature.

The liquid fraction, dissolved in ether, was separated on a Supelco DEGS 50-ft × 16-inch capillary column (liquid loading 10%). The sample was prepared by drying over $MgSO_4$, adding para-bromo phenol as an internal standard and bringing it almost to dryness in a Danish Kaderna evaporator. A sample chromatogram is shown in Figure IV. Confirmation of the major components were made by an associated Hitachi Perkin-Elmer RMS-4 mass spectrometer. Additional analyses are detailed in Graef (46).

A limited number of char fraction analyses were conducted on a Beckman I.R.-4 infrared spectrometer. A sample size of 0.5 mg residual to 800 mg KBr was found to give the best spectra (41).

RESULTS

Helium Plasma Product Characterization

The microwave reactor parameters are expected to affect the product distribution and thus, the conditions in Table I will be considered baseline reaction conditions. The absorbed energy is read directly from power monitors and this particular power level was the lowest that provided a consistently stable discharge over an acceptable range of pressures and flow rates.

The gross distribution of products formed from lignin in a helium discharge are shown in Table I. The values in parenthesis are those calculated by excluding the residual fraction. The uncertainty suggested by the upper limit (<10%) given for the volatile fraction is a consequence of the deposition of fine particle material in the nitrogen trap. The inability to dissolve this substance in a variety of solvents suggests that it should be categorized with the polymerized fraction, thus the lower limit (>3%) given for the polymerized fraction.

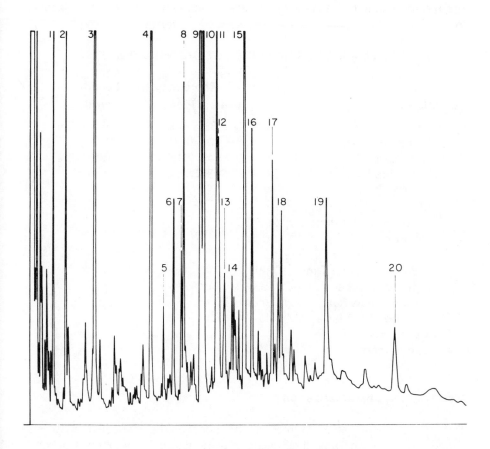

Figure 4. Gas chromatograph of condensable volatiles

The plasma pyrolysis values presented in Table I can be compared with the compilation by Allan and Mattila (17) of the thermal pyrolysis products of lignin under solvent-free conditions over a broad range of temperatures. The two product distributions are quite dissimilar. Thermal pyrolysis promotes liquefaction (78% on a residual-free basis) while plasma processing is primarily a gasification reaction (81% on a residual-free basis). As expected the plasma reactions cause a more severe degradation to lower molecular weight products.

Table I. OVERALL PRODUCT DISTRIBUTION

Reaction Product	Representative Conditions	
	Helium Plasma [1]	Pyrolysis [2]
Residual	33% (0%)	55% (0%)
Volatile Fraction	<10% (<15%)	35% (78%)
Permanent Gases [3]	54% (81%)	12% (22%)
Polymerized Products [4]	>3% (>4%)	— —

1. Baseline Reactor Parameters
   Helium Initial Pressure: 25 Torr
   Maximum Pressure: ~ 100 Torr
   Forward Power: 550 watts
   Batch Reaction Time: 10 min
   Total Absorbed Energy: 75 watt-hour
   Carrier Flow Rate: 86 cm^3/min
2. From Ref. 17, Allan and Matilla
3. $MW_{ave} \simeq 14$ g/mole
4. By difference

Gas Fraction Characterization

Detailed characterization of the gas fractions of the respective pyrolysis processes shown in Table II suggests that the dissimilarity extends beyond the distribution of the products.

Table II. COMPOSITION OF PERMANENT GASES[1]

	Helium Plasma	Thermal Pyrolysis[2]
Carbon Monoxide	44%	50%
Carbon Dioxide	2%	10%
Hydrogen	43%	None
Methane	2%	38%
Ethane	Trace	2%
Acetylene	14%	None
Higher Hydrocarbons	Trace	Trace
Sum	105%[3]	100%

1. Volume percent
2. Reference 17
3. Indicates error in measurements, ±5%

The analytical scheme for these studies precluded measurement of water. Both pyrolysis methods evolve carbon monoxide and carbon dioxide in comparable amounts. However, plasma processing produces 43% hydrogen and 14% acetylene on a volume basis while thermal pyrolysis gases contain neither component. Instead, the major hydrocarbon generated in the thermal pyrolysis system is methane (38%), while saturated hydrocarbons are minor components in the plasma process. These differences illustrate that the nature of conventional pyrolysis reactions is radically different from the microwave plasma pyrolysis reactions.

Condensible Liquid Characterization

The gas phase species were swept from the plasma zone by the carrier gas and a portion of them condensed in a liquid nitrogen cold trap. Some of the major condensible volatiles were identified with G.C.-Mass spectroscopy as shown in Figure IV. Quantitative determination of several of the larger peaks is given in Table III for the helium plasma reactor baseline conditions. Comparison of Table III with typical lignin pyrolysis products from Allan and Mattila (17) shown in Figure V reveals that while guaiacol and the cresols are present in both systems, a variety of other products, specifically the condensed aromatics, are not conventional pyrolysis products. The major components identified represent only 4% by volume of the volatile fraction while at least 50 additional compounds account for the remaining 96%. No attempt has yet been made to optimize or narrow the product distribution of the liquid fraction obtained in a helium plasma since the liquids represent only about 10 weight percent of the total products. Although current studies are addressing the issue of which experimental variables narrow the product

COMPOUND NAME	STRUCTURE	IDENTIFIED AS MAJOR COMPONENTS IN CONDENSIBLE VOLATILE FRACTION OF	
		HELIUM MICROWAVE PLASMA PYROLYSIS	THERMAL PYROLYSIS
phenol		✓	✓ ✓
o-cresol	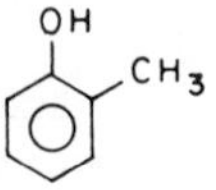	✓	✓ ✓
p-cresol		✓	✓ ✓
guaiacol	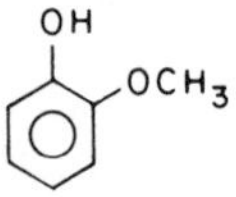	✓	✓ ✓
2,4 dimethyl phenol (xylenol)	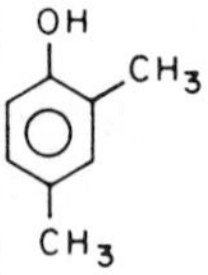		✓
2-methoxy,4-methyl phenol	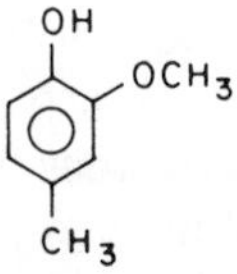		✓
2-methoxy,4-ethyl phenol	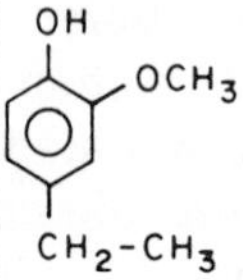		✓
2-methoxy,4-propyl phenol	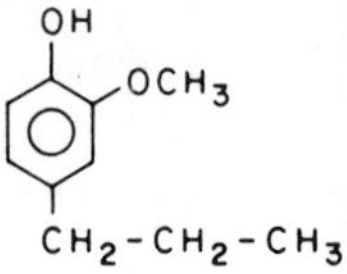		✓
p-vinyl phenol	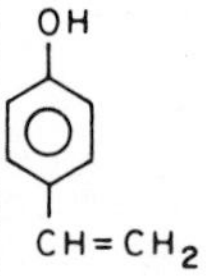		✓
napthalene	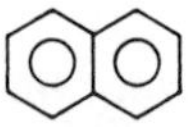	✓	
anthracene		✓	
styrene	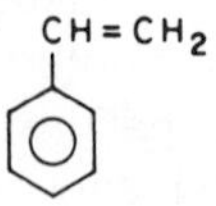	✓	
phenyl acetylene	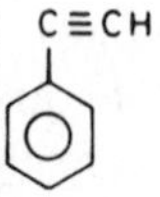	✓	
acenaphthcene	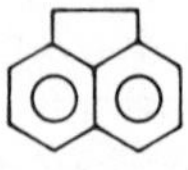	✓	

Figure 5. Contrast between plasma and thermal pyrolysis condensable volatile fraction (tars) produced in solvent-free pyrolysis of lignin

distribution (43,44), it is difficult to generalize about the reactions forming the condensible liquid fraction in this reactor geometry and quench zone configuration (44). Even if more favorable yields of the above compounds could be achieved, economic considerations suggest the decreased diversity of the gas fraction should take precedence and its composition should be optimized.

Table III. COMPOSITION OF VOLATILES
(Baseline Reactor Parameters)

Major Components	Volume Fraction of Tars, %
Styrene	0.5
Phenyl Acetylene	0.5
Napthalene	0.9
Guaiacol	0.2
O-Cresol	1.0
P-Cresol	0.4
Acenaphthene	0.1
Anthracene	0.5
	4.1
Other*	
Methylphenylacetylene	not
1,2-Dimethoxybenzene	quanti-
Other unidentified components, 50	tative
	100%

* Unconfirmed

Residual Analysis

An elemental analysis was performed (42) on the residual fraction. The following weight percents were obtained on the char: C, 83.92%; H, 2.09%; N, 0.45%; O, 8.5%; and remainder, 5.06%. The char is a black, porous material with a shape similar to an expanded pellet. Infrared spectra (41) of the char fraction showed virtually no features and none of the original lignin functional group absorption bands, substantiating that nearly complete reaction occurred. Although few residual fraction samples were analyzed spectroscopically, each was visually inspected and it is believed that the lack of functionality is representative of all char fractions. Certain runs not reported here showed unreacted lignin in a zone contiguous with the reactor wall. The pattern of the reacted and unreacted zones suggests that the pellet was misaligned in the field shielding a portion of the lignin from electron bombardment. This phenomenon, together with short duration runs showing a distinct shell of reacted lignin, lends support to the explanation that the lignin is transformed primarily by electron bombardment, a surface phenomenon, the rate of which depends on the electron concentration and energy.

Effects of Some System Parameters on Acetylene Production

In order to evaluate the importance of the major system parameters on the product distribution and composition, experiments were conducted in which power, carrier gas composition, flowrate, and time were varied (41). The variation of carrier composition and power are best described simultaneously since the effect of the input power is dependent on the carrier gas under consideration. This interaction exists since a change in either parameter alters the electron concentration and electron energy in the system (10).

The power, expressed as energy absorbed over the 10-minute run, was varied in several experiments and its effect on acetylene production is shown in Figure VI for three carrier gases, helium, argon, and hydrogen. The effects on other products are reported elsewhere (46). The higher concentration of H_2 when it serves as the carrier gas promotes an increase of roughly 1.5 times the acetylene produced in either inert carrier.

The mechanism for the homogeneous production of acetylene from the CO and H_2 in the plasma has been suggested by Mertz, et al. (32) as follows:

$$H_2 + e^- \rightarrow 2H\cdot + e^-$$
$$H\cdot + CO \rightarrow CH\cdot + O\cdot +$$
$$2CH\cdot \rightarrow C_2H_2$$

Although this mechanism can be expected to occur in H_2 carrier gas experiments, a considerable volume fraction, 43%, of the gaseous products is H_2 derived from the lignin itself (Table II). Thus the CH· radical is being generated directly from lignin and complex hydrocarbon fragmentation. Several studies (13,26,31) of coal plasma pyrolysis report acetylene production in somewhat smaller quantities than reported here consistent with the lower H to C ratio of coal. We suggest, however, that the acetylene production rate is a complex phenomenon dependent on the reactor geometry, local plasma temperature (47,48), plasma applicator configuration, quenching rate, and mass transfer limitations. This matter is under continued investigation.

DISCUSSION

Devolatilization kinetics experiments (43,44) and pellet behavior in which the unreacted-reacted interface is sharply marked indicate that the primary reactions within the solid can be described by a shell-progressive model of the type discussed by Carberry (45) and others. However, the expected secondary reactions described briefly below are quite different for volatiles escaping to the gas phase plasma or volatiles remaining in the pellet or char-residual. The differences between plasma and thermal pyrolysis regarding product distributions shown in Figure V and Tables II and III arise from the nature of the secondary reactions.

In the gas phase, the plasma reactor environment contains electrons generated and sustained by the microwave field (10). The electron, by virtue of its charge and minute mass, is the predominant "carrier" by which this transfer of electromagnetic energy to kinetic energy is achieved. Specifically, the electron, accelerated by the rapidly oscillating field, develops sufficient kinetic energy (1-2 KEV) or pseudo-temperatures (on the order of 10^4 °K) to dissociate, excite, or ionize other molecules present in the gas. The energetic electron also fragments the surface lignin and other hydrocarbons upon collision. Because of the relatively high concentration of free radicals and other energetic species, the plasma gas reactions are characterized as high temperature reactions occurring at rapid rates and producing other energetic species. The production of acetylene in a microwave plasma, as contrasted to ethane or methane in conventional pyrolysis, supports this concept of the plasma gas as a high energy zone, since as Figure VII from Baddour and Timmins (10) shows, acetylene is thermodynamically stable at higher temperature while methane and ethane are not. Although thermodynamic arguments cannot strictly be used in a microwave plasma as a consequence of the inequality of the electron temperature and the temperature of the ions and molecules, the production of acetylene tends to suggest that the electron concentration and velocity determine the energetics of the gas phase. Further details of the plasma gas reactions are discussed in Graef (41,46).

Although electron bombardment is rapid and the predominant form of heat transfer in the gas phase, transport processes within the pellet are quite different and much slower. Increased char yield and condensed aromatics found in this study are consistent with the following description of the processes occurring within the pellet. Upon collision with the pellet surface, the flux of electrons relases large amounts of heat which volatilizes and cracks the polymeric lignin. Depending on the gas composition as in Figure VI, the stoichiometry (or C/O/H ratios) of the biomass, and the mass transport situation, an amount of residual or char forms inward from the pellet surface, while the volatiles outflow increases the gas pressure near the pellet.

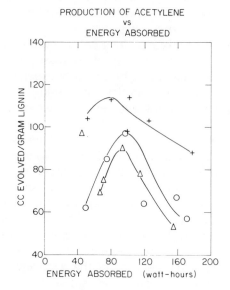

PRODUCTION OF ACETYLENE
vs
ENERGY ABSORBED

Figure 6. Acetylene evolved (standard cc) as a function of tubegrabed power absorbed over a 10-min experiment: (0) He, ($\triangle$) Ar, and (+) H_2 carriers; line is trend only

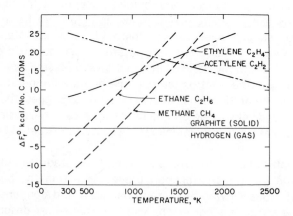

The Massachusetts Institute of Technology

Figure 7.   Thermodynamic stability of some hydrocarbons as a function of temperature (10)

Kinetic experiments show the char thickness progressively increases following the same rate as the weight loss (41,43,44) and this fact causes the rate and nature of further degradation reactions in the solid phase to be quite different from those occurring simultaneously in the gas phase for three reasons:

1. The hot char [300° to 1000°C as measured by infrared pyrometry (44)] and ash provide surfaces which can catalyze secondary thermal reactions of the reactive fragments of the escaping volatiles. The char can also be consumed by them as suggested by Lewellyn et al. (8) However, both processes occur at temperatures lower than the effective plasma temperature and degradation is not as severe.

2. The rate of heat penetration through the char layer is in proportion to the thermal properties and porosity of the char (43,44) since the mode of heat transfer to the unreacted lignin/char interface is by ordinary conduction through the char rather than by electron bombardment. Thus the apparent rate of devolatilization as measured by weight loss kinetics is slower (k $>$ 1-10 min^{-1}) than one might predict from electron pseudo-temperatures in the gas plasma. After an initial period, these apparent devolatilization rates are consistent with rates for conduction in porous char (44,49).

3. The porous interior of the pellet provides a resistance to mass transfer, i.e., confines the volatiles, which increases reactive fragment concentrations. This promotes polymerization and condensation reactions which form the greater char fraction and condensed aromatics (Table III and Figure V) than reported for conventional pyrolysis.

Depending on the carrier gas flow rate, vacuum pump capacity, degree of cracking and local plasma temperature, the just-volatilized higher molecular weight gases can be either swept from the plasma zone or undergo secondary plasma reactions. These gas-phase secondary reactions (described previously) occur to varying extents due to the residence time distribution (laminar flow) and spatially nonuniform plasma pseudo-temperature (48). The condensible volatile fraction components such as guaiacol and cresol that reflect the original lignin structure are rapidly quenched volatiles subjected to short residence times or low-plasma temperatures, perhaps from the outermost pellet layer reacted.

The description of the secondary reactions in the gas phase is further complicated by the fact that electron concentration and average electron velocity or energy do not remain constant for the entire experiment because of nonconstant system pressure and composition. The electron concentration is inversely related to pressure and highly dependent on gas composition via the ionization potential of the components (47). The absorbed power also changes with gas composition, further coupling the variables. Thus, the volatiles outflow reduces the plasma heating rate as a function of the rate and amount of pressure increase. Experiments conducted with increasing initial power as shown in Figure VI show a complex dependence on power and experiments are being conducted to uncover the mechanisms. A strong effect of particle size is observed and description of the coupled transport and reaction processes is reported elsewhere (43,44).

SUMMARY

Rapid, severe degradation has been shown to narrow the reaction product distribution in plasma pyrolysis of the aromatic fraction of biomass, lignin. This paper reports only five compounds with yields > 2% if the tar (3% to 10%) and char (33%) are considered two of the five. Primarily gasification occurs since yields of 51 weight percent gases are found. These gases, are 17 weight percent H_2 (43 volume percent) and C_2H_2 is 13 weight percent (14 volume percent). Char yield is reduced as suggested by Lewellyn because of consumption by increased radical concentrations, here, associated with the plasma. Condensed aromatics in addition to phenol-type compounds are found in the tar fraction, due presumably to mass transport limitations in the pellet. Despite the narrow product distribution and high heating value of the gas produced, energy consumption for the process is high due to the necessity of maintaining the plasma and to the large pellet size slowing the rate. Because of the potential of very fast reactions in the plasma, transport considerations, especially heat transfer, become increasingly important and point the way to further improvements.

ACKNOWLEDGEMENTS

This work was supported by the National Science Foundation under the Division of Advanced Energy and Resources Research and Technology Grant No. 7708979 and the NSF Engineering Initiation Grant Program. Martha Graef wishes to acknowledge financial assistance from the Department of Chemical Engineering, University of Washington. The helpful suggestions of K. V. Sarkanen, D. Hanson, D. Edelman, B. Hrutfiord, and R. Chan are also gratefully acknowledged.

REFERENCES

1. Sarkanen, K. V. *Science* **1976,** *191,* 773-76.

2. Goldstein, I. S. *Biotechnol Bioeng. Symp. Proc.* **1976,** (6), 293-301.

3. Rydholm, S. A. "Pulping Processes"; Interscience Publishers: New York, 1965.

4. Longwell, J. P. "Symposium Papers", 16th International Symposium on Combustion, sponsored by the Combustion Institute, Pittsburgh, 1976; Combustion Institute: Pittsburgh, 1976; 1-15.

5. Anthony, D.; Howard, J. P.; Hottel, H. C.; Meissner, H. P. "Symposium Papers", 15th International Symposium on Combustion, sponsored by the Combustion Institute: Pittsburgh, 1975; Combustion Institute, Pittsburgh, 1975; 1303.

6. Anthony, D. B.; Howard, J. B.; Hottel, H.; Meissner, H. P. *Fuel* **1976,** *55,* 121-28.

7. Suuberg, F. M. "Rapid Pyrolysis and Hydropyrolysis of Coal", Ph.D. Dissertation, Massachusetts Institute of Technology, Cambridge, Mass., 1977.

8. Lewellen, P.D.; Peters, W. A.; Howard, J. B. "Symposium Papers", 16th International Symposium on Combustion, sponsored by the Combustion Institute, Pittsburgh, 1977; Combustion Institute: Pittsburgh, 1977; 1471.

9. Goheen, D.W.; Henderson, J. T. *Cellulose Chem. and Tech.* **1978,** *13* (3), 363-72,.

10. Baddour, R. F.; Timmins, R. S. "The Application of Plasmas to Chemical Processing"; M.I.T. Press: Cambridge, Mass., 1967.

11. Hollahan, J.R.; Bell, A. T. "Techniques and Applications of Plasma Chemistry"; John Wiley: New York, 1974.

12. Bonet, C. *Chem. Eng. Prog.* **1976,** *72* (12), 63-69 .

13. Nicholson, R.; Littlewood, K. *Nature* **1972,** *236,* **397.**

14. Collins, J., U.S. Dept. of Energy, Washington, D.C., personal communication, 1978.

15. Shafizadah, F.; Sarkanen, K. V.; Tillman, D. A., eds. "Thermal Uses and Properties of Carbohydrates and Lignins"; Academic Press: New York, 1976.

16. Pearl, I.A. "The Chemistry of Lignin"; Marcel Dekker: New York, 1967; pp 276-83.

17. Allan, G. G.; Mattila, T. In "Lignins"; Sarkanen, K. V.; Ludwig, C. H., Eds.; Wiley-Interscience Publishers: New York, 1971; p 575.

18. Shafizadah, F.; Fu, Y. L. *Carbohydr. Res.* **1973,** *29,* 113.

19. Shafizadeh, F.; McIntyre, C.; Lundstrom, H.; Fu, Y. L. *Proc. Mont. Acad. Sci.* **1973,** *33,* 65-96.

20. Tran, D. Q.; Rai, C. *Fuel* **1978,** *57,* 293-98.

21. Tillman, D. A.; Sarkanen, K. V.; Anderson, L. L. "Fuels and Energy From Renewable Resources"; Academic Press: New York, 1977.

22. Shafizadeh, F.; Furneaux, R. H.; Cochran, T. G.; Scholl, J. P.; Sakai, Y., *J. Appl. Polym. Sci.* **1979,** 23 (12), 3525-39.

23. Bradbury, A.G.W.; Sakai, Y.; Shafidzdeh, R. *J. Appl. Polym. Sci.* **1979,** *23* (11), 3271-80.

24. Fairbridge, C.; Ross, R. A.; Sood, S. P. *J. Appl. Polym. Sci.,* **1979** 22, 497-510.

25. Goldstein, I.S. *Appl. Polym. Symp.* **1975,** (28), 259-67.

26. Che, S.C.L. Ph.D. Dissertation, University of Utah, Provo, Utah, 1974.

27. Bittman, R. Ph.D. Dissertation, University of California, Berkeley, Calif., 1966.

28. Fu, Y. C.; Blaustein, B.D. *Chem. Ind.* **1967,** 1257 (London).

29. Fu, Y. C.; Blaustein, B. D. *Fuel* **1968,** *47,* 463.

30.   Fu, Y. C.; Blaustein, B. D. *Ind. Eng. Chem. Process Design Develop* **1969,** *8,* 257.

31.   Fu, Y. C.; Blaustein, B. D.; Wender, I. In Flinn, J., Ed.; *Chem. Eng. Prog. Symp. Ser.* **1971,** *67* (112), 47-54.

32.   Mertz, S. F.; Asmussen, J.; Hawley, M. C. *IEEE Trans. Plasma Sci.* **1975,** *25* (Dec.), 297.

33.   Streitwieser, A.; Ward, H. E. *J. Amer. Chem. Soc.* **1962,** *84,* 1065.

34.   Bosisio, R. G. *J. Phys. E.* **1973,** *6,* 628.

35.   Gehrling Moore, Inc. Palo Alto, California.

36.   Knapp, E. M.; Ellis, W. T. U.S. Patent 3 560 347, 1971; U.S. Patent 3 449 213, 1969.

37.   Grannen, E. A.; Robinson, L. U.S. Patent 3 843 457, 1974.

38.   Bailin, L. J.; Sibert, M.; Jonas, L. A.; Bell, A. T. *Environ. Sci. Technol.* **1975,** *9,* (3), 254.

39.   Zaitsev, V. M.; Piyalkin, V. N.; Isyganov, E. A. *Gidroliz. Lesokhim. Promst.* **1975,** *3,* 10-12.

40.   Work, D. W. M.S. Thesis, University of Washington, Seattle, Wash., 1977.

41.   Graef, M. G. M.S. Thesis, University of Washington, Seattle, Wash., 1978.

42.   Swarzkopf Analytical Labs, New York, 1978.

43.   Chan, R. C. M.S. Thesis, University of Washington, Seattle, Wash., 1979.

44.   Wiggins, D. M.S. Thesis, University of Washington, Seattle, Wash., 1979.

45.   Carberry, J. J. "Chemical & Catalytic Reaction Engineering"; McGraw Hill: New York, 1977.

46.   Graef, M. K; Krieger, B. B., in preparation, 1979.

47.   MacDonald, A. D. "Microwave Breakdown in Gases"; John Wiley: New York, 1966.

48.   Bonet, C.; Bell, A. T. "Plasma Chemistry — 2 Transport Phenomena in Thermal Plasmas"; Pergamon Press: Elmsford, N.Y., 1975.

49.   Russel, W.; Saville, D.; Greene, M. I. "A Model for Short Residence Time Hydropyrolysis of Single Coal Particles"; *Amer. Inst. of Chem. Eng. J.* **1979,** *25* (Jan.), 65-80.

RECEIVED JUNE 18, 1980.

The Effects of Residence Time, Temperature, and Pressure on the Steam Gasification of Biomass

MICHAEL J. ANTAL, JR.

Department of Mechanical and Aerospace Engineering, Princeton University, Princeton, NJ 08540

The underlying science of thermochemical conversion of biomass materials to useful gaseous fuels is poorly understood. Recent experimental research in the U.S.A. (1) and Sweden (2) has offered new and important insights into the gasification process. The two research teams independently conclude that biomass gasification occurs in three steps: 1) pyrolysis, producing volatile matter and char; 2) secondary reactions of the evolved volatile matter in the gas phase; and 3) char gasification. Detailed understanding of the rates and products of these three steps offers important guidance for the improved design of biomass gasifiers.

Pyrolysis of biomass materials occurs under normal conditions at relatively low temperatures (300° to 500°C), producing volatile matter and char. Very rapid heating causes pyrolytic weight loss to occur at somewhat higher temperatures. In general, the volatile matter content of cellulosic materials approximates 90% of the dry weight of the initial feedstock. Woody materials contain between 70% and 80% volatile matter, and manures contain 60% volatile matter. However, it is known (3) that cellulosic materials can be completely volatilized when subject to very rapid heating (>10,000°C/sec). Several relatively complete reviews of the mechanisms and kinetics of cellulose pyrolysis are available in the literature (4-7).

0097-6156/81/0144-0313$05.50/0
© 1981 American Chemical Society

Volatile matter produced by pyrolysis of the biomass begins to participate in secondary, gas-phase reactions at temperatures exceeding 600°C. These reactions occur very rapidly and yield a hydrocarbon-rich syngas product. As recognized by Diebold (8), these reactions resemble the hydrocarbon cracking reactions employed in the manufacture of ethylene and propylene by the petrochemical industry (9,10). The secondary gas-phase reactions dominate the gasification chemistry of biomass.

At still higher temperatures (>700°C), pyrolytic char reacts with steam to produce hydrogen, carbon monoxide and carbon dioxide. Rates of gasification of biomass-derived chars are known to be higher than coal-derived chars (2); however, much higher temperatures are required to achieve char gasification than were initially required for the pyrolysis reactions. Catalysis of char gasification has been reported (11,12) with limited success.

Research described in this paper focuses on the second step of the gasification process, and details the effects of temperature and residence time on product gas formation. Cellulose is used as a feedstock for pyrolytic volatiles formation. Earlier papers (13,14) have discussed the effect of steam on cellulose pyrolysis kinetics. Two recent papers (15,16) presented early results on pelletized red alder wood pyrolysis/gasification in steam. Future papers will discuss results using other woody materials, crop residues, and manures (17,18). Research to date indicates that all biomass materials produce qualitatively similar results in the gasification reactor described in the following section of this paper. Effects of pressure on the heat of pyrolysis of cellulose are also discussed as a prelude to future papers detailing the more general effects of pressure on reaction rates and product slates.

EFFECTS OF TEMPERATURE AND RESIDENCE TIME ON THE SECONDARY, GAS-PHASE REACTIONS

Experimental Procedure

For the experiments described below, dry Whatman No.1 filter paper stored in a desiccant bottle was used as feedstock material. The use of an oven to obtain "bone dry" material was found to be futile due to the hygroscopic nature of the cellulose. The cellulose was assumed to have the chemical composition 0.444 C, 0.062 H, 0.494 0 on a mass fraction basis, and the char composition was determined to be 0.7835 C, 0.04 H, and 0.1765 by an independent laboratory.

A specially designed quartz, tubular, plug-flow reactor was fabricated to study the gas-phase reactions. Rates of gas formation by species can be measured using the reactor either in a differential or an integral mode. Results described here emphasize the integral aspects of the tubular reactor since they are the easiest to interpret.

A schematic of the experimental apparatus is given in Figure I. A typical experimental procedure was:

1) With all three furnaces cold, a small (0.1 to 0.5 g) sample of the material to be pyrolyzed is placed in the center of the pyrolysis reactor.

2) An inert gas is bled through ports D and E to cool the sample and purge the reactor, while furnaces 1 and 3 bring the steam superheater and the gas-phase reactor to the desired temperature.

3) The peristaltic pump is actuated and pumps water into the steam generator at a measured rate. Concurrently, a small amount of inert tracer gas (argon) is continuously injected through port A into the rear of the reactor.

4) When condensed water first begins to appear in the pyrolysis reactor, furnace 2 (which was preheated to the desired pyrolysis temperature) is moved into place around the pyrolysis zone of the reactor.

5) When pyrolysis temperatures are reached, the six-port Valco valve is switched and the 34-port Valco valve automatically takes 15 samples of the gas stream for later analysis in the Hewlett-Packard 5834a Gas Chromatograph (HPGC). Unsampled gas is collected in a Teflon bag for later analysis.

6) When all 15 samples have been taken, the six-port valve is switched again and the samples are automatically analyzed by the HPGC. Gases collected in the Teflon bag are sampled using a gas tight syringe and analyzed by the HPGC.

7) The char and tars produced during the experiment are collected and weighed. Water collected in the condenser is also weighed.

Temperatures within the reactor are controlled by various temperature controllers and monitored by Type K thermocouples with continuous recording on chart recorders. Measured temperature variations along the length of the gas-phase reactor have been described in an earlier publication (1).

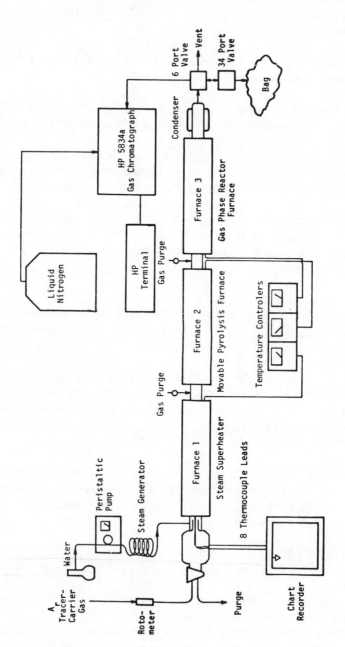

Figure 1.   Schematic of the tubular quartz reactor experiment

The evolved gas composition was observed to undergo considerable variation during the course of the experiment; consequently ten gas standards were acquired to calibrate the HPGC for quantitative analysis of the following gases: Ar, N_2, H_2, CO, CO_2, CH_4, C_2H_4, C_2H_6, C_3H_6, C_4H_8, C_4H_{10}, C_5H_{12}, and C_6H_{14}. Identification of the higher hydrocarbons ($>C_3$) is obscured by the fact that some other pyrolysis products have similar retention times. Analyses given in this paper for light hydrocarbons ($\leq C_3$) have been checked using a mass spectrometer. The HPGC uses a Poropak QS column in series with a Porosil column operating between -50 °C (cryogenic) and 200°C for gas analysis with a thermal conductivity detector (TCD). The carrier gas is an 8.5% H_2 $-$ 91.5% He mixture. A typical gas analysis takes 14 minutes.

The complete recovery of moisture and tars from the reactor sometimes poses difficulties. The moisture is absorbed on dry paper towels and weighed; whereas the tars condense on a rolled piece of aluminum foil inserted in the condenser. Mass balances are always better than 0.8, but can be misleading because much more water is used during the course of an experiment than solid reactant. The carbon balance is a better measure of the experiment's quality, and customarily ranges between 0.7 and 1.0 for the results reported here. Our inability to close the carbon balance in part reflects the formation of water-soluble carbonaceous compounds which are not subject to analysis by our existing instrumentation. Their presence is manifested by the color and odor of the collected water, which ranges from clear with an odor resembling automobile exhaust, to deep amber with a stronger, more noxious odor.

As designed, the reactor bears some resemblence to a dilute-phase transport reactor in that the solids and volatile pyrolysis products are present only in low concentrations in the steam reactant. During pyrolysis, the composition of gas in the gas-phase reactor using the lowest steam flow and a 0.1 g sample is nominally 68% steam, 28% volatiles, and 4% argon carrier (on a volume percent basis). Somewhat larger samples, leading to an increase in volatile concentrations, do not markedly affect the results reported here.

Rates of gas production can be measured using the reactor in either a differential or an integral mode. The differential mode employs the Valco valve system to obtain fifteen 0.6-ml samples of gas evolved during the course of the experiment. With Ar tracer gas injected at a measured rate, the dilution of the tracer gas sample can be directly related to the "instantaneous" rate of volatile gas production in the reactor. For example, with a tracer gas flow of 5 ml per min, a dilution of 50% in the gas sample would correspond to an "instantaneous" volatile gas production rate of 5 ml per

min. Unfortunately, departures from true plug flow within the reactor
(primarily due to the effect of the condenser on gas flow) make the
differential mode experimental data more difficult to interpret than indicated
above. Research reported here emphasizes the integral aspects of the reactor
design.

When used in the integral mode, total gas production by species is measured
using teflon bags to collect all the reactor effluent. The dependence of total
gas production on gas-phase residence time in the gas-phase zone of the
reactor is determined using the combined data of many experiments. This
data can be used to infer rates of gas production within the gas-phase
reactor. Kinetic models of gaseous species formation can be obtained
through a study of the effects of both temperature and residence time on
species production.

Kinetic Interpretation of Reactor Data

Consider the pyrolysis of a small sample of organic material in the pyrolysis
zone of the tubular reactor system. At any time t, the rate of evolution of
gaseous volatile matter $(\dot{m}_v)$ from the pyrolyzing sample is given by:

$$\dot{m}_v(t) = m\frac{dV}{dt} = mk[V^* - V(t)]^n; \; k = A \exp(-E/RT) \tag{1}$$

where m is the time dependent sample mass, m_i is the initial sample mass,
m_f is the final sample mass, $V^* = (m_i - m_f)/m_i$, $V = (m_i - m)/m_i$. A is the
pre-exponential constant, E is the apparent activation energy, R the Universal
Gas Constant, T is the time-dependent absolute temperature, and n is the
apparent order of the reaction. The concentration of volatile matter (C_v) in the
flowing stream is given by:

$$C_v(t) = \frac{\dot{m}_v}{\dot{m}_s/\rho_s + \dot{m}_v/\rho_v} \tag{2}$$

where $\dot{m}_s$ is the mass flow of steam in the tubular reactor and ρ_s and ρ_v
are the densities of steam and volatile matter (respectively). Assuming plug
flow, the volatile matter evolved at time t enters the gas-phase reactor at time
$t + \tau_i$ and leaves the gas-phase reactor at time $t + \tau_0$. The residence time
$\theta = \tau_0 - \tau_i$ of the volatiles in the gas phase is given by:

$$\theta = L^2/\int_0^L v(x)dx \approx \sigma L/\dot{V} \tag{3}$$

where L is the length of the gas-phase reactor, v is the spatially-dependent
gas velocity in the reactor, $\dot{V}$ is the volumetric flow of volatiles plus steam in
the reactor at the gas-phase reactor temperature T, and σ is the reactor's
effective cross sectional area.

Suppose that the rate of disappearance of condensible volatiles (due to cracking, reforming, etc. and treating the condensible volatiles as a single chemical species) in a differential volume element $\bar{v}\delta\tau$ of gas moving with average velocity $\bar{v} = L/\theta$ satisfies a first order rate law:

$$\frac{dC_V}{d\tau} = -C_V r_V \tag{4}$$

where the rate constant r_V is given by the Arrhenius expression:

$$r_V = A_V \exp(-E_V/RT) \tag{5}$$

The time variable τ can be thought to "track" the position of the differential volume element with initial volatile concentration $C_V(t)$ as it moves through the gas-phase reactor. The production of permanent gases produced by cracking/reforming reactions is also assumed to satisfy a first order rate law:

$$\frac{dC_j}{d\tau} = C_V r_j \tag{6}$$

where r_j is the rate constant associated with the jth permanent gas and:

$$r_j = A_j \exp(-E_j/RT) \tag{7}$$

As a first approximation, the gas-phase zone can be treated as an isothermal reactor (however, see the following section), leading to the expressions:

$$C_V(t + \tau_0) = C_V(t) \exp[-r_V (\tau_0 - \tau_i)] \tag{8}$$

$$C_j(t + \tau_0) - C_j(t) = \int_{\tau_i}^{\tau_0} C_V(t) r_j \exp[-r_V(\tau - \tau_i)]d\tau$$

$$= C_V(t)(r_j/r_V)\{1 - \exp[-r_V(\tau_0 - \tau_i)]\} \tag{9}$$

where r_V and r_J are constant (by assumption of constant T), and it is also assumed that the temperature of the pyrolysis zone is sufficiently "cold" so that the cracking/reforming reactions do not commence until the differential volume element enters the gas-phase reactor.

The mass of species j in the differential volume element emerging from the reactor at time $t + \tau_0$ is given by:

$$\sigma v\delta t C_j (t + \tau_0) \tag{10}$$

and the total mass m_j of species j produced during the experiment is given by:

$$m_j = \sigma \int v(t)C_j(t + \tau_0)dt$$

$$\simeq \bar{v}\sigma \int C_j(t + \tau_0)dt = [\bar{v}\sigma(r_j/r_v)]\{1-$$

$$\exp [-r_v(\tau_0 - \tau_i)]\}\int C_v(t)dt$$

$$+ \bar{v}\sigma \int C_j (t)dt \tag{11}$$

For short residence times $[r_v (\tau_0 - \tau_i) < < 1]$ Equation (11) becomes:

$$m_j \simeq \bar{v}\sigma r_j(\tau_0 - \tau_i)\int C_v(t)dt + \bar{v}\sigma \int C_j(t)dt \tag{12}$$

A plot of m_j vs. residence time ($\tau_0 - \tau_i$) should yield a straight line with temperature dependent slope $\bar{v} \sigma r_j \int C_v (t)dt = k_j$. A subsequent plot of $\ell n(k_j)$ vs. T^{-1} should yield straight lines whose slopes are the apparent activation energy E_j associated with the rate of production of species j. Thus, the data obtrained from the tubular reactor are susceptible to kinetic interpretation.

It should be noted that this analysis only provides an insight into the initial rates of the cracking reactions. Tertiary gas-phase reactions tend to obscure the interpretation of the kinetic data derived from the reactor. Moreover, the initial rate measurements for the cracking reactions are meaningful only for short residence times $[r_v (\tau_0 - \tau_i) < < 1]$. Because the cracking reactions occur rapidly, this constraint is difficult to satisfy.

Departures From an Ideal Isothermal Reactor

The tubular quartz reactor was designed anticipating the need to provide for long (5 seconds or more) gas-phase residence times in order to reform the oily volatile matter. Surprisingly, the observed reaction rates were so high that residence times on the order of 0.5 sec or less were needed to satisfy the condition $r_v(\tau_0 - \tau_i) < 1$. To obtain such short residence times, large mass flows were required which caused the reactor to deviate from its intended use as an ideal isothermal system. In addition, the Lindburg furnaces were observed to be less uniform in temperature than expected. Consequently, heat transfer plays a critical role in determining the residence time of the volatiles at temperature. The following paragraphs outline our "first order" approach towards recognizing the affects of heat transfer on the kinetic interpretation of the experimental data.

A simple energy balance for laminar fluid flow in a long tube leads to the equation:

$$\frac{dT}{dx} + \alpha T = \alpha T_w \tag{13}$$

where $\alpha = \pi Dh/\dot{m}c_p$, and $T(x)$ is the bulk gas temperature along the length x of the tube, T_w is the constant wall-temperature, D is the tube's diameter, h is the heat transfer coefficient, $\dot{m}$ is the mass flow of the gas and c_p is the specific heat of the gas. Equation (13) can be solved to determine the distance ℓ required for the gas to reach temperature T, with the result:

$$\ell = \left(-\frac{\dot{m}c_p}{\pi k N_{Nu_D}}\right)\ell n\left(\frac{T_w - T}{T_w - T_i}\right) \tag{14}$$

where $h = kN_{Nu_D}/D$, k is the thermal conductivity of the gas, and the Nusselt number $N_{Nu_D} = 3.7$.

In order to obtain an approximate residence time ($\tau_o - \tau_i$) for the volatiles in the gas-phase reactor, the length ℓ was calculated assuming $T_w - T = 5°C$. An error of 5°C in the gas-phase temperature measurement gives rise to about a 10% error in the determination of E_i. The residence time of the volatiles at temperature was then calculated using the formula:

$$\tau_0 - \tau_i = \frac{(L - \ell)\sigma}{\dot{m}_S/\rho_S + \dot{m}_V/\rho_V} \tag{15}$$

where L is the total length of the gas-phase section of the reactor, and σ is the apparent cross sectional area of the reactor.

Table I lists values of ℓ as a function of T_w for a steam flow of 0.34 g/min (used in the short residence time kinetic experiments). From this data, it is apparent that at the higher gas-phase temperatures, the volatiles spend about 50% of their time in the gas-phase reactor being heated to isothermal conditions. Consequently, kinetic measurements at the higher temperatures represent integrated values of the rates at lower temperatures, in addition to the (assumed) constant rate at T_w. This result clouds the kinetic interpretation of the reactor data. Although more effort could be made to explicitly account for the warmup time in the kinetic model for the reactor data, we have chosen to focus our effort on the design of a "second generation reactor" which will provide sufficient heat transfer rates to ensure nearly isothermal conditions for the shortest residence time experiments.

Table I. PARAMETERS USED TO CALCULATE
GAS-PHASE RESIDENCE TIME

T_w, °C	ℓ, cm	Effective Reactor Volume, cm^3	Effective Insert Volume, cm^3
750	13.4	46.4	30.2
700	13.9	48.1	31.3
675	14.2	49.2	32.0
650	14.5	50.2	32.6
625	14.8	51.3	33.3
600	15.2	52.7	34.2
575	15.6	54.0	35.1
550	16.1	55.8	36.2
500	17.2	59.6	38.7

$L = 29.2$ cm
$\sigma = 3.474$ cm^2
Bulk volume of insert $\cong 70$ cm^3
Length of insert $\cong 31.1$ cm

Some efforts have also been made to experimentally measure the temperature rise of the steam entering the gas-phase reactor. Qualitative agreement with the results of the heat transfer calculations was found; however, a brief calculation of the effect of radiation on the thermocouple's measurement of the gas temperature pointed to a significant error in the measurement. For example, with a steam flow of 0.12 g/min and a gas-phase temperature of 600°C, the thermocouple temperature was calculated to exceed that of the gas by 13°C. Carefully constructed radiation shields are needed to eliminate this effect. Because our research effort in this area was drawing to a close, it was decided to use the methodology described earlier (Equation 15) to estimate the residence time of the volatiles at temperature.

Departures From An Ideal Plug-Flow Reactor

Both turbulence and molecular diffusion cause tubular reactors to depart from ideal plug-flow behavior. It is standard practice to account for the two effects by a single dimensionless parameter D/vL, called the vessel dispersion number. This number is usually determined experimentally, and its magnitude indicates the degree of departure of the reactor from plug flow (D/vL) <0.01 for plug flow). A good discussion of the effects of dispersed plug flow on the kinetic interpretation of reactor data is given by Levenspiel (19).

The value $D/vL = 0.122$ was measured by T. Mattocks (17). Although this value suggests significant departures from plug flow, we believe most of the dispersion occurs in the condenser without affecting the kinetic measurements presented here.

Results and Discussion

Figures II, III, and IV display the dependence of gas production (g gas per g cellulose or % conversion) by species on gas-phase residence time for various gas-phase reactor temperatures. For these experiments, the steam super-heater was maintained at 350°C, and the pyrolysis furnace at 500°C. This latter setting gave rise to a measured sample heating rate of 100°C/min. Residence times were altered by varying the peristaltic pump's water flow rate between 0.06 and 0.34 g/min, and by inserting a closed quartz cylinder into the gas-phase reactor to reduce its apparent volume.

Data points reported in Figures II-V were accumulated over a period of eight months using experimental techniques which evolved and improved during that time period. Data points with residence times of two to three seconds were obtained using a water flow rate of 0.34 g/min for a 0.25 g sample.

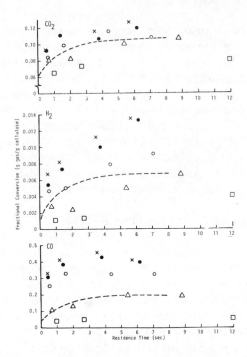

*Figure 2. Nonhydrocarbon gas production vs. residence time for various gas—
phase temperatures: (□) 500°, (△) 600°, (○) 650°, (●) 700°, (✕) 750°C*

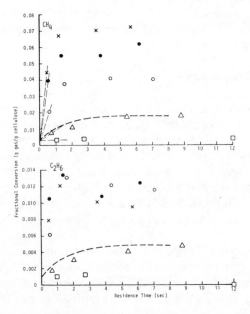

*Figure 3. Paraffinic hydrocarbon gas production vs. residence time for various
gas—phase temperatures: (□) 500°, (△) 600°, (○) 650°, (●) 700°, (✕) 750°C*

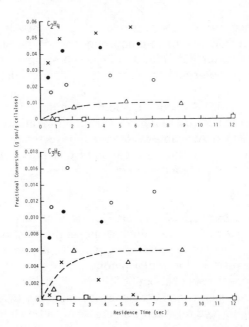

Figure 4. Olefinic hydrocarbon gas production vs. residence time for various gas—phase temperatures: (☐) 500°, (△) 600°, (○) 650°, (●) 700°, (✕) 750°C

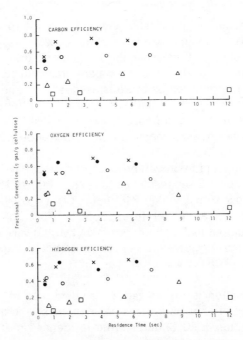

Figure 5. Carbon, hydrocarbon, and oxygen efficiency vs. residence time for various gas—phase temperatures: (☐) 500°, (△) 600°, (○) 650°, (●) 700°, (✕) 750°C

Shorter residence times were obtained using a quartz insert to reduce the gas-phase reactor's apparent volume. Longer residence times were obtained by reducing the water flow rate and cellulose sample size proportionately. Thus, all the data represents the same steam flow/cellulose weight ratio.

Since cellulose pyrolysis occurs in about one minute with a heating rate of 100°C/min, the data correspond to a steam dilution ratio of about 1.4 g steam per 1 g cellulose feed. Available evidence suggests that higher steam dilution ratios have little affect on the gasification results. Efforts are presently being made to more fully elucidate the effects of dilution ratio on steam gasification products.

Of the various gases represented in Figures II-IV, the behavior of carbon dioxide is simplest to interpret, since it shows the least dependence on gas-phase residence time or temperature. Apparently, the primary mechanism for CO_2 formation rests in the initial pyrolysis process. Secondary gas-phase reactions at temperatures of about 500°C contribute less to CO_2 formation. In order to increase gasification efficiency by reducing CO_2 formation (each molecule of CO_2 formed represents a net loss of carbon from the combustible products of the process), the conditions affecting the pyrolysis step of gasification must be carefully examined. For example, the use of high heating rate may reduce CO_2 formation.

Methane formation is also relatively easy to interpret. Increasing temperatures and increasing residence times result in increased methane formation. The slope of the dashed lines in Figure III gives the apparent rate of methane production at the various temperatures studied. The dependence of this production rate on temperature is used later in this section to estimate the activation energy for methane formation. Efforts to elucidate the mechanism of methane formation (most probably the pyrolysis/hydrogenation of higher hydrocarbons) are presently underway.

Carbon monoxide and hydrogen production data behave similarly, and reach a maximum at about 5 sec residence time and 700° to 750°C. Data for C_2H_6 production show some similarity to that of CH_4; however, C_2H_6 production reaches a maximum at temperatures of 650° to 700°C and residence times of about 2 sec. Competitive rates of formation by pyrolysis and consumption by pyrolysis or dehydrogenation reactions probably explain this observed behavior.

Ethylene production is maximized at temperatures of 700° to 750°C and residence times of about 6 sec, whereas propylene formation is favored by lower temperatures (650°C) and shorter residence times (2 sec).

In general, these results indicate that the gas-phase reaction temperature is the most significant parameter. The role of primary pyrolysis conditions and gas-phase residence times are much less significant. Moreover, for temperatures above 650°C, the initial rates of species formation are very high, so that much of the gas formation is complete in less than 0.5 sec. These very high rates of gas formation due to secondary reactions are of great significance for reactor design.

The preceding conclusions are substantiated by Figure V, which shows the effect of gas-phase temperature and residence time on the carbon, hydrogen and oxygen gasification efficiencies (carbon efficiency = carbon in gas ÷ feedstock carbon). Again, the gas-phase reactor temperature most signficantly affects the carbon and hydrogen efficiencies of the system.

Under the best conditions examined to date, 83% of the feedstock's energy was retained by the gaseous products of the process, and tar production was reduced to 3% of the feedstock weight. The gas had a heating value of 490 Btu/SCF. Other pertinent statistics are given in Table II.

Initial experiments in flowing argon with no steam present yield essentially the same results as the comparable steam runs. From this, it appears that the gasification process is dominated by cracking reactions and not steam reforming reactions. Variations in heating rate of the cellulose from 50°C/min to 200°C/min do not markedly affect results.

Using gas-phase residence times of 0.46 to 0.97 sec at temperatures of 750° and 500°C, respectively, apparent rates of production were measured for seven gaseous species: CO_2, H_2, CO, CH_4, C_2H_4, C_2H_6 and C_3H_6. Figures VI and VII are graphs of log (k_j/m_i) vs. T^{-1}, where k_j is the experimentally determined rate of production of gas species j. As indicated in Figures VI and VII, the slope of the lines connecting the values of log (k_j/m_i) gives the apparent activation energy E_j associated with the rate of production of each species j. Values for each E_j are given in Table III.

Several conclusions can be drawn from Figures VI and VII. At high temperatures (and short residence times), the rate of conversion of volatile matter to ethylene is second only to carbon monoxide and methane. Thus, large yields of ethylene can be expected from a well designed biomass gasifier. In contrast to ethylene, the rate of production of propylene peaks at 675°C and rapidly declines at higher temperatures. As expected, the rates of production of hydrogen and carbon monoxide are favored by high temperatures.

Table II. SELECTED GASIFICATION RESULTS FOR CELLULOSE

Steam Superheater Temperature	350°C
Pyrolysis Reactor Temperature	500°C
Gas-Phase Reactor Temperature	700°C
Gas-Phase Reactor Residence Time	3.5 sec
Sample Weight	0.125 g
Char Residue Weight	0.012 g
Char Residue Weight Percent	10%
Tar Residue Weight	0.003 g
Tar Residue Weight Percent	2%
Gas Volume Produced	84 ml
Gas Heating Value	490 Btu/SCF
Calorific Value of Gases	13.7 MM Btu/ton
Calorific Value of Char	2.8 MM Btu/ton
Calorific Value of Tars	0.5 MM Btu/ton
Mass Balance	0.84
Carbon Balance	0.96
Gas Analysis (Vol %)	
CO	52
H_2	18
CO_2	8
CH_4	14
C_2H_4	6
C_2H_6	1
C_3H_6	0.1
Other	0.9

Table III. APPARENT LOW TEMPERATURE (500°C ≤ T ≤ 675°C) ACTIVATION ENERGY (E_j) FOR VARIOUS GAS SPECIES

Gas Species	E_j (kcal/gmol)
CO_2	21
H_2	35
C_2H_6	38
C_2H_4	55
C_3H_6	55
CO	60
CH_4	67

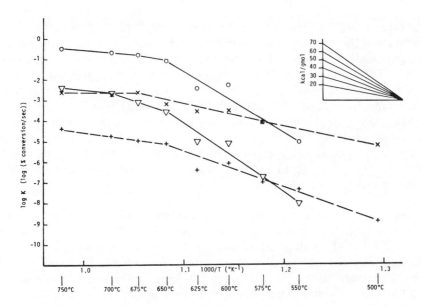

Figure 6.   Arrhenius plot of the gas—phase production rate for various gas species:
(×) CO₂, (+) H₂, (○) CO, (▽) CH₄

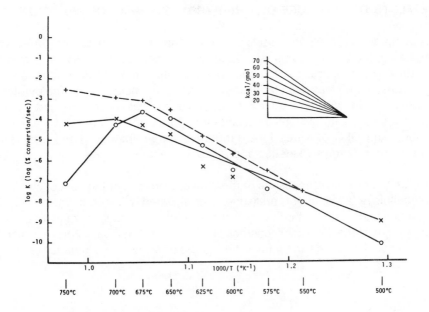

Figure 7.   Arrhenius plot of the gas—phase production rate for various gas species:
(×) C₂H₆, (+) C₂H₄, (○) C₃H₆

The rate curves for methane and ethylene "track" each other closely, suggesting that the same mechanism may be responsible for the formation of the two gases. All the rates exhibit a break at about 675°C, with lower apparent E_j above 675°C. This may indicate a change in the cracking mechanism, or it may be an artifact of poor heat transfer in the gas-phase reactor. A more explicit mechanistic interpretation of the data is made difficult by the effects of heat transfer and non-ideal plug flow on the kinetic interpretation of the experimental data. Nevertheless, the data are quite useful for engineering design purposes, and suggests criteria to be used for the design of second generation reactors intended to provide more accurate measurements of gas-phase cracking rates.

Finally, Figures II-IV also exhibit the fit (dashed lines) of Equation 11 to the experimental data for a gas-phase temperature of 600°C. Values for $\bar{v}\sigma \int c_j$ (t) dt and $\bar{v}\sigma \, (r_j/r_v) \int c_v$(t)dt at 600°C used in Equation 11 are listed in Table IV. The relatively good agreement of the model embodied in Equation 11 with the experimental data given in Figures II-IV is not entirely fortuitous, but less good at higher temperatures where the constraint $r_v \, (\tau_0 - \tau_i) << 1$ is not satisfied. A mechanistic interpretation of gas-phase phenomena is needed to improve our ability to mathematically predict the products of gasification under a variety of conditions.

EFFECTS OF PRESSURE ON THE PYROLYSIS HEAT OF REACTION

A comprehensive experimental research program to investigate the effects of pressure on the products of steam gasification of biomass is currently underway. A stainless steel, tubular microreactor similar to the quartz reactor described earlier has been fabricated for the experimental work. The pyrolysis furnace used with the quartz reactor system has been replaced in the pressurized steam system by a Setaram Differential Scanning Calorimeter (DSC). The DSC provides for quantitative determination of the effects of pressure on pyrolysis kinetics and heats of reaction.

Figure VIII presents the results of three measurements of the heat of pyrolysis of cellulose at different pressures. At elevated pressures, the pyrolysis reaction becomes exothermic, and char production increases from about 12% by weight of the cellulose feedstock at 1 bar to 16% at 6 bars. Future research is expected to refine this initial data and extend it over a broader range of pressures.

Table IV. NUMERICAL VALUES USED IN THE GAS-PHASE KINETIC MODEL

	$\bar{v}\sigma \int c_j(t)dt$	$\bar{v}\sigma (r_j/r_v) \int c_v(t)dt\vert_{T=600°C}$
CO_2	0.061	0.046
H_2	0.001	0.0057
CO	0.045	0.155
CH_4	0.0035	0.0145
C_2H_6	0.0009	0.0039
C_2H_4	0.0	0.010
C_3H_6	0.0002	0.0058

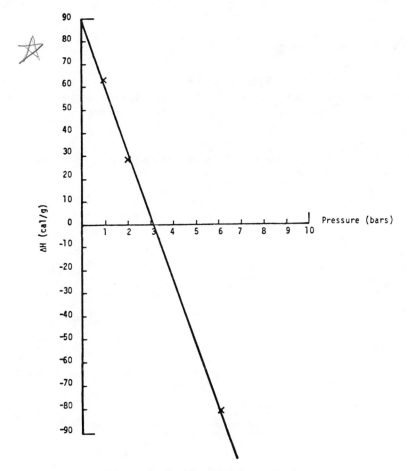

Figure 8.   Cellulose ΔH *pyrolysis* vs. *pressure*

CONCLUSIONS

Gas-phase, steam cracking reactions dominate the chemistry of biomass gasification. At temperatures above 650°C, these reactions proceed very rapidly and generate a hydrocarbon rich syngas containing commercially interesting amounts of ethylene, propylene, and methane. Increased pressure appears to inhibit the gasification process.

These results indicate that biomass gasifiers should be designed to provide for high heating rates and short residence time with gas-phase temperatures exceeding 650°C. Transport reactors, characterized by large throughputs, high heating rates, modest pressures, and short residence times appear to be ideally suited for this purpose. Future biomass gasifiers should rely on steam cracking to produce fuels and chemicals.

ACKNOWLEDGEMENTS

The assistance of Mr. W. Edwards and Mr. T. Mattocks in performing the tubular plug-flow reactor experiments is gratefully acknowledged. The measurement of the heat of pyrolysis of cellulose at six bars was made by Dr. P. Leparlouer while the author visited Setaram Laboratory in Lyon, France. The assistance of Setaram in this research is also gratefully acknowledged. This project was financed by the U.S. Environmental Protection Agency under Grant No. R804836010.

REFERENCES

1. Antal, M.J. "Symposium Papers", Energy From Biomass and Wastes, Symposium sponsored by the Institute of Gas Technology, Washington, D.C., August 1978; Institute of Gas Technology: Chicago, 1978.

2. Rensfelt, E.; Blomkvist, G.; Ekstrom, C.; Engstrom, S.; Espenas, B.G.; Liinanki, L. "Symposium Papers", Energy From Biomass and Wastes, Symposium sponsored by the Institute of Gas Technology, Washington, D.C., August 1978; Institute of Gas Technology: Chicago, 1978.

3. Lewellen, P.C.; Peters, W.A.; Howard, J.B. "Cellulose Pyrolysis Kinetics and Char Formation Mechanism", 16th International Symposium on Combustion, Cambridge, Mass., 1976.

4. Mackay, G.D.M. Canada Department of Forestry and Rural Development, Publ. 1201, Ottawa, Ont., 1967.

5. Roberts, A.F. *Combust. Flame* **1970,** *14,* 261.

6. Welker, J.R. *J. Fire Flammability* **1970,** *1,* 12.

7. Beall, F.C.; Eickner, H.W. U.S. Forest Service, 1970, FPL-130.

8. Diebold, J; Smith, G. Naval Weapons Center, NWC Technical Publication 6022, April 1978.

9. Hatch, L.F.; Matar, S. *Hydrocarbon Process.* **1978,** March, 129-139, Part 8.

10. Hatch, L.F.; Matar, S. *Hydrocarbon Process.* **1978,** March, 129-139, Part 9.

11. Feber, R.C.; Antal, M.J. U.S. Environmental Protection Agency, Report EPA-600/2-77-147, Cincinnati, Ohio, 1977.

12. Appell, H.R.; Pantages, P. in "Thermal Uses and Properties of Carbohydrates and Lignin", Shafizadeh, F.; Sarkanen, K.; Tillman, D., Eds. Academic Press: New York, 1976.

13. Antal, M.J.; Friedman, H.L.; Rogers, F.E. "Kinetic Rates of Cellulose Pyrolysis in Nitrogen and Steam", Eastern Section, The Combustion Institute Fall Meeting, Hartford, Conn., 1977.

14.   Antal, M.J.; Friedman, H.L.; Rogers, F.E. *Combust. Sci. Technol.,* in press.

15.   Reed, T.B.; Antal, M.J. "Preprints", 176th National Meeting of the American Chemical Society, Miami, Fla., 1978.

16.   Antal, M.J.; Reed, T.B. "Preprints", 176th National Meeting of the American Chemical Society, Miami, Fla., 1978.

17.   Mattocks, T. M.S.E. Thesis, Princeton University, Princeton, N.J., 1979.

18.   Antal, M.J.; Edwards, W.E.; Friedman, H.L.; Rogers, F.E. "A Study of the Steam Gasification of Organic Wastes", Final Progress Report to the U.S. Environmental Protection Agency, Princeton University, 1979.

19.   Levenspiel, O. "Chemical Reaction Engineering"; J. Wiley and Sons: New York, 1972.

RECEIVED JULY 28, 1980.

Gasification of Oak Sawdust, Mesquite, Corn Stover, and Cotton Gin Trash in a Countercurrent Fluidized Bed Pilot Reactor

STEVEN R. BECK, MAW JONG WANG, and JAMES A. HIGHTOWER

Texas Tech University, Box 4679, Lubbock, TX 79409

Research on pyrolyzing manure and wood using a pilot scale, counter-current, fluidized bed reactor to produce ammonia synthesis gas and hydrocarbons has been conducted by the Department of Chemical Engineering at Texas Tech University since 1970. Results have been encouraging and justify further study.

This paper reports the results of a comparison of the gasification of various biomass residues in the Synthesis Gas From Manure (SGFM) pilot plant. The residues evaluated include oak sawdust, mesquite, corn stover, and cotton gin trash. The SGFM process is based on a countercurrent, fluidized bed reactor. In this system, biomass is fed to the top of the reactor. As a result, the fresh feed is partially dried by direct contact with hot product gas prior to entering the reaction zone. This process has been described in detail by various researchers (1-4).

In the SGFM reactor, fresh feed enters the pyrolysis zone of the reactor prior to encountering an oxidizing atmosphere. As a result, significant amounts of olefinic compounds are formed and exit the reactor before they can decompose. This also results in the formation of tars which present problems in downstream processing.

0097-6156/81/0144-0335$05.00/0
© 1981 American Chemical Society

SGFM PILOT PLANT

The SGFM pilot plant was designed and constructed in 1975 to test the SGFM process in a continuous system and provide data for design and evaluation of a commercial facility. A schematic drawing of the SGFM pilot plant is shown in Figure I. The heart of the pilot plant is the reactor itself. The reactor is 15.2 cm in the lower 1.5 m and 20.3 m in the upper 1 m or disengaging zone. The solids are fed to the top of the reactor through a screw feeder, which controls the feed rate, and fall by gravity into the reactor itself. The air-steam mixture which enters the bottom of the reactor is preheated in an 8-m length of tubing that serves as a resistance heater. The char is removed from the reactor through a centerport opening in the bottom distributor plate. A hydraulic ram is used to prevent any bridging of the char in the discharge line. The gases exit the top of the reactor and pass into a cyclone that is operated at approximately 350°C. The cyclone is heated to prevent condensation of any of the reaction products and adequately removes most of the entrained solids. The gases leaving the cyclone then pass through a 3-stage impinger sequence which is operated at about 110° – 140°C. This serves to condense the tar but maintains the water in a vapor state. Following the impingers, the water is condensed in a double-pipe heat exchanger and collected in the downstream impinger section. The product gases are then passed through a turbine meter for flow rate measurements and vented to the atmosphere. Using this arrangement, good material balance data has been obtained, but there are some problems in operation of the pilot plant. With the screw feeder, very few problems have been encountered in feeding the biomass. The tar collection system is currently the major problem. Tar produced from biomass feedstocks is a very viscous material and tends to condense on all piping and also plug the impingers. For this reason, the errors that are apparent in material balance are primarily due to the inability to collect and measure all the tar produced. This is a relatively minor error because of the fact that the tar product is only about 5% of the raw feedstock weight.

FEEDSTOCKS

The corn stover used in this study was acquired from area farms in Lubbock County, and was ground in a hammermill such that it would pass a 1/4-inch screen. Because of the fibrous nature of the corn stover, all of the particles were not smaller than 1/4 inch. Some of the particles were greater than 1 inch long but only about 1/16 inch in diameter. Figure II gives the particle size distribution of the ground corn stover.

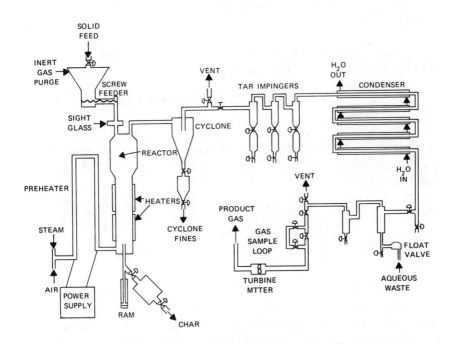

Figure 1. Flowsheet of SGFM pilot plant

The oak sawdust used in this study was obtained from Missouri. Figure III gives the particle size distribution of the sawdust.

Cotton gin trash was obtained in pelletized form from American Cotton Growers, Crosbyton Gin Division in Crosbyton, Texas. The gin trash was pelletized with no binder added. Figure IV gives the particle size distribution of the cotton gin trash.

The mesquite used in this study was obtained from Lubbock County. It was ground in a hammermill and then pulverized in a micro-pulverizer such that it would pass a 2-millimeter screen. Figure V gives the particle size distribution of the mesquite.

The moisture and ash content of the feedstocks are shown in Table I along with the heating value of each material.

Table I. FEEDSTOCK PROPERTIES

Feedstock	Moisture Content, %	Ash Content, %	High Heating Value, as-received, Btu/lb
Corn Stover	6.1	5.0	6,550
Oak Sawdust	35	0.9	4,842
Cotton Gin Trash	11.2	7.9	6,886
Mesquite	6.9	5.6	10,195

DISCUSSION

Gas Yield

The objective of this study was to determine the total gas yield and gas composition from the various feedstocks as a function of reactor temperature, air-to-feed ratio, and steam-to-feed ratio. The gas components of greatest interest are hydrogen, carbon monoxide, methane, and ethylene. These components contribute not only to the gas heating value, but also to the value of the gas as chemical synthesis feedstock.

The gas yields as a function of average reactor temperature are shown in Figure VI for dry sawdust, green sawdust, corn stover, cotton gin trash and mesquite. For comparison, the gas yields from cattle feedlot manure are also shown in Figure VI (2). The gas yields for all the feedstocks are greater than for manure. It is assumed that much of the hemicellulose and cellulose fed to the cattle was consumed during digestion. As a result, the manure is low in cellulose and high in lignin, protein, and fat. The biomass residues evaluated

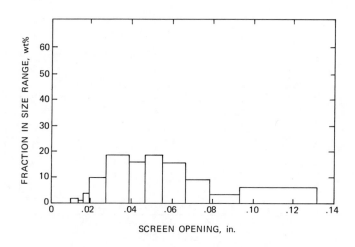

Figure 2. Particle size distribution of corn stover

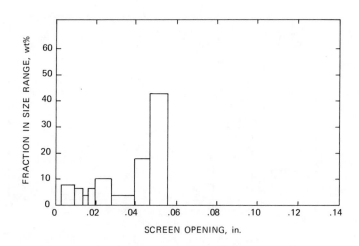

Figure 3. Particle size distribution of oak sawdust

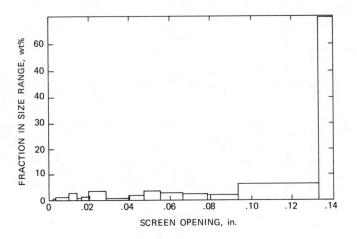

Figure 4.   Particle size distribution of cotton gin trash

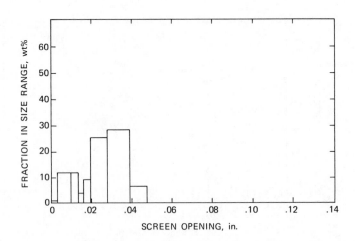

Figure 5.   Particle size distribution of mesquite

contain relatively high amounts of hemicellulose and cellulose. Previous studies reported by Stamm (5) show that hemicellulose and cellulose gasify at a much higher rate than does lignin. Therefore, unaltered plant matter, which is high in cellulose, should produce more gas than cattle manure which is low in cellulose.

The four runs with green sawdust are very interesting. Two runs, 56 and 57, were made with no steam injected into the reactor. This resulted in a higher gas yield than runs 54 and 55 which were made with steam injection. It is felt that the steam served as a heat sink and reduced the reaction temperature of the particles. These results also indicate that the steam-char reaction did not occur. Previous studies (6, 7) indicate that the reaction temperature must exceed 860°C in order for the steam-char reaction to be significant. This conclusion is also supported by the gas composition which will be discussed later. The gas yields for corn stover are very similar to those obtained for the dry sawdust. Corn stover poses some very difficult operating problems however. These will be discussed in the section on operating difficulties. The pelleted cotton gin trash behaved somewhat differently than the other feedstocks. The total gas yield was not greatly different, but it appeared that the steam-char reaction did occur. The gin trash was composed of some large pellets and a significant fraction of fines and broken pellets. This greatly affected fluidization in the reactor. The whole pellets probably did not fluidize until they were mostly ash. Consequently, they were able to lay on the distributor and undergo steam gasification at low temperature (<800°C) and long residence time. The char exiting the bottom of the reactor was light grey while the cyclone fines were black. With all other feedstocks, both the bottom char and cyclone fines were black.

The ash content of the char and cyclone fines supports the conclusion that the gin trash char underwent steam gasification. Table II shows the ash content from the various runs. The high ash content of the char from cotton gin trash indicates a high degree of conversion to gas.

The corn stover also showed a high level of conversion. This is probably due to the very small diameter of the particles which allows for rapid heat transfer. This results in high gasification rates.

Table II. ASH CONTENT OF CHAR AND CYCLONE FINES

Run No.	Feedstock	Raw Feed	Ash Content, %	
			Char	Cyclone Fines
51	Corn Stover	5.0	88.8	57.8
60	Corn Stover	5.0	89.5	51.7
71	Corn Stover	5.0	69.4	56.7
72	Corn Stover	5.0	46.6	72.8
73	Corn Stover	5.0	70.5	46.1
74	Corn Stover	5.0	86.5	28.9
76	Mesquite	5.6	19.6	29.9
77	Mesquite	5.6	17.3	31.4
78	Mesquite	5.6	24.0	34.1
79	Mesquite	5.6	18.6	20.5
67	Cotton Gin Trash	7.9	67.7	31.9
68	Cotton Gin Trash	7.9	82.8	28.7
69	Cotton Gin Trash	7.9	84.3	39.1
70	Cotton Gin Trash	7.9	90.3	29.3
54	Green Oak Sawdust	0.9	11.1	29.8
55	Green Oak Sawdust	0.9	9.0	28.4
56	Green Oak Sawdust	0.9	8.7	31.6
57	Green Oak Sawdust	0.9	2.2	19.6
64	Green Oak Sawdust	0.9	— —*	5.6
65	Green Oak Sawdust	0.9	— —*	7.1

* No char was obtained.

Additional evidence indicating lack of fluidization of the large gin trash pellets appeared when the bottom flange of the reactor was removed. With manure, sawdust, mesquite, and corn stover, the reactor was empty at the conclusion of each run. With pelleted cotton gin trash, a buildup of ash was found on the distributor plate. This means that the pelleted gin trash could not be used in the SGFM reactor on a continuous basis.

Gas Composition

The yields of hydrogen, carbon monoxide and ethylene are important when considering use of biomass as a chemical feedstock. Hydrogen is a valuable product for chemical synthesis and upgrading low quality fuels such as coal and heavy oils.

Shown in Figure VII is the ultimate hydrogen yield from the various feedstocks. This is calculated as the sum of the hydrogen yield and carbon

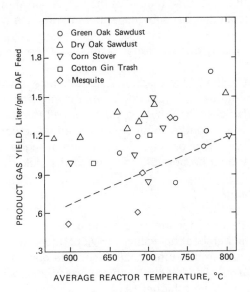

Figure 6. Total dry gas yield

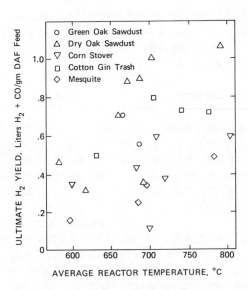

Figure 7. Ultimate hydrogen yield

monoxide yield. It is assumed that all of the CO can be shifted to H_2 and CO_2. The cotton gin trash and oak sawdust appear to produce the most hydrogen. In the case of gin trash, this is due to the steam-char reaction. For oak sawdust, it is probably due to higher reactivity of the material.

A unique aspect of the SGFM reactor is that ethylene is present in the gas at significant concentrations. Figure VIII shows the ethylene yield obtained from the various feedstocks. The ethylene yield from all the feedstocks studied is lower than for cattle manure. In all cases, the ethylene content is below the limit where recovery is economic. By operating at higher temperatures and gas velocity, it should be possible to increase the yield of ethylene.

Gas Heating Value

If the gas is to be used for fuel, the heating value of the gas is critical. In Figure IX, the higher heating value (HHV) of the raw gas is shown as a function of reactor temperature. In all cases, the gas has an HHV between 200 and 400 Btu/SCF. Feedstocks, such as oak sawdust, which produce more hydrocarbon gases, have a higher HHV.

Thermal Conversion

A relative measure of the thermal efficiency of the process is shown in Figure X. This shows the total heating value of the gas as a percentage of the total heating value of the feed. The relative values depend primarily on gas heating value and total gas yield.

Comparison of Feedstocks

Of the four feedstocks used in this study, oak sawdust proved to be the easiest to work with as no bridging occurred within the feed hopper and no plugging occurred between the screw feeder and the reactor inlet. The problem of tar removal and buildup in the impingers and downstream lines was no greater than that created by the other feedstocks.

Mesquite and corn stover were the second and third most desirable feedstocks in terms of ease of handling. Mesquite required additional preparation in that it had to be pulverized to prevent bridging within the feed hopper and clogging between the screw feeder and the reactor inlet. Tar removal problems were comparable to those of oak sawdust. Feeding difficulties were encountered with corn stover, as it tended to bridge within the feeder and the reactor inlet. Carry-over of fines in the product gas from corn stover increased tar buildup in the impingers and downstream lines. It is

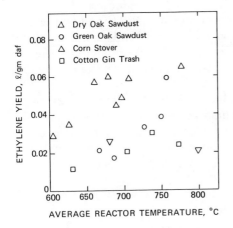

Figure 8. *Ethylene yield as a function of temperature*

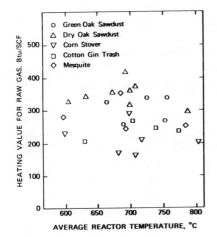

Figure 9. *Heating value of raw gas*

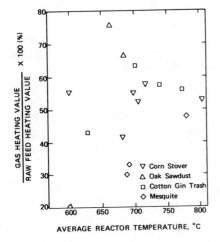

Figure 10. *Percentage conversion of raw feed heating value to gas heating value as a function of average reactor temperature*

the authors' opinion that difficulties encountered in the use of corn stover could be eliminated if the feedstock was pulverized to reduce the length-to-diameter ratio of the particles and an electrostatic precipitator was used to knock out all solid fines in the product gas stream as it left the reactor.

Cotton gin trash was the least favorable of the feedstocks in terms of handling. The formation of clinkers within the reactor is a serious problem, one which cannot be solved in the present system. The bottom flange of the reactor must be taken off in order to remove clinkers from the reactor. This situation is not conducive to any continuous operation process.

Product gas yields for the four feedstocks all increased as average reactor temperature increased. A high air-to-feed ratio was found to increase product gas yield for oak sawdust, corn stover, the value being 1.5 ℓ/g DAF feed, and the lowest from mesquite, the value being 0.51 ℓ/g DAF feed. The percentage of corn stover and cotton gin trash converted to product gas was higher than that of oak sawdust and mesquite. Ash contents of char and cyclone fines for corn stover were between 46 and 89% and between 46 and 73%, respectively, while those for cotton gin trash were between 67 and 90% and between 28 and 39%, respectively. Ash contents of char and cyclone fines from mesquite ranged from 17.3 to 24% and from 20 to 34%, respectively, and ash contents of cyclone fines from oak sawdust were 5.6 and 7.1%. The difference in percentage conversion is due to either a difference in gasification rates of the feedstocks, or a difference in residence time. It is believed that the high percentage conversion of corn stover was caused by a high pyrolysis and gasification rate. Corn stover feed rates were low, but high in terms of volume of feed because of the low bulk density of the feedstock. This situation allowed for a greater heat transfer area per pound of feed, thereby increasing conversion rate. As mentioned previously, char from cotton gin trash collected in the bottom of the reactor, giving the feedstock a long residence time.

The effect of average reactor temperature on product gas yield from cotton gin trash is not as pronounced as that for the other feedstocks. Char buildup in the bottom of the reactor creates this effect, as a long residence time allows for increased heat transfer and as a result, increased conversion of the feedstock to product gas. Low product gas yields from mesquite at low temperatures are probably a result of a low gasification rate. As temperature is increased, product gas rate increases sharply. The sharp rise in product gas rate at elevated temperature is probably due to the breakup of lignin which was not converted to product gas at lower temperatures.

Product gas yields from oak sawdust were higher than those from corn stover and mesquite and approximately the same as those from cotton gin trash at the same average reactor temperatures. Although gas yield is high, ash content of cyclone fines is relatively low in comparison to the other feedstocks. This indicates that high gas yields are probably a result of the composition of the oak sawdust. Ash content would have been greater if the high product gas yield had been caused by a long residence time or high gasification and pyrolysis rates. It is postulated that at increased heat transfer rates, (i.e., elevated temperatures), product gas yields from oak sawdust and mesquite would continue to increase while product gas yields from cotton gin trash and corn stover will level off and remain constant.

Increased heat transfer could be accomplished by a longer residence time in the reactor, higher temperatures or smaller particles. However, particle size must be large enough to prevent entrainment. If the reactor was operated in the temperature range covered in this study, char and cyclone fines from mesquite and oak sawdust could be recycled, increasing total gas yield per pound of dry, ash-free feed. Very little additional gas could be obtained with the recycle of char and cyclone fines from corn stover and cotton gin trash. A semiquantitative rating system for the various feedstocks was developed. This was based on operating considerations and product yields. This will serve as a guideline for future work. Table III presents a weighted comparison of the four feedstocks used in this study.

Table III. Weighted Comparison of Corn Stover, Oak Sawdust Cotton Gin Trash, and Mesquite

	A	B	C	D	E	F	G	H	I
1. Corn Stover	?	6	3	3	2	3	6	6	29
2. Oak Sawdust	?	15	4	3	3	6	10	10	51
3. Cotton Gin Trash	?	0	3	2	3	4	8	6	26
4. Mesquite	?	8	2	4	1	6	8	3	32

A = Availability and Price
B = Operability (1-15)
C = Product Gas Yield (0—4)
E = Ultimate Hydrogen Yield (0—4)
F = Gas Quality (0—6)
G = Calorific Value of Gas/lb Feed (0—10)
H = Percentage Conversion of Raw Feed Heating Value to Gas Heating Value (0—10)
I = Sum of A through H

The availability and price of the feedstocks is a major factor in determining the feasibility of a particular feedstock. Values are not presented because they will vary depending on location. Operability is a major concern and is therefore weighted heavily. Oak sawdust is the best feedstock as to operability because it feeds easily and requires very little treatment prior to feeding. Mesquite is rated lower than oak sawdust because it must be pulverized before feeding, while corn stover is rated even lower because it must be ground and has a tendency to plug and bridge within the hopper. Fine particles from corn stover are also blown out of the reactor because of its low bulk density, causing plugging in downstream lines. Cotton gin trash is rated at zero because of clinker formation within the reactor. This will not allow continuous operation in a system such as the one used in this study.

Product gas yield, total hydrocarbon yield, and ultimate hydrogen yield are rated on a scale of 1 to 4 because differences between feedstocks are not great enough to warrant a large deviation in values. Gas quality is considered to be of more importance than C, D, and E and is therefore rated on a scale from 1 to 6. It is considered more important because a higher gas quality will lower gas shipping costs.

The calorific value of the gas per pound of feed and the percentage conversion of raw feed heating value to gas heating value are both of major importance and are rated on a scale from 1 to 10. The calorific value of gas produced from mesquite is lower than other feedstocks at low temperatures, but increases to high values at high temperatures. Mesquite was given a high rating because the potential for higher yields of energy per pound of feed at elevated temperatures exists. The percentage conversion of energy stored in the feed to energy in the gas is rated heavily because low percentage conversions might indicate that higher energy yields could be obtained in an alternate process. The practicality of fluidized bed gasification is greatly reduced if a higher energy yield can be obtained from some other process.

SUMMARY

The Synthesis Gas From Manure (SGFM) process was designed to convert cattle feedlot manure to ammonia synthesis gas. Current work is aimed at using any biomass feedstock to produce either medium-Btu gas or chemical feedstocks. This paper presents a comparison of the experimental results compiled on gasification of oak sawdust, corn stover, mesquite, and cotton gin trash in the SGFM pilot plant. A weighted comparison of the product gas, hydrocarbon, and hydrogen yields, gas quality, calorific value of product gas, percentage conversion of raw feed heating value to gas heating value, and operability of each feed indicated that oak sawdust was the best feedstock.

REFERENCES

1. Beck, S. R. 3rd Annual Biomass Energy Systems Conference, Golden, Colo., June 5-7, 1979.

2. Beck, S. R.; Huffman, W. J.; Landeene, B. C.; Halligan, J. E. *Ind. Eng. Chem., Proc. Des. Dev.* **1979** *18*, 328.

3. Halligan, J. E.; Herzog, K. L.; Parker, H. W. *Ind. Eng. Chem., Proc. Des. Dev.* **1975** *14* (1), 64-69.

4. Halligan, J. E.; Sweazy, R. M. 72nd National Meeting of the American Institute of Chemical Engineers, St. Louis, May 21-24, 1972.

5. Stamm, A. J. *Ind. Eng. Chem.* **1956** *48* (3), 413-17.

6. Garrett, D. E. 70th Annual Meeting of the American Institute of Chemical Engineers, New York, November 16, 1977.

7. Rensfelt, E.; Blomkuist, G.; Ekstrom, S.; Espenas, B. G.; Liinanki, L. Conference on Energy from Biomass and Wastes Sponsored by the Institute of Gas Technology, Washington, D.C., 1978; Institute of Gas Technology, Chicago, 1978; paper No. 27.

RECEIVED JULY 7, 1980.

Thermochemical Gasification of Woody Biomass

H. F. FELDMANN, P. S. CHOI, H. N. CONKLE, and S. P. CHAUHAN

Battelle Columbus Laboratories, 505 King Avenue, Columbus, OH 43201

It is generally agreed that oil and gas supply problems can be partially met by the use of renewable resources such as forest products, agricultural materials, and urban wastes. The main advantage of renewable energy sources other than their renewable nature is that they are clean, i.e., low in sulfur and ash, and are highly reactive and nonagglomerating compared with fossil fuels. Gasification of renewable feedstock offers great potential particularly for retrofit of existing gas and oil-fired industrial boilers which represent about 84 percent of the boilers sold between 1963 and 1975 (1). Gasification can also be used for industrial dryers and furnaces and, on a larger scale, can be coupled in a combined-cycle system for power generation or used for the synthesis of transportation fuels. Although various types of gasification systems have been explored, there is a lack of steady-state data on the effect of various operating parameters including the use of catalysts to improve gasification reactivities, product distribution, and yields.

The objective of this paper is to present experimental data on effects of operating parameters, including the catalytic effects of wood ash, calcium oxide, and calcium carbonate, on wood gasification in a continuous reactor. These results will be utilized to guide operation of a multi-solid fluid-bed (MSFB) wood gasification process which is being developed by Battelle to improve the economics of producing a medium-Btu gas or synthesis gas from wood and other biomass (2).

0097-6156/81/0144-0351$06.25/0
© 1981 American Chemical Society

EXPERIMENTAL SYSTEM AND PROCEDURES

The catalytic wood gasification experiments were carried out in a 2.8-inch I.D. pressurized continuous reactor system. The experimental system is shown schematically in Figure I and consists of the following sections.

1. Hydrogen and steam feeding,
2. Wood feeding,
3. Gasifier,
4. Char withdrawal,
5. Liquid product collection, and
6. Gas metering and analysis.

Wood pellets are charged to the feed hopper and then the system is sealed and pressurized with hydrogen or nitrogen. The gasifier is then brought to the desired run temperature. After establishing the desired gas flow rates, wood feeding is initiated.

The gasifier is 12 feet in overall height with 8 feet within the heated zone. It operates with the flow of gas countercurrent to the downward-moving wood. Char is continually removed from the bottom of the gasifier to maintain a constant bed height and stored in the pressurized char receiver. Hot gases exiting the reactor are cooled in a condenser where the liquid products are collected. After removal of the liquid products, the gas is filtered, reduced in pressure, metered, and finally analyzed by a gas chromatograph and a continuous methane analyzer.

DISCUSSION

Experimental Results

Ultimate and proximate analyses of typical wood feed materials are given in Table I. Ranges of the operating conditions are:

Feed Type:	Wood pellets
Feed Rate:	4-12 lb/hr
Gasifier Temperature:	1150-1600 °F
Pressure:	10-216 psig
Residence Time:	30-130 min (nominal)
Steam/Wood Ratio:	0.10-0.56 lb/lb MAF
Feed Hydrogen/Wood Carbon Ratio:	0.24-0.58 lbM/lbM
Catalysts:	Wood ash, CaO and $CaCO_3$ incorporated into wood pellets

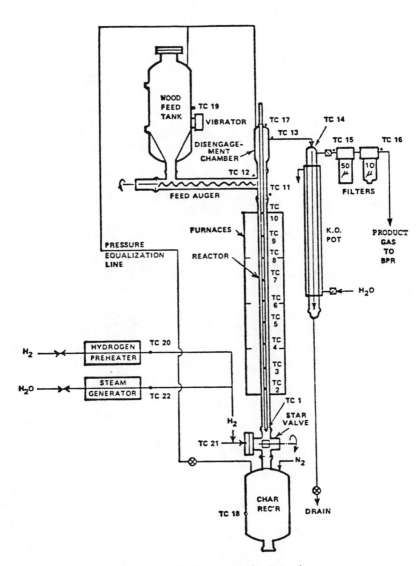

Figure 1.   Continuous gasification unit

Table I. ANALYSIS OF RAW AND CATALYZED WOOD FEED MATERIALS

Item	California Pellet	BCL Pellet	BCL Pellet	BCL Pellet	BCL Pellet	BCL Pellet	BCL Pellet	Tenn Pellet
Type of Wood	Hard, White	Hard, White	Hard, White	Hard, White	Hard, White	Hard, White	Hard, White	Hard, Mixed
Catalyst	None	None	1.5% Ash	5% CaO	9% $CaCO_3$	5% Ash	20% Ash	None
Ultimate Analysis, %								
C	46.78	47.41	45.96	44.56	44.81	45.56	37.22	40.36
H	5.75	5.86	5.62	5.60	5.45	5.62	4.68	4.70
N	0.64	0.11	0.14	0.11	0.08	0.08	0.14	0.22
S	0.14	0.02	0.05	0.07	0.09	0.08	0.15	0.13
Cl	0.02	0.03	0.02	0.02	0.03	0.01	0.03	0.04
Ash	0.66	0.66	2.22	4.95	6.08	5.28	15.22	10.46
Moisture	5.64	6.20	7.10	6.24	5.99	5.74	9.33	11.80
O (bal)	40.37	39.71	38.89	38.45	37.47	37.63	33.23	39.29
Volatile Matter, %	79.56	81.13	78.05	74.57	74.86	75.98	61.60	66.74
Fixed Carbon, %	14.14	12.01	12.63	14.24	13.07	13.00	13.85	11.00
Heating Value, Btu/lb	7897	8182	7855	7515	7275	7437	6222	6881

Table II. SUMMARY OF CONTINUOUS WOOD GASIFICATION TEST EXPERIMENTS

Item	11	11	11	12
Type of Feed[a]	CALF.P.	CALF.P.	CALF.P.	TENN.P.
Catalyst	No	No	No	No
Feed Rate, lb/hr	10.39	9.90	10.91	11.10
lb-MAF/hr	9.73	9.28	10.22	9.08
Bed Temperature, °F	1525.0	1504.0	1498.0	1580.6
Pressure, psig	212.1	212.1	212.1	214.3
Residence Time, [b] min	53.74	56.34	51.14	35.82
Steam Feed Rate, lb/hr	0.00	2.20	2.20	0.00
Hydrogen Feed Rate, SCFH	51.10	51.10	0.00	53.81
Steam/Wood Ratio, lb/lb-MAF	0.00	0.24	0.22	0.00
H_2/Wood-C Ratio, lbM/lbM	0.33	0.35	0.00	0.35
H_2 in Feed Gas, vol %	100.00	52.45	0.00	100.00
Prod. Gas Flow Rate, [c] SCFH	116.0	127.5	112.9	111.1
SCF/lb MAF	—	—	11.05	—
Carbon Conversion, wt%				
Gas Products	57.87	59.85	56.12	54.24
Liquid Products	2.65	3.06	2.82	2.71
Net Btu Output,[d] Btu/lb-MAF Wood				
To Gas Products	5170.87	5410.10	5378.95	4951.20
To Total	5206.01	5445.66	5412.98	5145.83
Gas Yield, SCF/lb-MAF Feed				
CH_4	4.80	4.54	3.41	4.87
CO	2.64	3.19	3.18	2.20
CO_2	1.41	1.51	2.05	1.84
H_2	2.93	4.40	2.30	3.15
Gas Composition (dry/wet)				
CH_4	40.24/26.64	33.07/21.70	30.87/19.39	39.77/25.61
CO	22.11/14.63	23.23/15.25	28.79/18.08	17.99/11.59
CO_2	11.85/7.84	10.99/7.21	18.55/11.65	15.02/19.67
H_2	24.61/16.29	31.99/20.99	20.79/13.06	25.75/16.58
C_2H_4	.38/.25	.19/.12	.40/.25	.38/.24
C_2H_6	.80/.53	.51/.33	.57/.36	.75/.48
N_2	0.00/0.00	0.00/0.00	0.00/0.00	.34/.22
H_2S	.02/.01	.02/.01	.02/.01	.01/.01
H_2O	0.00/33.18	0.00/34.37	0.00/37.19	0.00/35.60
Gas Heating Value,[e] Btu/SCF	576.44	523.46	487.21	561.12

Table II. CONTINUED

Item	12	12	12	12	14	14	15	15
Type of Feed[a]	TENN.P.	TENN.P.	TENN.P.	TENN.P.	TENN.P.	TENN.P.	TENN.P.	TENN.P.
Catalyst	No	No	No	No	No	No	No	No
Feed Rate, lb/hr	10.90	10.59	10.86	11.32	13.47	12.96	9.87	9.93
lb-MAF/hr	8.91	8.66	8.88	9.26	11.02	10.60	8.07	8.12
Bed Temperature, °F	1517.8	1495.1	1485.4	1488.2	1514.0	1403.0	1560.3	1459.2
Pressure, psig	214.3	214.8	214.3	214.3	69.9	74.1	20.9	20.9
Residence Time,[b] min	36.50	37.55	36.61	35.13	29.52	30.69	40.29	40.05
Steam Feed Rate, lb/hr	1.37	2.40	3.12	4.61	0.00	0.90	0.00	3.30
Hydrogen Feed Rate, SCFH	0.00	0.00	0.00	0.00	45.67	0.00	43.76	0.00
Steam/Wood Ratio, lb/lb-MAF	0.15	0.28	0.35	0.50	0.00	0.08	0.00	0.41
H_2/Wood-C Ratio, lbM/lbM	0.00	0.00	0.00	0.00	0.24	0.00	0.33	0.00
H_2 in Feed Gas, vol %	0.00	0.00	0.00	0.00	100.00	0.00	100.00	0.00
Prod. Gas Flow Rate,[c] SCFH	102.2	105.7	125.8	129.9	119.2	109.1	108.4	133.6
SCF/lb MAF	11.47	12.21	14.17	14.03	—	1029	—	17.09
Carbon Conversion, wt %								
Gas Products	53.35	58.71	64.89	62.37	47.62	48.66	54.70	76.35
Liquid Products	3.07	4.02	4.40	4.78	6.58	5.68	1.78	5.11
Net Btu Output,[d] Btu/lb-MAF Wood								
To Gas Products	5516.68	5707.55	6417.00	6187.47	4396.37	4756.21	4994.96	7330.58
To Total	5731.49	5988.79	6710.84	6500.77	4974.15	5207.82	5117.85	7754.02
Gas Yield, SCF/lb-MAF Feed								
CH_4	3.48	3.46	3.63	3.46	3.20	2.34	3.05	2.53
CO	2.94	3.39	3.64	3.25	2.63	3.23	3.86	6.63
CO_2	2.33	2.69	3.16	3.32	1.44	1.97	1.15	2.43
H_2	2.56	2.45	3.44	3.72	3.13	2.40	4.74	4.80
Gas Composition (dry/wet)								
CH_4	30.36/18.50	28.38/15.90	25.66/13.83	24.65/12.38	29.56/19.77	22.73/14.86	22.69/16.96	14.80/10.15
CO	25.61/15.61	27.81/15.58	25.73/13.86	23.18/11.64	24.33/16.27	31.34/20.49	28.77/21.51	38.84/26.64
CO_2	20.35/12.40	22.04/12.35	22.28/12.00	23.64/11.88	13.31/8.90	19.13/12.51	8.58/6.41	14.21/9.75
H_2	22.27/13.37	20.08/11.25	24.29/13.09	26.49/13.31	28.92/19.34	23.30/15.23	35.30/26.39	28.13/19.30
C_2H_4	.47/.29	.56/.31	.61/.33	.62/.31	2.29/1.53	1.96/1.28	3.30/2.47	2.89/1.98
C_2H_6	.74/.45	1.05/.59	1.34/.72	1.28/.64	1.36/.91	1.45/.95	1.17/.87	1.05/.72
N_2	.18/.11	.07/.04	.08/.04	.13/.07	.22/.15	.07/.05	.18/.13	.07/.05
H_2S	.02/.01	.01/.01	.01/.01	.01/.01	.01/.01	.01/.01	.01/.01	.01/.01
H_2O	0.00/39.05	0.00/43.98	0.00/46.12	0.00/49.76	0.00/33.12	0.00/34.63	0.00/25.25	0.00/31.41
Gas Heating Value,[e] Btu/SCF	480.80	467.66	453.00	441.00	530.15	461.93	508.63	429.39

Table II. CONTINUED

Item	16	16	17	17	18	18	18	18
Type of Feed[a]	BCLP	BCLP	BCLP	BCLP	BCLP	BCLP	BCLP	BCLP
Catalyst	No	No	1.5% Ash	1.5% Ash	5% CaO	5% CaO	5% CaO	5% CaO
Feed Rate, lb/hr	6.63	6.63	6.63	4.72	12.16	12.07	12.22	12.67
lb-MAF/hr	6.18	6.17	9.48	4.28	10.80	10.72	10.85	11.25
Bed Temperature, °F	1340.0	1285.0	1327.0	1347.0	1344.0	1284.0	1182.0	1390.0
Pressure, psig	20.2	18.7	21.0	20.1	19.3	20.1	18.6	20.3
Residence Time,[b] min	69.70	69.77	44.25	98.04	38.04	38.32	37.85	36.50
Steam Feed Rate, lb/hr	0.00	1.47	0.00	1.35	0.00	3.60	3.30	3.60
Hydrogen Feed Rate, SCFH	45.53	0.00	50.29	0.00	50.44	50.44	0.00	0.00
Steam/Wood Ratio, lb/lb-MAF	0.00	0.24	0.00	0.32	0.00	0.34	0.30	0.32
H_2/Wood-C Ratio, lbM/lbM	0.46	0.00	0.33	0.00	0.29	0.30	0.00	0.00
H_2 in Feed Gas, vol %	100.00	0.00	100.00	0.00	100.00	39.96	0.00	0.00
Prod. Gas Flow Rate,[c] SCFH	92.6	90.6	141.0	79.7	124.7	169.1	148.0	202.3
SCF/lb MAF	—	14.68	—	18.62	—	—	13.64	17.98
Carbon Conversion, wt %								
Gas Products	55.79	68.48	61.63	85.89	46.61	54.95	56.91	76.96
Liquid Products	5.06	9.11	3.05	12.83	2.66	4.41	4.21	4.68
Net Btu Output,[d] Btu/lb-MAF Wood								
To Gas Products	4796.32	6225.11	5435.62	7499.07	4011.44	4553.2	5140.76	7140.34
To Total	5126.94	6846.75	5592.89	8478.92	4778.14	4823.90	5408.00	7749.57
Gas Yield, SCF/lb-MAF Feed								
CH_4	2.78	2.24	2.99	2.85	2.20	1.69	1.56	2.07
CO	3.55	5.01	3.74	6.34	2.97	3.82	3.94	6.57
CO_2	1.31	2.55	1.79	3.06	1.18	2.33	2.70	2.46
H_2	6.59	4.21	5.63	5.61	4.61	7.49	3.01	6.32
Gas Composition (dry/wet)								
CH_4	18.54/13.68	15.26/9.72	20.13/16.36	15.31/10.34	19.07/14.00	10.72/7.41	11.45/7.85	11.50/8.50
CO	23.67/17.47	34.16/21.75	25.17/20.46	34.02/22.98	25.70/18.87	24.20/16.72	28.89/19.79	36.55/27.02
CO_2	8.77/6.47	17.37/11.06	12.02/9.77	16.41/11.08	10.21/7.50	14.77/10.21	19.79/13.56	13.69/10.12
H_2	44.03/32.49	28.68/18.26	37.87/30.78	30.10/20.33	39.95/29.34	47.47/32.81	36.77/25.19	35.13/25.37
C_2H_4	2.64/1.95	2.71/1.73	2.78/2.26	2.61/1.76	2.86/2.10	1.59/1.10	1.85/1.27	1.97/1.46
C_2H_6	1.81/1.34	1.40/.89	1.72/1.40	1.41/.95	1.63/1.20	1.16/.80	1.15/.79	1.08/.80
N_2	.50/.37	.40/.25	.30/.24	.13/.09	.58/.43	.07/.05	.09/.06	.08/.06
H_2S	.01/.01	.01/.01	.01/.01	.01/.01	.01/.01	.01/.01	.01/.01	.01/.01
H_2O	0.00/26.21	0.00/36.32	0.00/18.72	0.00/32.46	0.00/26.56	0.00/30.89	0.00/31.48	0.00/26.08
Gas Heating Value,[e] Btu/SCF	479.42	424.17	480.82	402.40	478.25	385.13	376.93	397.20

Table II. CONTINUED

Item	19	19	19	20	20	20	22	22
Type of Feed[a]	BCL. P.	BCL. P.	BCL. P.	BCL. P.	BCL. P.	BCL. P.	TENN. P.	TENN. P.
Catalyst	9% CaCO$_3$	9% CaCO$_3$	9% CaCO$_3$	5% Ash Spray	5% Ash Spray	Ash Spray	No	No
Feed Rate, lb/hr	10.82	10.61	11.18	12.12	12.18	12.58	10.30	10.15
lb-MAF/hr	9.51	9.33	9.83	10.78	10.84	11.19	8.54	8.41
Bed Temperature, °F	1393.0	1383.0	1393.0	1375.0	1341.0	1379.0	1398.2	1454.5
Pressure, psig	175	19.8	23.8	21.1	22.8	22.3	7.7	10.2
Residence Time,[b] min	44.71	45.60	43.28	43.90	43.69	2.30	41.24	41.84
Steam Feed Rate, lb/hr	0.00	3.00	3.00	0.00	3.10	2.90	0.00	2.93
Hydrogen Feed Rate, SCFH	50.57	50.57	0.00	46.10	46.10	0.00	49.40	49.40
Steam/Wood Ratio, lb/lb-MAF	0.00	0.32	0.31	0.00	0.29	0.26	0.00	0.35
H$_2$/Wood-C Ratio, lbM/lbM	0.31	0.34	0.00	0.26	0.26	0.00	0.36	0.37
H$_2$ in Feed Gas, vol %	100.00	44.46	0.00	100.00	41.39	0.00	100.00	44.47
Prod. Gas Flow Rate,[c] SCFH	107.6	185.2	158.6	119.2	204.1	167.4	111.0	154.3
SCF/lb MAF	—	—	16.13	—	—	14.96	—	—
Carbon Conversion, wt%								
Gas Products	47.67	71.42	69.80	44.70	58.25	63.26	48.90	61.55
Liquid Products	7.35	12.04	9.48	5.63	10.55	12.80	4.44	5.64
Net Btu Output,[d] Btu/lb-MAF Wood								
To Gas Products	3681.23	5983.82	6325.61	3866.76	5721.33	5985.22	3995.97	5214.99
To Total	4255.82	6722.53	6954.75	4156.56	6225.93	6620.77	4277.04	5555.50
Gas Yield, SCF/lb-MAF Feed								
CH$_4$	2.25	2.18	1.96	2.24	1.79	1.86	2.06	1.84
CO	3.19	5.63	5.93	2.64	4.94	5.40	3.52	4.88
CO$_2$	1.27	2.63	2.47	1.41	1.99	2.11	1.20	2.12
H$_2$	4.06	8.82	5.32	4.22	9.66	5.14	5.66	8.97
Gas Composition (dry/wet)								
CH$_4$	19.92/12.82	10.97/7.36	12.16/8.76	20.22/12.74	9.48/6.60	12.44/7.47	15.88/10.34	10.05/6.57
CO	28.24/18.17	28.33/19.02	36.73/26.49	23.89/15.05	26.23/18.26	36.11/21.68	27.11/17.65	26.63/17.41
CO$_2$	11.26/7.25	13.24/8.89	15.39/11.04	12.74/8.03	10.57/7.36	14.10/8.47	9.21/6.00	11.56/7.56
H$_2$	35.92/23.11	44.42/29.82	32.95/23.74	28.21/24.08	51.27/35.68	34.38/20.64	43.56/28.36	48.92/31.98
C$_2$H$_4$	2.67/1.72	1.59/1.07	1.56/1.12	2.68/1.69	1.03/.76	1.77/1.06	2.73/1.78	1.76/1.15
C$_2$H$_6$	1.54/.99	1.08/.72	1.14/.82	1.59/1.00	.78/.54	1.11/.67	1.12/.73	.85/.56
N$_2$	.40/.26	.37/.25	.13/.09	.66/.44	.56/.39	.09/.05	.37/.24	.23/.15
H$_2$S	.01/.01	.01/.01	.01/.01	.01/.01	.01/.01	.01/.01	.01/.01	.01/.01
H$_2$O	0.00/35.56	0.00/32.88	0.00/27.94	0.00/36.99	0.00/30.40	0.00/39.96	0.00/34.89	0.00/34.62
Gas Heating Value,[e] Btu/SCF	477.57	389.65	391.92	474.99	376.89	400.24	451.43	388.04

Table II. CONTINUED

Item	22	23	23	23	24	24	24	25
Type of Feed[a]	TENN. P.	TENN. P.	TENN. P.	TENN. P.	BCL. P.	BCL. P.	BCL. P.	TENN. P.
Catalyst	No	No	No	No	1.5% Ash	1.5% Ash	1.5% Ash	No
Feed Rate, lb/hr	9.91	10.65	10.69	10.67	12.36	12.40	12.59	3.88
lb-MAF/hr	8.22	8.82	8.85	8.84	11.28	11.31	11.49	3.02
Bed Temperature, °F	1396.5	1598.0	1557.1	1555.7	1364.0	1338.0	1404.0	1412.8
Pressure, psig	9.1	8.8	11.9	11.6	22.5	22.5	22.4	19.7
Residence Time,[b] min	42.86	39.88	39.75	39.79	43.04	42.92	42.25	129.84
Steam Feed Rate, lb/hr	3.00	0.00	2.60	3.30	0.00	0.00	3.05	0.00
Hydrogen Feed Rate, SCFH	0.00	51.00	51.00	0.00	47.00	100.00	47.00	28.80
Steam/Wood Ratio, lb/lb-MAF	0.37	0.00	0.29	0.37	0.00	0.00	0.27	0.00
H_2/Wood-C Ratio, lbM/lbM	0.00	0.26	0.36	0.00	0.26	0.55	0.25	0.58
H_2 in Feed Gas, vol %	0.00	100.00	48.23	0.00	100.00	100.00	42.26	100.00
Prod. Gas Flow Rate,[c] SCFH	135.9	131.6	217.3	185.8	128.6	175.2	218.8	54.4
SCF/lb MAF	16.53	—	—	21.01	—	—	—	—
Carbon Conversion, wt %								
Gas Products	66.38	60.23	86.36	89.41	49.30	46.10	67.57	52.87
Liquid Products	4.78	1.96	3.27	2.62	3.71	7.12	6.37	1.17
Net Btu Output,[d] Btu/lb-MAF Wood								
To Gas Products	6036.04	5068.78	7915.54	8379.39	4228.08	4092.65	6576.33	5653.96
To Total	6347.38	5166.01	8091.09	8513.18	4483.85	4635.66	7020.87	5677.83
Gas Yield, SCF/lb-MAF Feed								
Gas Composition (dry/wet)								
CH_4	8.88/6.24	17.35/13.28	10.26/7.99	11.34/8.84	20.18/14.13	15.16/10.64	12.42/8.70	10.55/8.47
CO	34.68/24.35	29.57/22.64	32.08/24.99	37.86/29.51	26.70/18.70	17.87/12.54	26.73/18.73	20.89/16.76
CO_2	15.51/10.89	9.50/7.27	8.91/6.94	12.95/10.09	12.31/8.62	7.79/5.47	11.30/7.92	8.56/6.87
H_2	38.68/27.16	38.99/29.85	45.90/35.75	34.64/27.00	35.20/24.65	55.50/38.95	45.95/32.20	46.91/37.64
C_2H_4	1.49/1.05	3.26/2.50	1.83/1.43	2.35/1.83	3.29/2.30	1.88/1.32	2.02/1.42	2.08/1.67
C_2H_6	.93/.65	.98/.75	.82/.64	.77/.60	1.91/1.34	1.66/1.16	1.31/.92	1.52/1.22
N_2	.14/.10	.33/.25	.19/.15	.08/.06	.39/.27	.13/.09	.26/.18	.47/.38
H_2S	.01/.01	.01/.01	.01/.01	.01/.01	.01/.01	.01/.01	.01/.01	.01/.01
H_2O	0.00/29.77	0.00/23.43	0.00/22.10	0.00/22.06	0.00/29.96	0.00/29.82	0.00/29.92	0.00/19.76
Gas Heating Value,[e] Btu/SCF	364.90	465.43	398.50	398.84	489.27	449.05	415.00	476.06

Table II. CONTINUED

Item	25	25	25	26	26	27	27
Type of Feed[a]	TENN.P.	TENN.P.	TENN.P.	TENN.P.	TENN.P.	BCL.P. 20% Ash	BCL.P. 20% Ash
Catalyst	No	No	No	No	No	No	No
Feed Rate, lb/hr	3.96	6.83	11.97	11.09	11.15	8.12	8.11
lb-MAF/hr	3.08	5.31	9.31	8.62	8.67	6.13	6.12
Bed Temperature, °F	1432.7	1403.8	1370.0	1617.0	1580.0	1402.0	1403.0
Pressure, psig	19.8	18.1	23.0	12.5	14.5	21.2	23.1
Residence Time, [b] min	127.21	73.76	42.07	44.76	44.51	72.21	72.20
Steam Feed Rate, lb/hr	1.20	2.16	2.40	0.00	2.64	0.00	3.40
Hydrogen Feed Rate, lb/lb-MAF	0.00	0.00	0.00	52.30	0.00	51.00	0.00
Steam/Wood Ratio, lb/lb-MAF	.39	0.41	0.26	0.00	0.30	0.00	0.56
H_2/Wood-C Ratio, lbM/lbM	0.00	0.00	0.00	0.37	0.00	0.54	0.00
H_2 in Feed Gas, vol %	0.00	0.00	0.00	100.00	0.00	100.00	0.00
Prod. Gas Flow Rate, [c] SCFH	73.4	102.7	160.9	123.1	164.7	93.6	168.1
SCF/lb MAF	23.83	19.34	17.28	—	19.00	—	27.47
Carbon Conversion, wt %							
Gas Products	97.24	77.64	73.90	52.94	80.90	51.95	102.48
Liquid Products	3.89	5.37	4.62	2.23	3.45	1.51	1.66
Net Btu Output, [d] Btu/lb-MAF Wood							
To Gas Products	9502.94	7519.23	7049.17	4668.99	7708.67	4166.59	10052.12
To Total	9734.68	7843.66	7312.11	4787.99	7916.09	4196.95	10091.93
Gas Yield, SCF/lb-MAF Feed							
CH_4	2.90	2.26	2.25	2.46	2.26	2.62	2.21
CO	8.41	6.18	6.27	3.77	7.53	2.66	9.93
CO_2	3.26	3.14	2.48	1.28	2.28	1.73	2.87
H_2	8.49	7.12	5.70	6.14	6.31	7.67	11.90
Gas Composition (dry/wet)							
CH_4	12.18/9.29	11.69/7.70	13.02/8.41	17.21/12.74	11.88/8.86	17.19/12.52	8.03/6.61
CO	35.28/26.90	31.96/21.05	36.27/23.43	26.42/19.56	39.65/29.56	17.44/12.71	36.14/29.76
CO_2	13.67/10.42	16.23/10.69	14.33/9.26	8.97/6.64	12.02/8.96	11.35/8.27	10.43/8.59
H_2	35.59/10.42	36.83/24.26	32.96/21.29	42.98/31.82	33.22/24.27	50.20/36.57	43.30/35.65
C_2H_4	1.89/1.44	1.93/1.27	2.14/1.38	3.07/2.27	2.33/1.74	2.07/1.51	1.16/.96
C_2H_6	.98/.75	1.04/.68	1.10/.71	1.03/.76	.80/.60	1.43/1.04	.60/.96
N_2	.40/.31	.31/.20	.17/.11	.31/.23	.09/.07	.30/.22	.34/.28
H_2S	.01/.01	.01/.01	.01/.01	.01/.01	.01/.01	.01/.01	.01/.01
H_2O	0.00/23.74	0.00/34.14	0.00/35.40	0.00/25.97	0.00/25.44	0.00/27.15	0.00/17.66
Gas Heating Value, [e] Btu/SCF	398.54	388.74	407.80	464.50	405.70	450.10	365.90

(a) Cal P = pellet supplied by California Pellet Mill, hardwood
    Tenn P = pellet supplied by Tennessee Woodex, hardwood and bark
    BCL P = pellet made at Battelle-Columbus, hardwood

(b) Based on volumetric rate of feed material.

(c) Monitored at 70°F.

(d) Net Btu Input = $\dfrac{\text{Gross Btu output in Product Gas} - \text{Btu Input in Feed Gas}}{\text{Wood Feed Rate (MAF)}}$

(e) For raw dry gas.

Results obtained in the continuous unit are summarized in Table II. The runs carried out included hydrogasification and steam gasification, as well as gasification with hydrogen and steam mixtures. In order to identify the optimum operating conditions and most effective catalyst application, often more than one parameter was changed simultaneously. Thus the correlations presented here must be considered empirical and more experimental work should be carried out to isolate individual parameter effects.

This study established that the following parameters had the most significant effect on gasification.

- Type of gasifying agent — whether steam or hydrogen.
- Feed gas/wood ratio.
- Temperature.
- Solids residence time.
- Presence of CaO and/or recycled wood ash catalyst.

The effect of reactor pressure was mainly to shift the methane concentration in the gas. However, the heat content of the gas generated (net gas yield/lb maf wood x   heating value of gas) was, if anything, increased by lowering the reactor pressure.

The discussion that follows describes the results of steam gasification and hydrogasification separately since the type of gasifying agent had perhaps the greatest effect on product yield and composition. Despite the simultaneous changing of parameters for these experiments, the results are summarized in a series of plots to illustrate trends and avoid forcing the casual reader to wade through the tables. The effects of gasification parameters will for the most part be illustrated by their effect on the net gaseous Btu yield. This is probably the single most important output parameter because it measures the saleable commodity of the process. A cursory comparison of some of the results presented in these tables for steam gasification with data generated by Professor Ingemor Bjerle's group at the University of Lund-Sweden in an approximately 1.5-ft I.D. fluid bed indicated that at similar conditions, the gas compositions were remarkably similar.

From a process point of view, one of the surprising and important implications of this study is that steam gasification is much more effective than hydrogasification even if a methane-rich gas is the desired end product.

Steam Gasification

In these experiments, pure steam was injected into the gasifier distributor. The effect of temperature on the net Btu product gas yield is shown in Figure II. This figure shows that:

● Increases in temperature increase the net Btu yield.
● CaO added to the pellets in a 5 weight percent ratio reduces the gasification temperature required to achieve a given Btu yield by about 100°F.
● CaCO$_3$, at least at the 9 weight percent level, did not improve gasification.
● Wood ash at a 5 weight percent ratio did not affect gasification.

Figure III shows the effect of solids residence time on net Btu yield. There is a question as to whether the high Btu yield shown for the pellets containing 20 percent recycled wood ash is due to the catalytic effect of the wood ash or the relatively high steam/wood ratio used for that run. Further experiments are needed to isolate the catalytic effects from the effects of other process parameters.

The steam/wood ratio affected both the net Btu yield of the gas and the fraction of carbon converted to liquid products. The increase in gaseous net Btu yield with increasing steam/wood ratio is shown in Figure IV. Figure V shows a significant increase in both carbon conversion to gas and liquid products with increasing steam/wood ratio. One explanation for the increase in liquid products with increasing steam is that increasing the steam concentration reduces the amount of volatiles that crack to form residual carbon.

The direct formation of methane increases the product gas heating value and reduces the amount of heat required for gasification. On the other hand, if the gas is to be used for subsequent synthesis in a grass roots synthesis plant, methane is undesirable since it is a diluent that increases compression costs and the size of the synthesis portion of the plant. The operating parameters found to have the greatest effect on methane concentration were the temperature and pressure. From coal gasification experience, one would expect methane concentrations to be higher for hydrogasification than for steam gasification, but with wood it is not at all clear that the increased methane concentration in the gas is due to hydrogasification. For example, a change from pure hydrogen to pure steam during a run always resulted in a lowering of reactor temperature. For both steam gasification and hydro-gasification, the methane concentration in the product gas increased with increasing reactor temperature as shown in Figure VI. Also note that the

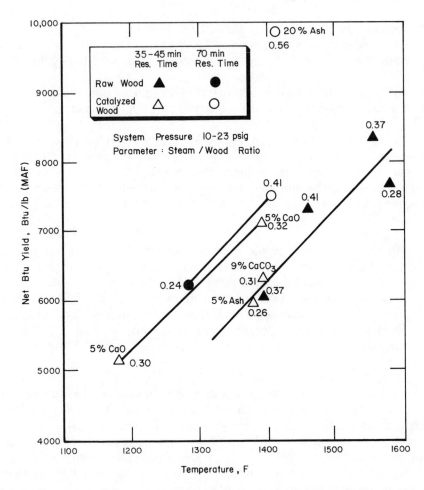

Figure 2.   Net BTU yield (gas) as a function of temperature in steam gasification of wood

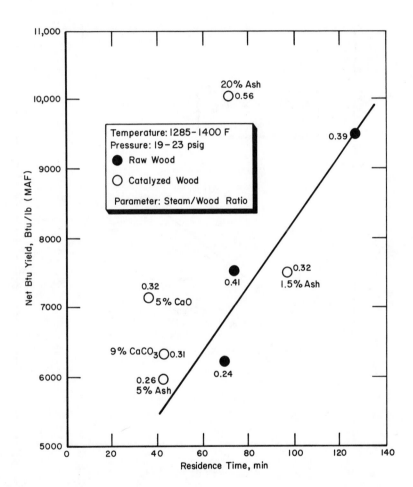

Figure 3. Effect of residence time on net BTU yield (gas) in steam gasification of wood

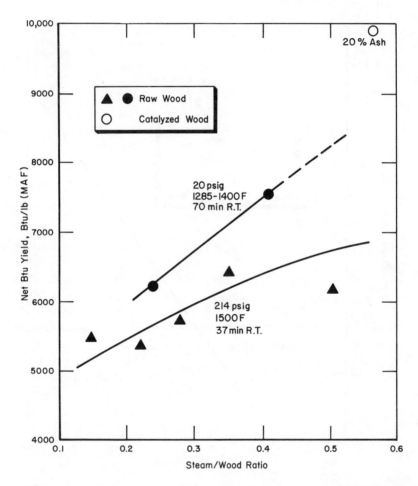

Figure 4.   Effect of steam/wood ratio on net BTU yield (gas) in steam gasification of wood

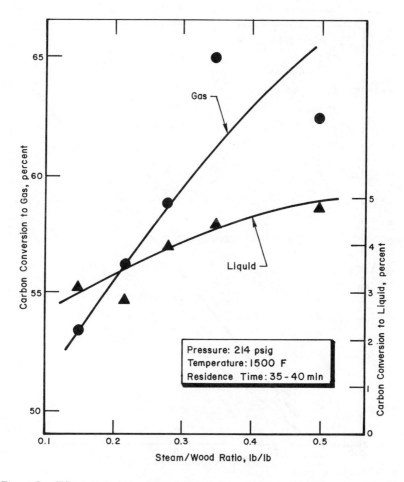

Figure 5.   Effect of steam/wood ratio on carbon conversion to gaseous and liquid products

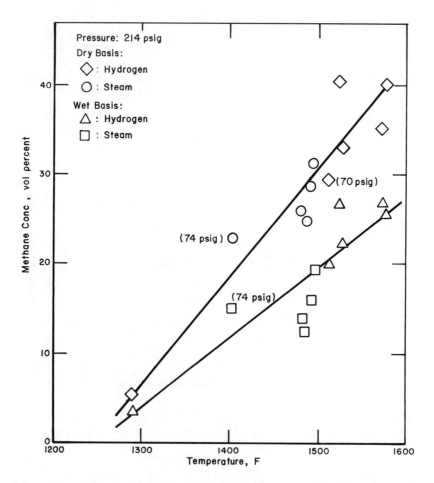

*Figure 6.   Methane concentration against temperature in gasification of raw wood
with steam and H$_2$*

methane concentrations measured for steam gasification extrapolate directly to those for hydrogasification at the higher temperature. Thus, Figure VI suggests that the higher methane concentrations observed for hydrogasification could simply be due to the higher temperatures achieved during the hydrogasification portion of the tests.

Data at lower pressures than the 200 psig used to illustrate the importance of temperature in Figure VI suggest that the relative effects of hydrogen versus steam on methane concentration could be greater at lower pressures. For example, several methane concentrations for lower pressure tests for both catalyzed and uncatalyzed wood are plotted against temperature in Figure VII. These results suggest that hydrogen is more effective than steam in increasing methane concentrations at lower pressures than at higher pressures where temperature rather than gasifying agent has the greatest effect.

With either steam or hydrogen, increases in the total system pressure increase methane concentration as shown in Figure VIII. While, as discussed above, the relative effects of hydrogen versus steam on methane concentration at the higher system pressures are probably due to temperature rather than the different gasifying agents, the effect of system pressure seems pronounced for both hydrogen and steam over the pressure range from atmospheric to 100 psig. Above this pressure range, increases in pressure do not greatly increase the methane concentration.

With only one exception, the gas compositions from both steam and hydrogasification and that from an experiment with a very short steady-state period were well in excess of methanation equilibria. This is shown in Figure IX where the equilibrium constant for the reaction $CO + 3H_2 \rightleftharpoons CH_4 + H_2O$ and the experimental values of the partial pressure ratio $(P_{CH_4} P_{H_2O})/(P_{CO} P^3_{H_2})$ are plotted against temperature. Since the experimental ratios are well above the equilibrium constant, introducing a methanation catalyst would decrease the methane yield and thereby increase the heat required to carry out the gasification reaction.

Direct Hydrogasification

In the direct hydrogasification experiments, hydrogen is fed into the distributor at the bottom of the gasifier. As the data in Table II show, there is no great difference in gas compositions between the tests in which pure steam is the feed gas and those in which pure hydrogen is used. However, because of the gas/solids contact in the 3-inch I.D. reactor, there is a considerable difference in the gas composition seen by the wood particles as

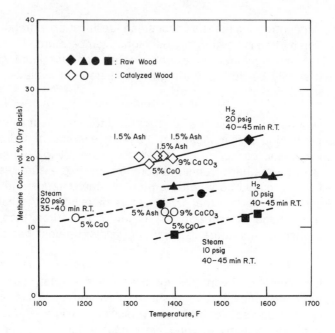

Figure 7. Methane concentration against temperature in gasification of wood with steam and H_2

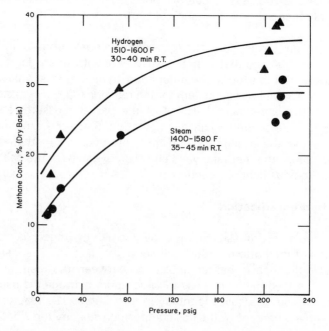

Figure 8. Effect of system pressure on methane concentration in raw product gas in gasification of wood

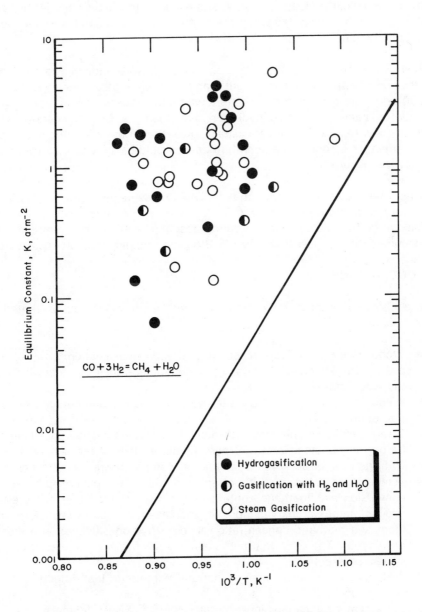

Figure 9. Comparison of theoretical and actual equilibrium constants

they move downward in plug flow through the reactor. In the case of pure hydrogen feed, the steam is evolved at the reactor top as the particles dry and the more highly converted wood char in the lower portion of the reactor sees only hydrogen.

Tests with mixtures of steam and hydrogen provide another example that steam is a better gasification agent than hydrogen at pressures at least up to 200 psig. This is illustrated in Figure X where the net Btu yield of product gas is plotted against fraction of hydrogen in the feed gas. Since temperatures for the tests with pure steam were usually lower than those with pure hydrogen, the effect, on a constant temperature basis, would be more pronounced than shown in Figure X.

Thus, these tests demonstrate that, for lower pressure gasification systems, it is very important for the char to "see" steam in order to achieve the maximum Btu yield in the product gas and that, at the temperatures examined thus far (less than 1600°F), Btu yields are higher at lower pressures.

CONCLUSIONS

The following tentative conclusions were reached from the data presented in this paper.

- Wood ash and CaO are reasonably effective gasification catalysts for pellets. The effectiveness apparently depends on their incorporation into the wood matrix.
- Wood ash, CaO, and $CaCO_3$ seemed to increase organic liquid product formation.
- Steam gasification proceeds at a much higher rate than hydrogasification. Carbon conversions 30 to 40 percent higher than those achieved with hydrogen can be achieved with steam at comparable residence times.
- Temperature is the most important parameter in affecting both carbon conversion and methane content in the raw product gas. Increases in gasifier temperature resulted in increased carbon conversion and increased methane concentration in both hydrogasification and steam gasification. The effect on methane concentration was greater at higher system pressures.
- Residence time also is an important parameter that affects carbon conversion.
- For steam gasification, lower total system pressure increased carbon conversion. However, methane concentration increased at higher system pressure.

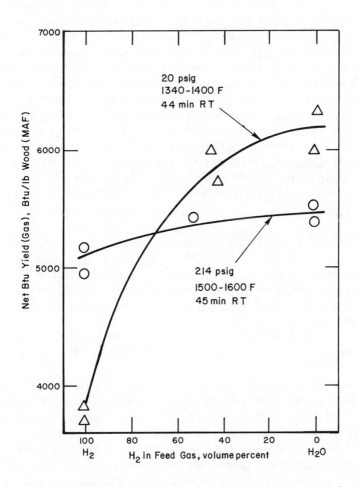

Figure 10.   Effect of gasification agent on net BTU yield in gas product

- Increasing the steam/wood ratios over the range of 0 to 0.45 lb/lb significantly increased carbon conversion.
- Higher pressures of about 200 psig are required for hydrogen to approach steam in its effectiveness in reacting with the wood char.

Because of a high "volatile matter" content of wood (approximately 80 weight percent), it seems reasonable that devolatilization plays a major role in the overall conversion process. Since devolatilization is probably suppressed with increasing system pressure, gasification and hydrogasification reactions with the solid phase are probably much more important at high pressure than at low pressure. At lower pressure, the major conversion to gas occurs through reactions of hydrogen and steam with the vapor products of devolatilization. Thus, the increased effectiveness of steam at the lower pressures suggests that steam cracking of the volatiles to gaseous products is probably more effective than hydrocracking. The additional conversion of solid-phase carbon requires high system pressures for hydrogasification than for steam gasification which is not nearly as dependent on pressure. These factors are the probable reasons for the increased gas yields at lower pressure and the increased effectiveness of steam compared to hydrogen.

The observations that at higher system pressures, temperature influences methane concentration in the gas phase much more than whether steam or hydrogen is the gasifying agent, and the much higher-than-equilibrium methane concentrations suggests that methane probably arises from the primary devolatilization process rather than subsequent solid phase reactions.

SUMMARY

The effects of gasification parameters for both catalyzed and raw wood in gasification experiments with hydrogen, hydrogen/steam, and steam are described. Calcium oxide, calcium carbonate, and wood ash were used as catalysts. Experimental results indicate that steam is a more effective gasification agent for wood than hydrogen. Steam gasification proceeds at a higher rate resulting in a greater net Btu recovery in the product gas. When incorporated into pelletized wood, wood ash and calcium oxide are both effective in increasing carbon conversion and net Btu recovery. Wood ash is effective for both hydrogen and steam gasification while calcium oxide seems more effective in steam than in hydrogen atmospheres. In addition, calcium carbonate was found to increase organic liquid product formation. Methane concentrations in excess of that predicted by thermodynamic equilibrium were achieved over the entire range of hydrogen/steam ratios and pressures studied.

ACKNOWLEDGEMENTS

This study was sponsored by the Biomass-Thermochemical Conversion Group of the Solar Energy Division, U.S. Department of Energy. The guidance and support of Mr. Nello DelGobbo of this group is gratefully acknowledged.

REFERENCES

1. Hall, E.H., *et al.* "Comparison of Fossil and Wood Fuels", Washington, D.C., March 1976, EPA-600/2-76-056.

2. Feldmann, H.F. "Symposium Papers", Energy from Biomass and Wastes, Symposium sponsored by the Institute of Gas Technology, Washington, D.C. Aug. 14-18, 1978; Institute of Gas Technology: Chicago, 1978.

RECEIVED JUNE 30, 1980.

ECONOMICS AND ENERGETICS

Comparative Economic Analysis of Chemicals and Synthetic Fuels from Biomass

FRED A. SCHOOLEY, RONALD L. DICKENSON, STEPHEN M. KOHAN,
JERRY L. JONES, PAUL C. MEAGHER, KENT R. ERNEST,
GWEN CROOKS, KATHERINE A. MILLER, and WING S. FONG
SRI International, 333 Ravenswood Avenue, Menlo Park, CA 94025

The study summarized in this paper was designed to examine numerous technological processes for producing useful fuels and chemicals from biomass.

The specific objectives of the study were:

● To determine those biomass missions most likely to result in energy market penetration in the years 1985, 2000, and 2020;
● To quantify the level of market penetration expected in those years;
● To provide R&D program recommendations for the U.S. Department of Energy Fuels from Biomass Systems Branch.

The details of this work are presented in the seven-volume final report to the U.S. Department of Energy. The titles of these volumes are:

I   Summary and Conclusions
II   Mission Selection, Market Penetration, Modeling, and Economic Analysis
III Feedstock Availability
IV Thermochemical Conversion of Biomass to Fuels and Chemicals
V   Biochemical Conversion of Biomass to Fuels and Chemicals
VI Mission Addendum
VII Program Recommendations

0097-6156/81/0144-0379$09.75
© 1981 American Chemical Society

METHODOLOGY*

Market Penetration Model

The methodology underlying the comparative economic assessment is based in part on our previous work in the field of energy market analysis. In particular, this investigation has drawn on previously developed analytical approaches to the problem of forecasting the expected market potential of newly introduced energy technologies and commodities. To aid in these past analyses, a computer model was developed and applied to the evaluation of market potentials for various solar technologies and synthetic fuels. However, the level of model detail necessary to investigate specific biomass missions has required further modeling effort. This work has focused on the development of a methodology that uses an iterative process to converge equilibrium biomass supply/demand/price conditions. The following paragraphs describe model inputs, summarize the methodology, and discuss details of the market penetration formulas contained within the model.

Input Data

The data required for operation of the model consist of the resource availabilities of the various types of biomass feedstocks, the process economics of biomass conversion options, a framework of energy demands and market prices, and a set of three parameters that are used to describe the interaction of the biomass-derived products with the markets in which they compete.

Biomass Resources

Because of the unique character of biomass resource availability, computer modeling requires a more sophisticated approach than that suitable for other types of resources. Unlike other energy sources, biomass is both a renewable and depletable feedstock at the same time. It is renewable over long periods and depletable in the short term because of restrictions such as length of the growing season and rate of residue generation.

Biomass availability is most readily described in a manner similar to that typically used for fossil fuels by the use of a curve that considers the resource quantity available as a function of price. Unlike fossil fuel resource curves, however, the biomass curves vary in time, reflecting changes in the expected future availability of biomass feedstocks from residues and energy farms.

---

* See Appendix A for definitions of the terms used in the paper, and Appendix B for product demand and price data sources.

Five major biomass feedstocks were considered in this analysis — low moisture plants, high moisture plants, woody crops, manure, and marine crops. For each of these categories, Mission Analysis Final Report, Volume III, describes a set of resource curves, corresponding to the years 1975, 1985, 2000, and 2020 — the time frame of the analysis. Prices for the intervening years are found by interpolation within the computer model.

Conversion Economics

The process economic data required for biomass conversion describe the costs and efficiencies required in producing an energy product from a biomass feedstock. This information determines the production price for any biomass product at any specific biomass feedstock cost.

Specifically, the computer model cost inputs consist of a specific capital cost (SCC) in units of dollars per million Btu of biomass-derived product per year and an operating and maintenance cost (M) in dollars per million Btu. The efficiency input parameter, e, specifies the amount of product energy obtainable per unit of energy contained in the biomass feedstock. These parameters are supplied for each feedstock-product mission.

In determining the overall product price, a capital recovery factor (CRF) is first applied to the SCC to obtain a capital charge in dollars per million Btu of product energy. The maintenance and feedstock costs are added to this, resulting in the following product cost equation:

Biomass product production price = CRF X SCC + M + feedstock cost/e

Where:

$$
\begin{aligned}
CRF &= \text{capital recovery factor} \\
SCC &= \text{specific capital cost} \\
M &= \text{operating and maintenance cost} \\
e &= \text{efficiency factor.}
\end{aligned}
$$

Expected Energy Demands and Prices

In order to perform a market penetration analysis, a framework of energy product demands and market price projections in which the biomass-derived products compete must be assumed. For each of the markets under consideration, an overall energy demand sets the size of the available market open to the biomass products. A projection of the alternative (nonbiomass) fuel market price also is given as an estimate of the competitive environment that the biomass-derived product will face. Both the product demand and the market price projections are input as a function of time and region.

Market Penetration

The market penetration parameters are input to the model to characterize marketplace behavior. Three different parameters are used — market share, the behavioral lag half-life, and the behavioral lag response.

Model Utilization

The market penetration analysis uses an iterative process that converges toward equilibrium biomass supply/demand/price conditions. Supply/demand/price equilibrium is defined here as the situation in which the production price of the biomass feedstocks is at the levels necessary to match the supplies of feedstocks with the demands for them. Figure I is a flow diagram of the procedure.

Estimates are first made for the equilibrium marginal production prices of each major biomass feedstock as a function of time and region. Because of the relationships between feedstock price and quantity of feedstock available (as described by a set of biomass resource curves), this also determines estimates for the equilibrium supply quantities of each major feedstock.

Using the initial feedstock production prices, the model calculates the equilibrium marginal production prices of each biomass-derived product as a function of time. (The equation shown earlier is used to find these prices.) Product penetration estimates can then be made for each biomass product (shown later in this section), and the level of demand for each of the feedstocks can be evaluated for the estimated feedstock prices.

To test for equilibrium conditions, the demand estimates for each major feedstock are compared with the supply quantity estimate obtained earlier. On the first iteration through this procedure, supply-demand mismatches possibly will occur for various feedstocks and at various time periods. These discrepancies indicate that the equilibrium feedstock prices have not been accurately estimated and that adjustments must be made. If the calculated demand for a particular feedstock exceeds the quantity available at the estimated feedstock price level, for example, then the estimated price is too low and should be adjusted upward. On the next iteration through the procedure, one of the effects of this adjustment will be to make a larger feedstock supply available, as determined by the appropriate biomass resource curve. Another effect will be to push the biomass product prices

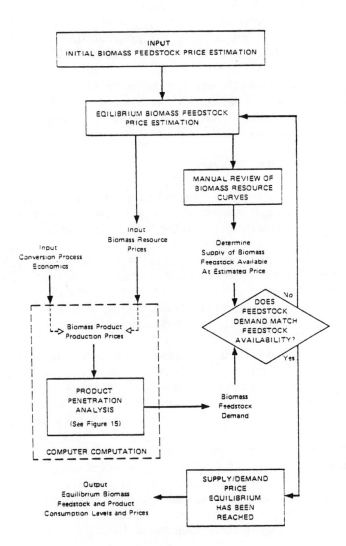

Figure 1.   Overall methodology of biomass comparative analysis

upward, causing a reduction in demand as calculated by the product penetration analysis. These two effects will push the biomass feedstock supply and demand levels toward closer agreement. At this point, the direction and size of the mismatch is again evaluated, feedstock prices are again adjusted, and another iteration of the process is performed. After several such iterations, the equilibrium supply/demand/price condition will be reached for each feedstock and for each time period chosen.

The results obtained from the final equilibrium iteration will be those that are useful to the overall biomass mission analysis. Most importantly, they will provide estimates of regional demand for each major biomass feedstock and each major biomass-derived product.

Product Penetration Analysis

The basic steps followed in the product penetration assessment are presented in Figure II. This procedure first entails a static economic analysis based on the competition between the biomass-derived product price and the market price. The result is a steady-state market share, reflecting a situation that would be expected to exist after a period during which the competitive economic forces remained constant. Steady-state conditions do not hold soon after the introduction of a new, cost-competitive mission, technology, or product. The marketplace will be in a state of flux as the newcomer gains wider recognition and acceptance.

This dynamic market behavior is modeled as a "behavioral lag" constraint, which reduces the rate at which a new mission or technology may be introduced. With this information a dynamic biomass product market share can be generated and applied to the product demand forecast to obtain an estimate of the potential biomass mission penetration as a function of time.

In this methodology, the results produced are not interactive with the basic framework of energy demands and alternative prices in which the technologies compete. Thus, the assumed market demands and prices are not directly perturbed by the biomass mission market penetration, a valid assumption as long as the potential biomass mission demand does not become too large a share of the total demand.

Steady-State Market Share

As an idealization, the share of a particular market that a single new technology or product can attain at any particular time under steady-state

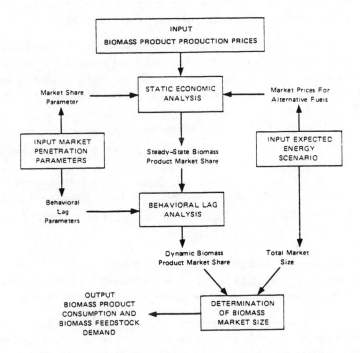

Figure 2.   Logic used in biomass product penetration analysis

conditions can be represented by the curves shown in Figure III and given by:

$$\text{Steady-State Market Share to Biomass-Derived Product} = \frac{1}{1 + \left(\frac{P_b}{P_m}\right)^{\gamma}}$$

where P_b and P_m are the marginal price of the biomass product and the marginal market price, respectively.* This static representation says that when P_b and P_m are equal and under steady-state conditions, the market will be shared equally. The market share parameter (γ) is a measure of market imperfections, price variations, and consumer preferences.

When two or more new technologies are competing for a share of the same market, a more general market share formula must be used. For example, if N different biomass missions all produce the same product (such as SNG), then their respective market shares would be represented by the following equation:

$$\text{Steady-State Share to N}^{th}\text{ Biomass Product} = \frac{1}{\left(\frac{P_{B_N}}{P_{B_1}}\right)^{\gamma} + \left(\frac{P_{B_N}}{P_{B_2}}\right)^{\gamma} + \ldots + \left(\frac{P_{B_N}}{P_{B_N}}\right)^{\gamma} + \left(\frac{P_{B_N}}{P_M}\right)^{\gamma}}$$

where P_{B_I} through P_{B_N} represents the prices of the first through the Nth biomass products and P_M represents the expected market price. If all of the prices P_{B_I} through P_{B_N} and P_M were equal, each biomass product would receive $1/(N + 1)$ of the market.

In this representation, a single representative marginal price is used for each biomass product and alternative fuel. Actually, significant individual variations from these representative prices do exist. For example, continued

---

* The market penetration model converts the average market prices into marginal prices before use in the steady-state market share analysis. The equation used for this purpose is:

$$P_{m_t} = \frac{0.25 \, (P_t + P_{t+1}) \, (D_t + D_{t+1}) - 0.25 \, (P_t + P_{t-1}) \, (D_t + D_{t-1}) \, (1 - \Delta T / X)}{0.5 \, [D_{t+1} - D_{t-1} + (D_t + D_{t-1})(\Delta T / X)]}$$

where:

P_m is the marginal market price.
t   denotes the time period under consideration.
P is the average market price.
D is the total demand for the market.
ΔT is the size of the time intervals used in the analysis (5 years).
X is the lifetime of the energy facility or production unit used in the market.

governmental regulations might introduce significant disparities into the future marginal prices of pipeline gas. Also, other sources such as LNG imports and various high-Btu synthetic products will be introduced at yet different prices, causing an even wider distribution of prices about our assumed representative marginal price. The market share parameter is used to model such price variations.

Decision-makers also influence market share. Even if a new technology is somewhat more expensive than the alternative, some fraction of purchasers will choose it, perhaps because of environmental or "energy independence" considerations. Alternatively, some fraction of purchasers will continue to use their familiar fuel source even if economic considerations dictate a change to a new one. Imperfect price information also affects the market share curve. These various factors are aggregated into the one market share parameter. In a perfect market with high price elasticity and none of these real world effects, γ would be infinite, and the product with even a very slight economic advantage would obtain a 100 percent steady-state market share.

Based on previous work with energy commodities, we have found that the characteristic response patterns of various markets can be modeled by a suitable choice of γ. Large industrial and utility markets, for example, could generally be modeled with a γ value in the range of 25 to 35. These high values reflect the strong response to price variations by industrial and utility consumers who deal with large quantities of energy and are acutely aware of economic considerations. The γ value used for this market penetration analysis was 20, reflecting the somewhat larger degree of market imperfection that would be expected to obtain in markets that are available to biomass-derived products. Even lower γ values would be applicable for smaller scale energy consumers, who typically behave in a less strictly economic fashion. Factors such as aesthetics and novelty, for example, would be expected to play a much larger role in the decision-making process of a residential consumer than an industrial buyer.

Dynamic Market Response

The market share curve in Figure III is only a static representation. To assess the dynamics of market penetration, a "dynamic market response curve" can be used to describe how fast the current market will move toward the static price-determined market share curve as a result of real-world behavioral response. This is called the behavioral lag effect.

The dynamic market response curve, given by:

$$\frac{1}{1 +\left(\dfrac{h}{n}\right)^{\alpha}}$$

is shown in Figure IV where:

h = behavioral lag half-life (time required for one-half of the market to respond to the entrance of a new product).

n = years since new product introduction.

α = behavioral lag response parameter.

This curve tends to slow the introduction of a new technology based on the time that it takes for market decision makers to accept and switch to the new product.

The behavioral lag parameters, h and α, provide a means of quantifying the dynamic market response. A half-life of 10 years was chosen for this analysis, allowing for a maximum of 50 percent market penetration at a point 10 years from the date assumed for commercialization. The second behavioral lag parameter, α, fixes the relative shape (curvature) of the dynamic market response curve once the half-life parameter has been chosen; a value of 4 was chosen for this parameter.

To find the share of the open market captured by a new fuel in any particular year, the equilibrium market share and dynamic market response curve are multiplied. This is done on an annual basis, resulting in a dynamic market share for the new product that varies with time. The actual biomass product demand is then found as a function of time by applying the dynamic market share to estimates of the size of the market that is available to the new product.

Outputs

The market penetration calculations are performed at five-year intervals over the time frame of the analysis. However, particular emphasis is placed on the analysis of three periods - the year term (1985), the intermediate term (2000), and the long term (2020).

The results of the penetration analysis provide estimates of the expected market penetration (in quads) of each mission by region and time interval. The expected demands for each type of feedstock and the equilibrium marginal prices for biomass feedstocks and products are also determined.

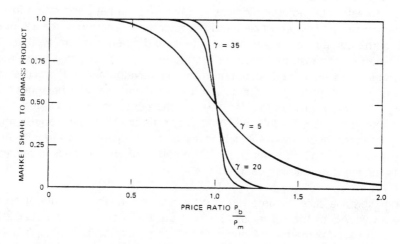

Figure 3. Steady-state market share

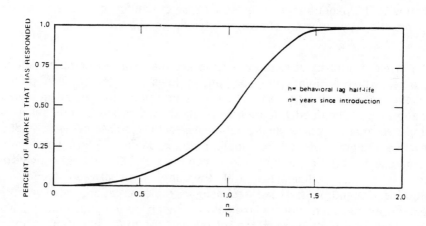

Figure 4. Dynamic market response

RESULTS

The study involved the projection of biomass feedstock availability by market price within U.S. census regions and the development of a computerized model to estimate regional biomass fuel product market penetration in five-year increments. This effort required the regional projection of market prices and demands for ten conventional fuels and chemicals as well as the derivation of biomass fuel product production data on 53 missions. Fifteen missions were examined in detail with the development of process flow diagrams, descriptions, economics, and energy and material balances.* The costs of energy production for the 15 missions under regulated utility financing and a 65%-to-35% debt-to-equity ratio are shown in Tables I and II. The optimistic estimates reflect high by-product values and product yields for the biochemical missions and a 20 percent reduction in base case capital costs for the thermochemical missions.

Using the base-case assumptions for feedstock availability (without federal incentives), 15 of the 53 missions penetrate the market by the year 2020, producing approximately 5.4 quads of fuel and chemical products, including electricity and steam. Assuming federal incentives and optimistic but achievable feedstock availability, 17 of the 53 missions penetrate the market by year 2020, producing approximately 10.3 quads of fuel and chemical products. The penetrations expected by type of fuel for the "base case" and "optimistic" scenarios are shown in Table III.

The levels of market penetration shown are the result of computer model routines which simulate the competitive fuels environment and project usable energy demand as a function of energy price and historical market relationships. The model formulations allow a consistent comparison of projected market prices (marginal costs) with biomass-derived product revenue requirements and compute product demand levels considering numerous factors, including mission commercialization dates, conversion process thermal efficiencies, and the time lag between technology introduction and widespread technology implementation. However, the relationship between expected biomass product price (revenue required) and alternative fuel and chemical market price is the most important factor in determining annual market penetration levels.

---

* See Appendix C for the results of the sensitivity analyses for each of the 15 missions, showing the effects on costs of changes in operating factors, feedstock prices, facility life and facility size.

Table I. DETAILED MISSION ANALYSIS RESULTS:
LARGE THERMOCHEMICAL FACILITIES

Route	Conversion[a] Process	Revenue Required ($/MM Btu)[b]	
		Base Case	Optimistic[c]
Wood to:			
Heavy fuel oil	CL	5.37	4.76
Methanol	GOB	7.77	6.72
Ammonia	GOB	164.00	141.00
($/short ton)			
SNG	GOB	6.41	5.56
Steam	DC	3.00	2.73
Electricity	DC	16.38	14.40
Steam and	DC	3.42	3.06
Electricity			
Oil and char[d]	P	4.50	4.00

[a] Key: CL = catalytic liquefaction; GOB = gasification, oxygen-blown; DC = direct combustion; P = pyrolysis.

[b] 1977 dollars in year 1985. Data source, SRI Detailed Analysis — Regulated Utility Financing.

[c] Capital cost, 80% of base case.

[d] Char valued at $1.25 per million Btu.

Table II. DETAILED MISSION ANALYSIS RESULTS:
LARGE BIOCHEMICAL FACILITIES

Route	Conversion[a] Process	Revenue Required ($/MM Btu)[b]	
		Base Case	Optimistic[c]
Cattle manure to IBG[d]	AD	4.87	4.37
Cattle manure to SNG[d] 100,000 head environmental feedlot	AD	7.02	2.75
10,000 head environmental feedlot	AD	14.44	7.50
Wheat straw to ethanol	F	52.60	29.20
Sugarcane to ethanol	F	26.95	20.00
Kelp to SNG	AD	20.70	10.70
Algae to ethanol	F	26.90	17.40
Wheat straw to IBG (40% conversion)	AD	23.76	9.00

[a] Key: AD = anaerobic digestion; F = fermentation.
[b] 1977 dollars in year 1985. Data source, SRI Detailed Analysis — Regulated Utility Financing.
[c] High by-product values and product yields.
[d] SNG = Substitute natural gas; IBG = Intermediate-Btu gas.

Table III. MARKET PENETRATION — BIOMASS PRODUCTS

Year	Estimated Biomass-Derived Products[a] 10^15 Btu		
	1985	2000	2020
BASE CASE SCENARIO			
Gaseous products (SNG, IBG, LBG)[b]	0	0.13	0.29
Methanol/Ethanol	0	0	0
Ammonia[c]	0.02	0.40	0.56
Process steam or steam/electric	0.68	2.32	4.01
Pyrolytic fuel oils	0.04	0.61	0.53
Total Quads	0.74	3.46	5.39
OPTIMISTIC SCENARIO			
Gaseous products (SNG, IBG, LBG)[b]	0	0.21	0.45
Methanol/Ethanol	0	0	0
Ammonia[c]	0.13	0.41	0.58
Process steam or steam/electric	1.10	4.17	8.39
Pyrolytic fuel oils	0.04	0.87	0.83
Total Quads	1.27	5.66	10.25

a Excludes existing biomass energy products.
b SNG = Substitute natural gas; IBG = Intermediate-Btu gas; LBG = Low-Btu gas.
c Assumes 18.3 million Btu/ton of ammonia.

Mission Ranking

We assumed that a large development facility would be desirable to demonstrate the feasibility of each high potential mission and to assure the achievement of future commercial conversion operations. Table IV shows a figure of merit (cost-benefit ratio), calculated by dividing the program funding requirement for each mission (based on the cost of a large development facility operating for five years) by the annual quad penetration in 2020. The results of this analysis provide the ranking of missions shown in Table V.

The model outputs and market demand levels are extremely sensitive to feedstock scenario assumptions* as indicated by the doubling in product demand with a change from base case to optimistic scenario (see Table VI). However, separating SRI-generated mission input data from data generated by others resulted in only a slight change in model results (an increase in biomass fuel product penetration from 10.00 to 10.25 quads in the year 2020 and a change in total feedstock demand from 13.4 to 13.6 quads of biomass). As indicated by the ratio of fuel product demand to feedstock demand, the overall conversion efficiency percentage is about 75 percent, reflecting the strong influence of direct combustion mission penetration.

SUMMARY

Biomass offers a significant potential for reducing national dependence on imported fossil fuels through the conversion of a renewable energy source to useful liquid and gaseous fuels, electric power, process steam, and chemicals. Several previous studies have indicated that feasible national goals could be the production of about 5 quadrillion Btu (quads) of energy by the year 2000 and 10 quads of energy by the year 2020. The results of these studies indicate that these are realistic and achievable goals provided Federal funding levels for biomass development are increased and Federal incentives are successfully applied to increase biomass feedstock availability.

The study summarized in this paper involved the identification of over 1,100 possible missions (specific conversion routes from biomass feedstock to useful fuel and chemical products to end-use markets) prior to the selection of 15 missions for detailed analysis.

---

* As well as other factors such as end use demand projections and foreign oil prices.

Table IV. RESEARCH AND DEVELOPMENT COST-BENEFIT RATIO CALCULATIONS

Mission[a] Feedstock	Product	Commercialization Date[b]	Estimated Development Program Funding Required (millions of dollars)					Total Quads 2020	Development Cost-Benefit Ratio ($/10⁶ Btu)
			Feedstock Preparation and Production	Pilot Plant	Large Demo Plant[c]	Demo Operation (5 years)	Total $		
Direct gasification — oxygen blown									
26 High moisture	SNG	1985	5	—	170	96	266	0.05	5.32
32 High moisture	Ammonia	1985	5	—	86	59	150	0.47	0.32
34 High moisture	IBG	1985	5	—	151	113	264	0.03	8.80
35 Low moisture	IBG	1985	5	—	116	15	131	0.01	13.10
Subtotal							811	0.56	1.45
Direct gasification — air blown/staged									
31 Wood	Ammonia	1985	—	—	135	170	305	0.11	2.77
42 High moisture	LBG	1985	5	—	65	107	172	0.15	1.15
Subtotal							477	0.26	1.83
Pyrolysis — maximum gas yield									
28 Wood	SNG	1985	—	—	78	111	189	0.11	1.72
Pyrolysis — maximum liquids									
18 Wood	Oil and Char	1985	—	—	61	52	113	0.68	0.17
19 Low moisture	Oil and Char	1985	5	—	61	52	113	0.15	0.75
Subtotal							226	0.84	0.27
Direct combustion									
9 Wood	Steam	(earlier than) 1975	—	—	94	59	153	1.21	0.13
10 Low moisture	Steam	1975	5	—	94	59	153	1.01	0.15
24 Wood	Steam/electric	1975	—	—	109	60	169	3.36	0.05
25 Low moisture	Steam/electric	1975	5	—	109	60	169	2.81	0.06
Subtotal							644	8.39	0.08
Anaerobic digestion									
13 Manure	IBG	1985	—	—	11	10	21	0.15	0.14
14 Manure	SNG	1985	—	—	14	13	27	0.02	1.35
30 High moisture	SNG	1985	—	—	97	50	147	0.02	7.35
39 High moisture	IBG	1980	5	5	74	58	132	0.01	13.20
Subtotal							327	0.20	1.64

a Missions number 1 through 25 were evaluated by SRI
b Date estimated to reflect initial mission technical and economic feasibility for modeling purposes.
c Demonstration plant costs assume largest design practicable — 500 to 3,000 dry tons per day of feedstock. Source of cost data — Appendix A.

Table V. MISSION RANKING

Product	Feedstock	Process	Evaluated by
Highest ranking (mission with lowest cost-benefit ratios)			
Process steam with electrical by-product	Wood or low moisture	Combustion	SRI
Steam	Wood or low moisture	Combustion	SRI
IBG	Manure	Anaerobic digestion	SRI
Oil and charcoal	Wood or low moisture	Pyrolysis (maximum liquid yields)	SRI
Other missions showing penetration (higher cost-benefit ratios)			
Ammonia	Wood or high moisture	Gasification staged air/or oxygen blown	SRI
SNG	Manure	Anaerobic digestion	SRI
SNG	High moisture	Anaerobic digestion	SRI
IBG	Low moisture	Gasification (oxygen blown)	SRI
LBG	High moisture	Gasification (air blown)	Others
SNG	Wood	Pyrolysis (maximum gas yield)	Others
SNG	High moisture	Gasification (oxygen blown)	Others
IBG	High moisture	Gasification (oxygen blown)	Others
IBG	High moisture	Anaerobic digestion	Others

Table VI: FEEDSTOCK AVAILABILITY AND DEMAND IN YEAR 2020

Feedstock Availability (Model Input) Market Price ($/ton)	Millions of Dry Tons Base Case	Optimistic Case
10	200	200
20	326	390
30	484	792
40	621	1715
50	718	2783

Biomass Feedstock Demand (Model Results) Type	10^{15} Btu Base Case	Optimistic Case
Low moisture	3.5	5.1
High moisture	0.9	1.0
Woody	2.4	7.0
Manure	0.5	0.5
Aquatic	0	0
Total	7.3	13.6

Biomass Fuel Product Demand (Model Results) Product	10^{15} Btu Base Case	Optimistic Case
Gases (SNG, IBG, LBG)	0.29	0.45
Ammonia	0.56	0.58
Steam and electricity	3.91	8.39
Fuel oil (L.S.)	0.53	0.83
Total	5.39	10.25

Based only on market penetration projections, processes that appear to offer minor future contributions include marine crops, catalytic liquefaction, and fermentation to produce ethanol. Missions that appear to have greater near-term commercialization potential include:

- Gasification of wood and low moisture plants to produce LBG, IBG, SNG, and ammonia.
- Pyrolysis of wood and low-moisture plants to produce SNG, fuel oil and char.
- Combustion of wood and low-moisture plants to produce steam and steam with electricity as a by-product.
- Anaerobic digestion of manure and high-moisture terrestrial crops to produce IBG and SNG.

Overall, the success of biomass energy development (as measured by the levels of future quad production) can best be achieved by emphasis on production methods, procedures and policies designed to increase feedstock availability. Biomass fuel product market penetration is highly sensitive to feedstock scenario and costs. A doubling of feedstock availability at $30 per ton allows biomass energy production to approximately double. Therefore, financial incentives of various types, possiby including loan, tax credit, and subsidies may be desirable to: (a) increase the use of available crop and timber lands for the production of energy crops and combination energy, food/fiber crops, and (b) encourage construction and operation of biomass fuel and chemical product production facilities.

REFERENCE

1.   Schooley, F. A., *et al.* Dec 1978-Jan 1979, U.S. Department of Energy Final Report.

Appendix A

DEFINITIONS

Biomass product production price — The total of all costs necessary for the manufacture of a biomass-derived product, including manufacturer's profit.

Market price — The price against which the biomass-derived product must compete for market share.

Steady-state market share — The fraction of the product market that a biomass-derived product would supply at a point long after the biomass product production price and market price had become unchanging with time.

Market share parameter (γ) — A variable used to characterize the behavior of an energy marketplace. This variable is an estimate of the steady-state market share of a biomass product once the comparative economics are determined.

Behavioral lag — A means of describing the resistance of market decision-makers to the introduction of new energy products and technologies. This concept forms the basis for the dynamic market response curve.

Behavioral lag half-life (h) — The time required for one-half of a given market to respond to the availability of a new energy product or technology.

Behavioral lag response parameter (α) — This variable fixes the relative shape of the dynamic market response curve for a particular market once the behavioral lag half-life has been chosen.

Mission — A specific biomass feedstock-to-product conversion process or procedure directed to a designated fuel or petrochemical market.

Appendix B

PRODUCT DEMAND AND PRICE DATA SOURCES

1975 Product Demands and Prices were based upon the following sources:

Fuel	Source of Data
Coal	U.S. Bureau of Mines — Mineral Industry Surveys Quarterly and Annual Reports: Bituminous Coal and Lignite Distribution Quarterly, Pennsylvania Anthracite Annual, Coke and Coal Chemicals Annual
Natural Gas	U.S. Bureau of Mines — Mineral Industry Survey — Natural Gas Annual
Electric Power	Edison Electric Institute Statistical Yearbook — Data adjusted by Gulf Energy Model Outputs to yield base, intermediate, and peak load data
Gasoline	Federal Highway Administration's Highway Statistics Annual and the National Petroleum Factbook Platt's Oilgram Price Service
Residual Fuel Oil	U.S. Bureau of Mines — Mineral Industry Survey — Fuel Oil Sales Annual, Federal Power Commission — Annual Summary of Cost and Quality of Steam — Electric Plant Fuels
Crude Oil	U.S. Bureau of Mines — Mineral Industry Survey — Petroleum Statement Annual, Oil and Gas Journal — Annual Refining Issue, Federal Energy Administration — Monthly Energy Review
Methanol	Chemical Marketing Reporter
Ammonia	Several recent studies by the SRI Chemical Economics Department, SRI Chemical Economics Handbook, U.S. Bureau of Census Data on Ammonia Producers Shipments
IBG	Not applicable for 1975
Steam	Based on the conversion of coal at 85% efficiency to satisfy the Industrial Process Steam Market — Fuel and Energy Prices Forecasts, EPRI, EA-411 and EA-443, April and September 1977.

Fuel demand and price projections are based on the following sources:

- Brookhaven National Laboratory, *Regional Reference, Energy Systems,* EPRI EA-462 (June 1977).
- ERDA, *Market Oriented Program Planning Study* (September 1977).
- Foster Associates, Inc., *Fuel and Energy Price Forecasts,* EPRI EA-411 (April 1977).
- SRI International, *Fuel and Energy Price Forecasts: Quantities and Long-Term Marginal Prices,* EPRI EA-433 (September 1977).
- SRI International, "Assessing the Benefits of the Gas Research Institute's Research and Development Programs", SRI Project 6955 (March 1978):

Projections of ammonia demand were made by applying a consumption factor to the Obers projection of crop yields.

Appendix C. SUMMARY OF COST SENSITIVITY DATA
ON 15 MISSIONS

Mission

1	Wood to oil
2	Wood to methanol
3	Wood to ammonia
4	SNG from wood
5	Steam from wood
6	Electricity from wood
7	IBG from cattle manure
8	SNG from cattle manure
9	Wheat straw to IBG
10	Wheat straw to ethanol
11	High sugar content plant to ethanol
12	Wood to oil via pyrolysis (maximum liquid yield)
13	Kelp to SNG via anaerobic digestion
14	Kelp to alcohol via fermentation
15	Cogeneration of electricity and steam from wood

MISSION I — SELECTED SUMMARY DATA
WOOD TO OIL VIA CATALYTIC LIQUEFACTION

		Sensitivity To							
	Base Case	Feedstock Price			Operating Capacity (%)		Capital Investment		Project Life
A. Product									
Bbl/day	5268	5268	5268	5268	5268	5268	5268	5268	5268
10^9 Btu/day	30.6	30.6	30.6	30.6	30.6	30.6	30.6	30.6	30.6
B. Feedstock									
ODT/day[a]	3000	3000	3000	3000	3000	3000	3000	3000	3000
10^9 Btu/day	57.6	57.6	57.6	57.6	57.6	57.6	57.6	57.6	57.6
\$/10^6 Btu	1.0	2.0	1.5	0	1.0	1.0	1.0	1.0	1.0
C. Total Capital Investment							−20%	+30%	
10^6 dollars	144.9	146.5	145.7	143.4	144.7	144.5	116.5	189.6	144.9
\$/10^6 Btu[b]	1.95	1.98	1.97	1.93	2.22	2.52	1.57	2.52	1.72
D. Annual Cost of Feedstock									
10^6	18.8	37.7	28.3	0	16.8	14.7	18.8	18.8	18.8
\$/10^6 Btu	1.88	3.76	2.82	0	1.87	1.87	1.88	1.88	1.88
E. Annual Operating Cost[c]									
10^6	15.5	15.5	15.5	15.5	15.3	15.1	13.2	18.9	15.5
\$/10^6 Btu	1.54	1.54	1.54	1.54	1.70	1.93	1.31	1.88	1.54
% Operating Capacity	90	90	90	90	80	70	90	90	90
F. Revenue Requirements[d]									
Regulated Utility	31.2	42.2	36.7	20.1	33.6	36.6	27.6	36.5	29.8
\$/Bbl of oil \$$10^6$ Btu[e]	5.37	7.28	6.33	3.47	5.79	6.32	4.76	6.29	5.14
G. Plant Life Years	20	20	20	20	20	20	20	20	30

[a] Assumes a 50% moisture content feedstock and 19.2 million Btu/dry ton — process efficiency = 53%.
[b] Capital component of product cost.
[c] Excludes feedstock cost and plant depreciation.
[d] Calculated to yield a 15% rate of return on equity and a 9% return on debt (65% debt and 35% equity). Income tax = 52%.
[e] Assumes 5.8 × 10^6 Btu/bbl.

MISSION II. SELECTED SUMMARY DATA
WOOD TO METHANOL VIA GASIFICATION (OXYGEN-BLOWN REACTOR)

	Base Case		Feedstock Price			Operating Capacity (%)		Capital Investment		Project Life
								−20%	+30%	
A. Product										
10³ Gal/day	100	200	600	600	600	600	600	600	600	600
10⁹ Btu/day	5.5	11.0	33.2	33.2	33.2	33.2	33.2	33.2	33.2	33.2
B. Feedstock										
ODT/day[a]	500	1000	3000	3000	3000	3000	3000	3000	3000	3000
10⁹ Btu/day	9.6	19.2	57.6	57.6	57.6	57.6	57.6	57.6	57.6	57.6
$/10⁶ Btu	1.0	1.0	1.0	1.0	1.0	1.0	1.0	1.0	1.0	1.0
C. Total Capital Investment										
$/10⁶	58.13	100.76	270.26	269.48	267.16	268.28	268.50	215.61	348.36	268.71
$/10⁶ Btu[b]	4.33	3.51	3.37	3.36	3.32	4.31	3.76	2.69	4.34	2.95
D. Annual Cost of Feedstock										
$/10⁶	3.14	6.28	37.69	28.16	0	14.66	16.75	18.84	18.84	18.84
$/10⁶ Btu	1.73	1.73	3.47	2.60	0	1.73	1.73	1.73	1.73	1.73
E. Annual Operating Cost[c]										
$/10⁶	6.88	13.02	29.38	29.38	29.38	28.38	28.88	25.10	35.80	29.39
$/10⁶ Btu	3.81	3.61	2.69	2.69	2.69	3.33	2.98	2.30	3.28	2.69
% Operating Capacity	90	90	90	90	90	70	80	90	90	90
F. Revenue Requirements[d]										
Regulated Utility $/10⁶ Btu[e]	9.87	8.85	9.53	8.65	6.01	9.37	8.47	6.72	9.35	7.37
G. Plant Life Years	20	20	20	20	20	20	20	20	20	30

[a] Assumes a 50% moisture content feedstock and 19.2 million Btu/dry ton — process efficiency = 58%.

[b] Capital component of product cost.

[c] Excludes feedstock cost and plant depreciation.

[d] Calculated to yield a 15% rate of return on equity and a 9% return on debt (65% debt and 35% equity). Income tax = 52%.

[e] Assumes 55,610 Btu/gal.

MISSION III. SELECTED SUMMARY DATA

AMMONIA FROM WOOD VIA GASIFICATION WITH AN OXYGEN-BLOWN REACTOR

	Base Case	Feedstock Price				Sensitivity To							
						Operating Capacity (%)			Capital Investment			Project Life	
									−20%		+30%		
A. Product													
Ton/day	250	1542	1542	1542	1542	1542	1542	1542	1542	1542	1542	1542	1542
10^9 Btu/day[a]	4.6	28.2	28.2	28.2	28.2	28.2	28.2	28.2	28.2	28.2	28.2	28.2	28.2
B. Feedstock													
ODT/day	500	3000	3000	3000	3000	3000	3000	3000	3000	3000	3000	3000	3000
10^9 Btu/day	9.2	57.6	57.6	57.6	57.6	57.6	57.6	57.6	57.6	57.6	57.6	57.6	57.6
$/$10^6$ Btu	1.00	2.00	1.50	1.00	0	1.00	1.00	1.00	1.00	1.00	1.00	1.00	1.00
C. Total Capital Investment													
10^6	65.0	268.9	268.0	267.1	265.7	266.9	267.1	267.3	214.4	267.3	346.5	267.3	267.3
$/$10^6$ Btu[b]	5.96	4.74	4.73	4.71	4.68	6.57	5.52	4.71	3.86	4.71	6.00	4.24	4.24
$/ton	127.40	86.75	86.49	86.24	85.73	120.38	101.17	86.24	70.53	86.24	109.77	77.64	77.64
D. Annual Cost of Feedstock													
10^6	3.1	37.7	28.3	18.8	0	14.7	16.8	18.8	18.8	18.8	18.8	18.8	18.8
$/$10^6$ Btu	2.09	4.06	3.04	2.03	0	2.03	2.03	2.03	2.03	2.03	2.03	2.03	2.03
$/ton	38.27	74.28	55.71	37.14	0	37.14	37.14	37.14	37.14	37.14	37.14	37.14	37.14
E. Annual Operating Cost[c]													
10^6	5.7	20.6	20.6	20.6	20.6	20.6	20.6	20.6	20.6	20.6	20.6	20.6	20.6
$/$10^6$ Btu	3.78	2.22	2.22	2.22	2.22	2.22	2.22	2.22	2.22	2.22	2.22	2.22	2.22
$/ton	69.08	40.55	40.55	40.55	40.55	40.55	40.55	40.55	40.55	40.55	40.55	40.55	40.55
% Operating Capacity	90	90	90	90	90	70	80	90	90	90	90	90	90
F. Revenue Requirements[d]													
Regulated Utility													
$/$10^6$ Btu	12.82	11.02	9.99	9.99	6.90	11.52	10.82	11.23	7.74	11.23	10.79	8.49	8.49
$/ton	235.00	202.00	183.00	183.00	126.00	198.00	198.00	205.00	142.00	205.00	197.00	155.00	155.00
DCF													
$/$10^6$ Btu	18.93	15.67	14.64	16.39	11.52	16.79	15.00	16.39	11.46	16.39	16.81	13.60	13.60
$/ton	346.00	287.00	268.00	300.00	211.00	307.00	274.00	300.00	210.00	300.00	308.00	249.00	249.00
G. Plant Life Years	20	20	20	20	20	20	20	20	20	20	20	30	30

[a] Assumes 18.3 million Btu/ton — process efficiency = 49%.

[b] Capital component of product cost.

[c] Excludes feedstock cost and plant depreciation.

[d] Calculated to yield a 15% rate of return on equity and a 9% return on debt (65% debt and 35% equity). Income tax = 52%.

MISSION IV. SELECTED SUMMARY DATA
SNG PRODUCTION FROM WOOD VIA GASIFICATION (OXYGEN-BLOWN REACTOR)

	Base Case			Sensitivity To							
				30 Years Plant Life	Capital Investment		Feedstock Price			Operating Capacity (%)	
A. Product											
10^6 SCF/day	6.4	12.7	38.2	38.2	38.2	38.2	38.2	38.2	38.2	38.2	38.2
10^9 Btu/day	6.0	12.0	36.1	36.1	36.1	36.1	36.1	36.1	36.1	36.1	36.1
B. Feedstock											
ODT/day[a]	500	1000	3000	3000	3000	3000	3000	3000	3000	3000	3000
10^9 Btu/day	9.6	19.2	57.6	57.6	57.6	57.6	57.6	57.6	57.6	57.6	57.6
$/10^6$ Btu	1.00	1.00	1.00	1.00	1.00	1.00	1.50	2.00	0	1.00	1.00
C. Total Capital Investment					−20%	+30%					
$10^6	50.1	88.5	238.5	238.5	191.4	309.2	239.3	240.1	237.0	238.3	238.1
$/10^6$ Btu[b]	3.40	3.01	2.70	2.70	2.16	3.49	2.71	2.71	2.68	3.03	3.46
D. Annual Cost of Feedstock											
$10^6	3.14	6.3	18.8	18.8	18.8	18.8	28.3	37.7	0	16.8	14.7
$/10^6$ Btu	1.59	1.59	1.59	1.59	1.59	1.59	2.39	3.18	–	1.59	1.59
E. Annual Operating Cost[c]											
$10^6	5.9	10.3	25.2	21.2	21.4	30.9	25.1	25.2	25.1	24.9	24.5
$/10^6$ Btu	2.97	2.61	2.12	1.79	1.81	2.61	2.12	2.13	2.12	2.36	2.66
% Operating Capacity	90	90	90	90	90	90	90	90	90	80	70
F. Revenue Requirements[d]											
Regulated Utility $/10^6$ Btu	7.96	7.21	6.41	6.08	5.56	7.69	7.22	8.02	4.80	6.98	7.71

[a] Assumes a 50% moisture content feedstock and 19.2 million Btu/dry ton — process efficiency = 63%.

[b] Capital component of product cost.

[c] Excludes feedstock cost and plant depreciation.

[d] Calculated to yield a 15% rate of return on equity and an 9% return on debt (65% debt and 35% equity). Plant life = 20 years. Income tax = 52%.

MISSION V. SELECTED SUMMARY DATA
STEAM PRODUCTION FROM WOOD VIA DIRECT COMBUSTION

| | Base Case | | Sensitivity To | | | | | |
			Capital Investment		Feedstock Price		Operating Capacity	
A. Product								
10³ lb/hr	239	478	1434	1434	1434	1434	1434	1434
10⁹ Btu/day[a]	7.4	14.8	44.4	44.4	44.4	44.4	44.4	44.4
B. Feedstock								
ODT/day[b]	500	1000	3000	3000	3000	3000	3000	3000
10⁹ Btu/day	9.6	19.2	57.6	57.6	57.6	57.6	57.6	57.6
$/10⁶ Btu	1.00	1.00	1.00	1.00	1.50	0	1.00	1.00
C. Total Capital Investment			−20%	+30%				
$10⁶	17.4	32.3	76.0	121.3	94.9	92.6	93.7	93.3
$/10⁶ Btu[c]	0.98	0.90	0.71	1.13	0.89	0.86	1.12	1.56
D. Annual Cost of Feedstock								
$10⁶	3.1	6.3	18.9	18.9	28.3	0	14.7	10.5
$/10⁶ Btu	1.30	1.30	1.30	1.30	1.95	0	1.30	1.30
E. Annual Operating Cost[d]								
$10⁶	2.6	4.5	10.2	14.1	11.8	11.8	11.2	10.6
$/10⁶ Btu	1.08	0.95	0.72	0.98	0.82	0.82	1.00	1.33
% Operating Capacity	90	90	90	90	90	90	70	50
F. Revenue Requirements[e]								
Regulated Utility $/10⁶ Btu	3.36	3.15	2.73	3.41	3.66	1.68	3.42	4.19

a 450 psia — 810°F.
b Assumes a 50% moisture feedstock and 19.2 million Btu/dry ton — process efficiency = 77%.
c Capital component of product cost.
d Excludes feedstock cost and plant depreciation.
e Calculated to yield a 15% rate of return on equity and a 9% return on debt (65% debt and 35% equity). Plant life = 20 years. Income tax = 52%.

MISSION VI. SELECTED SUMMARY DATA
ELECTRICITY PRODUCTION FROM WOOD VIA DIRECT COMBUSTION

	Base Case			Sensitivity To						
				Capital Investment		Feedstock Price		Operating Capacity (%)		
				−20%	+30%					
A. Product										
Plant Size, MW	25	50	150	150	150	150	150	150	150	150
MWh/day	600	1200	3600	3600	3600	3600	3600	3600	3600	3600
10^9 Btu/day	2.04	4.08	12.24	12.24	12.24	12.24	12.24	12.24	12.24	12.24
B. Feedstock										
ODT/day[a]	500	1000	3000	3000	3000	3000	3000	3000	3000	3000
10^9 Btu/day	9.6	19.2	57.6	57.6	57.6	57.6	57.6	57.6	57.6	57.6
$/10^6 Btu	1.00	1.00	1.00	1.00	1.00	1.50	0	1.00	1.00	1.00
C. Total Capital Investment										
$10^6	30.7	58.2	165.6	133.3	214.2	166.3	164.3	165.4	165.1	164.8
$/10^6 Btu[b]	7.00	6.61	6.25	5.03	8.09	6.29	6.19	7.15	9.08	12.44
D. Annual Cost of Feedstock										
$10^6	2.8	5.6	16.8	16.8	16.8	25.1	0	14.7	11.5	8.4
$/10^6 Btu	4.70	4.70	4.70	4.70	4.70	7.05	—	4.70	4.70	4.70
E. Annual Operating Cost[c]										
$10^6	4.1	7.6	19.3	16.6	23.4	19.3	19.3	18.8	18.1	17.4
$/10^6 Btu	6.93	6.43	5.43	4.67	6.57	5.43	5.43	6.04	7.39	9.77
% Operating Capacity	80	80	80	80	80	80	80	70	55	40
F. Revenue Requirements[d]										
Regulated Utility										
$10^6 Btu	18.63	17.74	16.38	14.40	19.36	18.77	11.62	17.89	21.17	26.91

[a] Assumes a 50% moisture content feedstock and 19.2 million Btu/dry ton — process efficiency = 21%.

[b] Capital component of product cost.

[c] Excludes feedstock cost and plant depreciation.

[d] Calculated to yield a 15% rate of return on equity and a 9% return on debt (65% debt and 35% equity). Plant life = 20 years. Income tax = 52%.

MISSION VII. SELECTED SUMMARY DATA
IBG PRODUCTION FROM CATTLE MANURE VIA ANAEROBIC DIGESTION

	Base Case			Sensitivity To					
				Capital Investment		Feedstock Price		Operating Capacity (%)	
				−20%	+30%				
A. Product									
Head of Cattle	10,000	100,000	250,000	10,000	10,000	100,000	100,000	10,000	10,000
10^6 SCF/day (as CH_4)	0.226	2.26	5.65	0.226	0.226	2.26	2.26	0.226	0.226
10^6 Btu/day	226	2260	5650	226	226	2260	2260	226	226
B. Feedstock									
ODT/day[a]	45	450	1125	45	45	450	450	45	45
10^6 Btu/day	675	6750	16,875	675	675	6750	6750	675	675
$/$10^6$ Btu	0.33	0.33	0.33	0.33	0.33	0.67	0.165	0.33	0.33
C. Total Capital Investment									
10^6	1.83	10.82	23.91	1.48	2.34	10.82	10.82	1.83	1.83
$/$10^6$ Btu[b]	2.68	1.35	1.13	2.01	3.60	1.35	1.35	3.19	3.66
D. Annual Cost of Feedstock									
10^6	0.07	0.74	1.86	0.07	0.07	1.48	0.37	0.07	0.06
$/$10^6$ Btu	1.00	1.00	1.00	1.00	1.00	1.99	0.50	1.06	1.04
E. Annual Operating Cost[c]									
10^6	0.39	1.87	3.45	0.36	0.44	1.87	1.87	0.38	0.38
$/$10^6$ Btu	5.26	2.52	1.86	4.86	5.94	2.52	2.52	5.75	6.62
% Operating Capacity	90	90	90	90	90	90	90	80	70
F. Revenue Requirements[d]									
Regulated utility $/$10^6$ Btu	8.94	4.87	3.99	7.87	10.54	5.86	4.37	10.00	11.32

a Process efficiency = 33.5%.
b Capital component of product cost.
c Excludes feedstock costs and plant depreciation.
d Calculated to yield a 15% rate of return on equity and a 9% return on debt (65% debt and 35% equity). Plant life = 20 years. Income tax = 52%.

MISSION VIII. SELECTED SUMMARY DATA
SNG PRODUCTION FROM CATTLE MANURE VIA ANAEROBIC DIGESTION

	Base Case			Sensitivity To						
				Capital Investment		Feedstock Price			Operating Capacity (%)	
A. Product										
Head of Cattle	10,000	100,000	250,000	10,000	10,000	100,000	100,000	100,000	10.000	10.000
10^6 SCF/day (as CH_4)	0.204	2.04	5.10	0.204	0.204	2.04	2.04	2.04	0.204	0.204
10^6 Btu/day	204	2040	5100	204	204	2040	2040	2040	204	204
B. Feedstock										
ODT/day[a]	45	450	1125	45	45	450	450	450	45	45
10^6 Btu/day	675	6750	16,875	675	675	6750	6750	6750	675	675
$/$10^6$ Btu	0.33	0.33	0.33	0.33	0.33	0.165	0.67	0	0.33	0.33
C. Total Capital Investment				−20%	+30%					
10^6	2.4	13.6	29.3	1.9	3.1	23.8	23.8	23.8	2.4	2.4
$/$10^6$ Btu[b]	4.17	2.11	2.01	3.28	5.43	2.26	2.18	2.05	4.82	5.51
D. Annual Cost of Feedstock										
10^6	0.07	0.70	1.75	0.07	0.07	0.35	1.40	0	0.06	0.05
$/$10^6$ Btu	1.04	1.04	1.04	1.04	1.04	0.52	2.08	–	1.04	1.04
E. Annual Operating Cost[c]										
10^6	0.6	2.6	4.9	0.6	0.7	2.6	2.6	2.6	0.6	0.6
$/$10^6$ Btu	9.23	3.87	2.90	8.49	10.42	3.87	3.87	3.87	10.24	11.70
% Operating Capacity	90	90	90	90	90	90	90	90	80	70
F. Revenue Requirements[d]										
Regulated Utility $/$10^6$ Btu	14.44	7.02	5.95	12.81	16.89	6.65	8.13	5.92	16.10	18.25

[a] Process efficiency = 30.2%.
[b] Capital component of product cost.
[c] Excludes feedstock cost and plant depreciation.
[d] Calculated to yield a 15% return on equity and a 9% return on debt (65% debt and 35% equity). Plant life = 20 years. Income tax = 52%.

MISSION IX. SELECTED SUMMARY DATA
IBG FROM WHEAT STRAW VIA ANAEROBIC DIGESTION

60% COD Reduction — Plant: 5 × 10⁶ SCF/day

	Base Case	Sensitivity To Feedstock Price ($35/ton)	Sensitivity To Feedstock Price ($15/ton)	Sensitivity To Operating Capacity (80%)	Sensitivity To Operating Capacity (70%)	Sensitivity To Capital Investment (+30%)	Sensitivity To Capital Investment (−20%)
A. Product — 10^6 SCF/day (500 Btu/SCF)	5.0	5.0	5.0	5.0	5.0	5.0	5.0
10^9 Btu/day	2.5	2.5	2.5	2.5	2.5	2.5	2.5
B. Feedstock — ODT/day[a]	500	500	500	500	500	500	500
10^9 Btu/day	7.5	7.5	7.5	7.5	7.5	7.5	7.5
$/ton	25	35	15	25	25	25	25
C. Total Capital Investment — 10^6	12.6	12.6	12.6	12.6	12.6	15.7	10.6
$/10^6$ Btu[b]	2.10	2.10	2.10	2.36	2.70	2.61	1.76
D. Annual Cost of Feedstock — 10^6	4.1	5.8	2.5	3.6	3.2	4.1	4.1
$/10^6$ Btu	4.99	6.99	2.99	4.99	4.99	4.99	4.99
E. Annual Operating Cost[c] — 10^6	5.1	5.1	5.1	4.7	4.3	5.4	5.0
$/10^6$ Btu	6.21	6.21	6.21	6.48	6.80	6.52	6.01
% Operating Capacity	90	90	90	80	70	90	90
F. Revenue Requirements[d] — Regulated Utility $/10^6$ Btu	13.30	15.30	11.30	13.83	14.49	14.12	12.76

60% COD Reduction — Plant: 300 × 10⁶ SCF/day

	Base Case	Sensitivity To Feedstock Price ($35/ton)	Sensitivity To Feedstock Price ($15/ton)
A. Product — 10^6 SCF/day	300	300	300
10^9 Btu/day	15.0	15.0	15.0
B. Feedstock — ODT/day	3000	3000	3000
10^9 Btu/day	45.0	45.0	45.0
$/ton	25	35	15
C. Total Capital Investment — 10^6	65.7	65.6	65.5
$/10^6$ Btu	1.82	1.82	1.81
D. Annual Cost of Feedstock — 10^6	24.7	35.6	14.8
$/10^6$ Btu	4.99	6.99	2.99
E. Annual Operating Cost — 10^6	27.4	27.4	27.4
$/10^6$ Btu	5.55	5.55	5.55
% Operating Capacity	90	90	90
F. Revenue Requirements — Regulated Utility $/10^6$ Btu	12.36	14.36	10.36

40% COD Reduction — Plant: 2.8 × 10⁶ SCF/day

	Base Case	Sensitivity To Feedstock Price ($35/ton)	Sensitivity To Feedstock Price ($15/ton)	Sensitivity To Operating Capacity (80%)	Sensitivity To Operating Capacity (70%)	Sensitivity To Capital Investment (+30%)	Sensitivity To Capital Investment (−20%)
A. Product — 10^6 SCF/day	2.8	2.8	2.8	2.8	2.8	2.8	2.8
10^9 Btu/day	1.4	1.4	1.4	1.4	1.4	1.4	1.4
B. Feedstock — ODT/day	500	500	500	500	500	500	500
10^9 Btu/day	7.5	7.5	7.5	7.5	7.5	7.5	7.5
$/ton	25	35	15	25	25	25	25
C. Total Capital Investment — 10^6	12.6	12.6	12.6	12.6	12.6	15.7	10.6
$/10^6$ Btu	3.74	3.74	3.74	4.23	4.84	4.56	3.05
D. Annual Cost of Feedstock — 10^6	4.1	5.8	2.5	3.7	3.2	4.1	4.1
$/10^6$ Btu	8.92	12.50	5.35	8.92	8.92	8.92	8.92
E. Annual Operating Cost — 10^6	5.1	5.1	5.1	4.7	4.3	5.4	5.0
$/10^6$ Btu	11.10	11.10	11.10	11.56	12.13	11.74	10.73
% Operating Capacity	90	90	90	80	70	90	90
F. Revenue Requirements — Regulated Utility $/10^6$ Btu	23.76	27.34	20.19	24.71	25.89	25.22	22.79

40% COD Reduction — Plant: 16.8 × 10⁶ SCF/day

	Base Case	Sensitivity To Feedstock Price ($35/ton)	Sensitivity To Feedstock Price ($15/ton)
A. Product — 10^6 SCF/day	16.8	16.8	16.8
10^9 Btu/day	8.4	8.4	8.4
B. Feedstock — ODT/day	3000	3000	3000
10^9 Btu/day	45.0	45.0	45.0
$/ton	25	35	15
C. Total Capital Investment — 10^6	65.68	65.63	65.74
$/10^6$ Btu	3.25	3.25	3.25
D. Annual Cost of Feedstock — 10^6	24.75	34.65	14.85
$/10^6$ Btu	8.29	12.50	5.35
E. Annual Operating Cost — 10^6	27.4	27.4	27.4
$/10^6$ Btu	9.90	9.90	9.90
% Operating Capacity	90	90	90
F. Revenue Requirements — Regulated Utility $/10^6$ Btu	22.07	25.65	18.50

a Process efficiency = 33.3% for 60% COD reduction and 18.7% for 40% COD reduction.
b Capital component of product cost.
c Excludes feedstock cost and plant depreciation.
d Calculated to yield a 15% rate of return on equity and a 9% return on debt (65% debt and 35% equity). Plant life = 20 years. Income tax = 52%.

MISSION X. SELECTED SUMMARY DATA
WHEAT STRAW TO ETHANOL (4% SUGAR SOLUTION) VIA ENZYMATIC HYDROLYSIS AND FERMENTATION
25 Million Gallons Per Year of Ethanol
(Facility Daily Outputs: 500 Tons of Sugar and 75,768 Gallons of Ethanol)

	Base Case	Feedstock Price			Capital Investment		Operating Capacity (%)		
					+30%	−20%			
A. Product									
10³ Gal/day	75.8	75.8	75.8	75.8	75.8	75.8	75.8	75.8	75.8
10⁹ Btu/day	5.5	5.5	5.5	5.5	5.5	5.5	5.5	5.5	5.5
B. Feedstock									
ODT/day[a]	3270	3270	3270	3270	3270	3270	3270	3270	3270
10⁹ Btu/day	49.0	49.0	49.0	49.0	49.0	49.0	49.0	49.0	49.0
$/ton (dry)	15.0	9.0	0	30.0	15.0	15.0	15.0	15.0	15.0
C. Total Capital Investment									
Sugar plant ($10⁶)	94.9	94.6	94.2	95.5	121.8	76.9	94.3	94.6	94.7
Ethanol plant ($10⁶)	32.9	32.6	32.2	33.6	43.2	26.2	33.2	33.0	33.0
Total Capital Investment, $10⁶	127.8	127.2	126.4	129.1	165.0	103.1	127.5	127.5	127.7
$/10⁶ Btu[b]	8.90	8.86	8.79	9.00	11.50	7.18	11.38	9.98	9.52
D. Annual Cost of Feedstock									
$10⁶	16.2	9.7	0	32.4	16.2	16.2	12.5	14.3	15.2
$/10⁶ Btu	9.00	5.40	0	18.00	9.00	9.00	6.94	7.95	8.44
E. Annual Operating Cost[c]									
Sugar plant ($10⁶)	54.4	54.4	54.4	54.4	56.6	53.0	44.2	49.3	51.7
Ethanol plant ($10⁶)	8.2	8.2	8.2	8.2	8.2	8.2	6.7	7.4	7.8
Total Annual Operating Costs ($10⁶)	62.6	62.6	62.6	62.6	64.8	61.2	50.9	56.7	59.5
$/10⁶ Btu	34.70	34.70	34.70	34.70	36.00	33.92	36.15	35.31	35.29
% Operating Capacity	90	90	90	90	90	90	70	80	84
F. Revenue Requirements[d]									
Regulated Utility									
Sugar, $/lb	0.25	0.23	0.20	0.30	0.27	0.24	0.27	0.26	0.26
Ethanol, $/10⁶ Btu	52.60	48.96	43.49	61.70	56.50	50.10	54.47	53.24	53.25

[a] Process efficiency = 11%.
[b] Capital component of product cost.
[c] Excludes feedstock cost and plant depreciation.
[d] Calculated to yield a 15% rate of return on equity and a 9% return on debt (65% debt and 35% equity). Plant life = 20 years. Income tax = 52%.

MISSION XI. SELECTED SUMMARY DATA
SUGARCANE MILL
High Sugar Content Plant to Ethanol (10.7 % Sugar Solution) Via Fermentation
(Facility Daily Outputs, 500 Tons of Sugar and 75,768 Gallons of Ethanol)

	Base Case	Sensitivity To				
		Feedstock Price			**Capital Investment**	
A. Product						
10^3 Gal/day	75.8	75.8	75.8	75.8	75.8	75.8
10^9 Btu/day	5.5	5.5	5.5	5.5	5.5	5.5
B. Feedstock						
ODT/day [a]	2756	2756	2756	2756	2756	2756
10^9 Btu/day	41.3	41.3	41.3	41.3	41.3	41.3
$/ton (dry)	65.0	50.0	100	0	65.0	65.0
C. Total Capital Investment					+30%	−20%
Sugar plant (10^6)	49.2	49.2	49.2	49.1	63.8	39.5
Ethanol plant (10^6)	21.0	20.7	21.7	19.8	21.2	20.8
Total Capital Investment, 10^6	60.2	69.9	70.9	68.9	85.0	60.3
$/$10^6$ Btu [b]	4.17	4.84	4.91	4.77	5.82	4.18
D. Annual Cost of Feedstock						
10^6	29.56	22.74	45.47	0	29.56	29.56
$/$10^6$ Btu	16.28	12.53	25.05	0	16.29	16.29
E. Annual Operating Cost [c]						
Sugar plant (10^6)	6.8	6.8	6.8	6.8	8.4	5.8
Ethanol plant (10^6)	5.0	5.0	5.0	5.0	5.0	5.0
$/$10^6$ Btu	6.50	6.50	6.50	6.50	7.38	5.95
F. Revenue Requirements [d]						
Regulated Utility						
Sugar, $/lb	0.13	0.11	0.18	0.04	0.14	0.12
Ethanol, $/$10^6$ Btu	26.45	23.87	36.46	11.27	29.49	26.42

[a] Sugar plant operates 165 day/year at 1000 ton/day. Annual average = 500 ton/day. Operating percent sugar plant = 45%; Ethanol plant = 90%. Process efficiency = 13.3%.
[b] Capital component of product cost.
[c] Excludes feedstock cost and plant depreciation.
[d] Calculated to yield a 15% rate of return on equity and a 9% return on debt (65% debt and 35% equity). Plant life = 20 years. Income tax = 52%.

MISSION XII. SELECTED SUMMARY DATA
WOOD TO OIL FOR DIRECT COMBUSTION AND CHAR VIA PYROLYSIS
(MAXIMUM LIQUID YIELD)

	Base Case			Sensitivity To — Plant Size and Feedstock Price			Operating Capacity (%)	Capital Investment −20%	Capital Investment +30%	Project Life
	26 / 151 / 7.7	52 / 302 / 15.4	156 / 918 / 46.2	26 / 151 / 7.7	52 / 302 / 15.4	156 / 918 / 46.2	156 / 918 / 46.2	156 / 918 / 46.2	156 / 918 / 46.2	156 / 918 / 46.2
A. Product										
Oil, 10^3 gal/day[a]	26	52	156	26	52	156	156	156	156	156
Char, ton/day[b]	151	302	918	151	302	918	918	918	918	918
10^9 Btu/day	7.7	15.4	46.2	7.7	15.4	46.2	46.2	46.2	46.2	46.2
B. Feedstock										
ODT/day[c]	500	1000	3000	500	1000	3000	3000	3000	3000	3000
10^9 Btu/day	9.6	19.1	57.3	9.6	19.1	57.3	57.3	57.3	57.3	57.3
$/$10^6$ Btu	1.0	1.0	1.0	2.0	2.0	2.0	1.0	1.0	1.0	1.0
C. Total Capital Investment										
10^6	12.3	22.2	61.4	12.6	22.7	63.0	61.2	49.7	79.0	61.4
$/$10^6$ Btu[d]	0.7	0.7	0.6	0.7	0.7	0.7	0.7	0.5	0.7	0.5
D. Annual Cost of Feedstock										
10^6	3.1	6.3	18.8	6.3	12.6	37.7	16.8	18.8	18.8	18.8
$/$10^6$ Btu	1.4	1.4	1.4	2.7	2.7	2.7	1.4	1.4	1.4	1.4
E. Annual Operating Cost[e]										
10^6	3.0	5.3	10.3	3.0	5.3	10.3	10.3	9.0	12.4	10.3
$/$10^6$ Btu	1.3	1.1	0.7	1.3	1.1	0.7	0.8	0.6	0.9	0.7
% Operating Capacity	90	90	90	90	90	90	80	90	90	90
F. Revenue Requirements (Total Product Basis)[f]										
Regulated Utility $/$10^6$ Btu	3.4	3.2	2.7	4.7	4.5	4.1	2.9	2.5	3.0	2.6
$/$10^6$ Btu (Oil only)[g]	6.1	5.6	4.5	8.9	8.5	6.2	4.9	4.0	5.1	4.2
G. Plant Life Years	20	20	20	20	20	20	20	20	20	30

[a] Assumes 0.250 lb/lb dry wood or 52 gal of oil/dry ton.
[b] Assume 0.302 lb/lb dry wood.
[c] Assumes a 50% moisture content feedstock and 19.1 million Btu/dry ton — process efficiency = 80.6%.
[d] Capital component of product cost.
[e] Excludes feedstock cost and plant depreciation.
[f] Calculated to yield a 15% rate of return on equity and a 9% return on debt (65% debt and 35% equity). Income tax = 52%.
[g] Assumes char valued at $1.25/$10^6$ Btu and represents 55% of total output.

MISSION XIII. SELECTED SUMMARY DATA
KELP TO SNG VIA ANAEROBIC DIGESTION

		Plant Size			Sensitivity to Feedstock Price					
A.	**Product**									
	10^6 SCF/day (as CH_4)	5.6	16.8	33.6	33.6	33.6	33.6	16.8	16.8	16.8
	10^9 Btu/day	5.6	16.8	33.6	33.6	33.6	33.6	16.8	16.8	16.8
B.	**Feedstock**									
	DAFT/day [a]	1000	3000	6000	6000	6000	6000	3000	3000	3000
	10^9 Btu/day	16	48	96	96	96	96	48	48	48
	$/ton (DAF)	100	100	100	25	200	0	25	200	0
C.	**Total Capital Investment**									
	10^6	30.1	68.9	115.7	113.6	118.4	113.0	67.9	70.3	67.6
	$/10^6 Btu [b]	2.2	1.6	1.3	1.3	1.4	1.3	1.6	1.7	1.6
D.	**Annual Cost of Feedstock**									
	10^6	33.0	99.0	198.0	49.5	396.0	0	24.7	198.0	0
	$/10^6 Btu	17.9	17.9	17.9	4.5	35.7	0	4.5	35.7	0
E.	**Annual Operating Cost [c]**									
	10^6	4.3	9.3	16.0	16.0	16.0	16.0	9.3	9.3	9.3
	$/10^6 Btu	2.3	1.7	1.5	1.5	1.5	1.5	1.7	1.7	1.7
	% Operating Capacity	90	90	90	90	90	90	90	90	90
F.	**Revenue Requirements [d]**									
	Regulated Utility $/10^6 Btu	22.4	21.2	20.7	7.3	38.6	2.8	7.8	39.1	3.3

[a] Process efficiency = 35%. DAFT denotes dry ash-free tons.

[b] Capital component of product cost.

[c] Excludes feedstock costs and plant depreciation; assumes 16 million Btu/dry ash-free ton of feedstock.

[d] Calculated to yield a 15% rate of return on equity and a 9% return on debt (65% debt and 35% equity). Plant life = 20 years. Income tax = 52%.

MISSION XIV. SELECTED SUMMARY DATA
KELP TO ETHANOL VIA ACID HYDROLYSIS AND FERMENTATION
25 Million Gallons per Year

| | Base Case | 50% Sugar Conversion | | | | | | 80% Sugar Conversion Base Case |
		Sensitivity To Feedstock Prices		Capital Investment		Operating Capacity (%)		
A. Product				+30%	−20%			
Sugar, ton/day	500	500	500	500	500	500	500	500
Ethanol, 10^3 gal/day	76	76	76	76	76	76	76	76
10^9 Btu/day	5.7	5.7	5.7	5.7	5.7	5.7	5.7	5.7
B. Feedstock								
DAFT/day[a]	1126	1126	1126	1126	1126	1126	1126	703
10^9 Btu/day[b]	18.0	18.0	18.0	18.0	18.0	18.0	18.0	11.25
$/ton (dry)	75	100	0	75	75	75	75	75
C. Total Capital Investment								
Sugar plant (10^6)	27.9	28.0	27.7	36.3	22.3	27.9	27.8	19.5
Ethanol plant (10^6)	28.0	28.0	28.0	36.4	22.4	28.0	28.0	26.5
Total Capital Investment[c]	55.9	56.0	55.7	72.7	44.7	55.9	55.8	46.0
$/$10^6$ Btu[d]	4.5	4.5	4.5	5.9	3.6	5.1	5.8	3.7
D. Annual Cost of Feedstock								
10^6	27.9	37.2	0	27.9	27.9	24.7	21.6	17.4
$/$10^6$ Btu	14.9	19.8	0	14.9	14.9	14.9	14.9	9.2
E. Annual Operating Cost[d]								
Sugar plant (10^6)	5.2	5.2	5.2	5.9	4.7	5.0	4.7	3.4
Ethanol plant (10^6)	7.0	7.0	7.0	7.9	6.4	6.4	6.3	7.0
Total Annual Operating Cost (10^6)	12.2	12.2	12.2	13.8	11.1	11.4	11.0	10.4
$/$10^6$ Btu	6.5	6.5	6.5	7.4	5.9	6.8	7.5	5.5
% Operating Capacity	90	90	90	90	90	80	70	90
F. Revenue Requirements[e]								
Regulated Utility Sugar, $/lb	0.11	0.14	0.03	0.12	0.11	0.12	0.11	0.08
Ethanol, $/$10^6$ Btu	26.9	30.8	11.0	28.2	29.6	26.8	28.2	19.0

[a] Process efficiency = 32%. DAFT denotes dry ash-free tons.
[b] Assumes 16 million Btu/DAFT of feedstock, 80% carbohydrate.
[c] Excludes feedstock cost and plant depreciation.
[d] Capital component of product cost.
[e] Calculated to yield a 15% rate of return on equity and a 9% return on debt (65% debt and 35% equity). Plant life = 20 years. Income tax = 52%.

MISSION XV. SELECTED SUMMARY DATA
COGENERATION OF STEAM AND ELECTRICITY FROM WOOD VIA DIRECT COMBUSTION

	Base Case			Sensitivity To						
				Capital Investment		Feedstock Price		Operating Capacity (%)		
				-20%	+30%					
A. Product										
Plant capacity, MW	3.65	7.3	21.9	21.9	21.9	21.9	21.9	21.9	21.9	21.9
MWh/day	87.6	175	525	525	525	525	525	525	525	525
Steam, 10^3 lb/hr	210	420	1260	1260	1260	1260	1260	1260	1260	1260
10^9 Btu/day	7.2	14.5	43.5	43.5	43.5	43.5	43.5	43.5	43.5	43.5
B. Feedstock										
ODT/day[a]	500	1000	3000	3000	3000	3000	3000	3000	3000	3000
10^9 Btu/day	9.6	19.2	57.6	57.6	57.6	57.6	57.6	57.6	57.6	57.6
$/$10^6$ Btu	1.00	1.00	1.00	1.00	1.00	1.50	0	1.00	1.00	1.00
C. Total Capital Investment										
$10^6	21.6	40.3	109.1	87.9	140.8	109.8	107.7	108.9	108.6	108.3
$/$10^6$ Btu[b]	1.37	1.29	1.16	0.94	1.49	1.17	1.15	1.33	1.68	2.30
D. Annual Cost of Feedstock										
$10^6	2.8	5.6	16.8	16.8	16.8	25.1	0	14.7	11.5	9.4
$/$10^6$ Btu	1.32	1.32	1.32	1.32	1.32	1.98	—	1.32	1.32	1.32
E. Annual Operating Cost[c]										
$10^6	2.9	5.1	12.0	10.2	14.6	12.0	12.0	11.8	11.5	11.3
$/$10^6$ Btu	1.36	1.19	0.94	0.80	1.16	0.94	0.94	1.06	1.32	1.78
% Operating Capacity	80	80	80	80	80	80	80	70	55	40
F. Revenue Requirements[d]										
Regulated Utility $/$10^6$ Btu	4.05	3.80	3.42	3.06	3.97	4.09	2.09	3.71	4.32	5.40

[a] Assumes a 50% moisture content feedstock and 19.2 million Btu per dry ton — process efficiency = 75.5%
[b] The capital component of product costs.
[c] Excludes feedstock cost and plant depreciation.
[d] Calculated to yield a 15% return on equity and a 9% return on debt (65% debt and 35% equity). Plant life = 20 years.Income tax = 52%.

RECEIVED SEPTEMBER 9, 1980.

An Analysis of Gasohol Energetics

WILLIAM A. SCHELLER

Department of Chemical Engineering, University of Nebraska, Lincoln, NB 68588

The question of energy utilization and energy efficiency in the gasohol program has generated much discussion and considerable controversy. The purpose of this paper is to present a detailed energy balance associated with the components of the gasohol program, i.e., grain production, fuel alcohol production and the replacement of gasoline with ethanol to produce gasohol. The overall energy balance involves a comparison of a gasohol fuel economy with a gasoline fuel economy including the energy impact of the distillers dried grain which becomes available to the livestock feeding industry.

ENERGY FOR CORN PRODUCTION

Energy requirements for the production of agricultural products vary considerably from country to country. In less developed areas, the energy expended per unit of production is usually considerably lower than in highly developed areas of the world. On the other hand, the product production per unit of land is usually related to the energy expenditure and in those areas where less energy is expended, less product yield is obtained. Table I compares the energy consumption and product production in Mexico with hand labor and with oxen power with that for a modern U.S. farm. The corn production with hand labor is very energy efficient requiring only 35,000 Btu's per bushel, but the grain yield is only 31 bushels per acre. When the farmer adds an ox to assist in the corn production, not only does the energy expended per bushel of grain increase, but the net grain yield per acre is cut

0097-6156/81/0144-0419$05.75/0
© 1981 American Chemical Society

almost in half because of the need to feed the ox. On a modern U.S. farm, the energy expenditure per bushel of corn is about 3.5 times that for corn production with hand labor in Mexico. However, the yield of corn per acre is increased by a factor of 2.8. If large quantities of grain are to be produced, it is important that the production per unit of land be maximized.

Table I. COMPARISON OF ENERGY REQUIREMENTS FOR CORN PRODUCTION[a,b]

| | Mexico | | U.S.A. |
	Hand Labor	Oxen Power	Modern Farm
1000 Btu/Acre	1084	1572	10,510
Bushels/Acre	31	15	86
Btu/Bushel	35,000	104,800	122,200

[a] Excludes energy for manufacture of farm machinery.
[b] Data from Reference 1.

Table II shows the evolution of energy requirements for the production of corn in the United States between 1950 and 1975. Total energy consumption per acre including the energy for fuel, fertilizer, pesticides, herbicides, manpower, etc., but not including energy for the manufacture of the farm machinery, increased from 3.8 million Btu's per acre in 1975 to 10.51 million Btu's per acre in 1975 (a factor of 2.8). The yield per acre in turn increased from 38 bushels per acre to 86 bushels per acre or a factor of about 2.3. During this same period, the energy consumed per bushel of corn increased only 22.2%. In all cases, the energy consumption expressed as equivalent gallons of oil consumed per bushel of corn was less than 1.

Table II. ENERGY REQUIREMENTS FOR CORN PRODUCTION IN THE U.S.A.[a,b]

	1950	1959	1970	1975
1000 Btu/Acre	3800	6030	9760	10,510
Bushels/Acre	38	54	81	86
Btu/Bushel	100,000	111,600	120,500	122,200
Gal Equivalent Oil/ Bushel	0.67	0.74	0.80	0.81

[a] Excludes energy for the manufacture of farm machinery, 1 gal of equivalent oil equals 150,000 Btu.
[b] Data from References 1 and 2.

While detailed fuel, fertilizer, and chemical consumptions are not yet available for 1979, these will probably be somewhat higher than the 1975 figures. On the other hand, it is estimated that the average yield of corn in 1979 will exceed 100 bushels per acre. Because the figures in Table II are average figures for the nation, they include the energy requirements for an average amount of irrigation. Unirrigated corn will require less energy per acre for production while corn produced in arid areas will require more energy per unit of land area. For purposes of the energy comparisons contained in this paper, the energy figure for 1975 was used.

ENERGY REQUIREMENTS FOR FUEL ALCOHOL PRODUCTION

When corn is fermented to produce alcohol, a number of products are produced. The most desirable product is probably the grain alcohol (ethanol). However, small amounts of heavier alcohols known collectively as fusel oil are also produced in the fermentation. In the beverage industry, the fusel oil is considered to be undesirable and a large amount of energy is expended in removing it from the grain alcohol. When one is manufacturing a fuel grade alcohol for use in gasohol, the fusel oil is a desirable component and should remain mixed with the grain alcohol rather than being removed. Fusel oil production is about 0.5% of the ethanol.

Table III contains a material balance showing the composition of the corn used for the calculations and the products produced. In addition to fuel alcohol, a high-protein cattle feed called distillers dark grain (DDG) is also produced in the fermentation process. As will be seen later, inclusion of DDG in the cattle ration provides more weight gain than if the cattle had been fed the original corn which was used to produce the fuel alcohol and DDG. Carbon dioxide is a second byproduct produced in the fermentation, but for purposes of this analysis, it has been assumed to be vented to the atmosphere.

Table III. MATERIAL BALANCE PRODUCTION OF FUEL ALCOHOL

Corn Component, lb/bu		Product, unit/bu
Starch	34.07	Fuel Alcohol, 2.6133 gal
Protein	4.73	DDG (10% H_2O), 18.016 lb
Other	8.52	Carbon Dioxide, 16.886 lb
Moisture	8.68	Alcohol Losses, 0.0531 gal
Total	56.00	Remaining H_2O, 3.4492 lb

The recovery of alcohol by distillation is very efficient with a typical total loss of alcohol between the fermentor and the anhydrous product of only 2%. It is also interesting to note that there is sufficient moisture contained in the grain to supply the chemical needs for conversion of starch to alcohol and to account for the 10% moisture content in the distillers grains. All additional water added to the system is simply to supply appropriate concentrations and streams of appropriate fluidity.

In the last few years, there has been a dramatic reduction in the energy requirements for grain alcohol production mainly through more efficient heat recovery in the plant. Table IV compares the energy consumption in a beverage alcohol plant of 1973 with fuel alcohol plants designed in early 1978 and late 1979. The beverage alcohol plant consumed about 172,000 Btu's per gallon of alcohol produced. By early 1978, with the realization that there was no need to produce a highly purified neutral spirits for fuel alcohol and with the introduction of modest heat recovery facilities, the energy requirements dropped to 125,000 Btu's per gallon. In a late 1979 plant design which included use of furnace stack gases to dry the distillers grain and development of a pressure profile in the plant to increase potential heat recovery, the energy requirement dropped to 69,600 Btu's per gallon of alcohol produced. The high heat of combustion of one gallon of anhydrous fuel grade alcohol is about 84,200 Btu's.

Table IV. EVOLUTION OF GRAIN ALCOHOL PLANT ENERGY REQUIREMENTS

	Beverage Plant	Early 1978 Fuel Plant	Late 1979 Fuel Plant
190° Proof Spirits	109,000	68,000	52,900
Anhydrous Alcohol	None	14,000	Included
Subtotal	109,000	82,000	52,900
DDG Production	63,000	43,000	16,700
Total	172,000	125,000	69,600
Gal Equivalent Oil/ Gal Alc.	1.15	0.82	0.46

The energy consumptions in Table IV include the fossil fuel consumption associated with the generation of the electrical needs for the alcohol plant as well as the fossil fuel burned in generating steam for the plant. For the overall gasohol energy analysis, the late 1979 fuel alcohol plant energy consumption is used.

BASIS FOR ENERGY COMPARISON OF GASOLINE AND GASOHOL FUEL SYSTEMS

In making a consistent and valid comparison of the energy consumption in a gasohol fuel system and in a gasoline fuel system, there are a number of factors that must be considered. These factors include any difference in fuel consumption (miles per gallon) between the two systems, any difference in fuel octane number between the two systems, any change in fuel volume between the two systems which may result from blending of the fuel components, and any change in corn requirements that would be necessary to maintain equal quantities of beef production in both systems.

Research has been conducted to provide information about each of these factors. In Nebraska, a two million mile road test program was conducted over a 34-month period in which the fuel economy of unleaded gasoline was compared with the fuel economy of gasohol. Data from this test indicate the gasohol-fueled cars obtained on the average 6.7% more miles per gallon than the cars fueled on unleaded gasoline. For purposes of the energy comparison in this paper, it has been assumed that the gasohol cars would obtain only 3% more miles per gallon.

Measurements by independent laboratories have shown that when a mixture is prepared containing 10% anhydrous ethanol and 90% unleaded gasoline, the average octane, [(R + M)/2], is three numbers higher for gasohol than for the unleaded gasoline used as the base stock. In this paper, a three octane number increase is used in the energy comparison. In blending 10% anhydrous ethanol with 90% unleaded gasoline, laboratory measurements have shown that the total volume of the mixture is 0.23% greater than the sum of the volume of the components. For this paper, this excess volume of mixing has been assumed to be zero.

Finally, feeding trials involving distillers dark grain have shown that beef cattle receiving this component in their diet show increased weight gain over cattle not receiving this material. Tests carried out in Kentucky indicate that beef cattle receiving distillers dark grains from the fermentation of 20% of the corn fed to the animals gained 12.9% more weight than those cattle receiving the total ration of corn. Tests conducted in Nebraska support the conclusion that distillers dried grains are a better feed component than the whole corn from which they are produced. For purposes of this paper, it has been assumed that the increased weight gain associated with feeding distillers dark grains is 6% rather than the 12.9% reported from Kentucky.

The results of experimental observations and the evaluation basis used in this paper are contained in Table V.

**Table V. COMPARISON OF EXPERIMENTAL GASOHOL DATA
WITH ENERGY EVALUATION BASIS**

	Experimental Observation	Evaluation Basis
Increased MPG, %	6.7	3.0
Excess Volume of Mix, %	0.23	0
Increased Octane, (R + M)/2	3	3
Increased Beef Weight Gain With DDG, %	12.9	6

CRUDE OIL SAVINGS ASSOCIATED WITH GASOHOL BLENDING

When one gallon of anhydrous grain alcohol is blended with 9 gallons of gasoline to produce 10 gallons of gasohol, there will be a reduction in crude oil requirements for the manufacture of automotive fuel. The most obvious reason for this is that grain alcohol has replaced gasoline in the mixture. Other factors also affect the amount of crude oil used including the fact that the 9 gallons of unleaded gasoline can be produced at a lower octane number because the addition of grain alcohol will raise the octane number of the blend to meet the market specification. Furthermore, because a car will travel further on one gallon of gasohol than it will on one gallon of gasoline and since the automotive fuel market is a demand to drive a total number of miles, less gasoline will be needed to meet this demand.

The quantitative effect of these factors expressed as crude oil savings is shown in Table VI. By replacing one gallon of gasoline out of ten with one gallon of ethanol, we save not only one gallon of crude oil but also a slight amount more because energy is not required to refine that gallon of crude oil. This saving by replacement amounts to 1.014 gal of crude oil per gallon of grain alcohol. The requirements of a lower octane number for the 9 gallons of unleaded gasoline will save 0.286 gal of crude oil per gallon of grain alcohol, and obtaining 3% more miles per gallon with gasohol will save 0.300 gallons of crude oil per gallon of grain alcohol for a total crude oil saving of 1.6 gallons. Information related to these crude oil savings was obtained from refinery simulation studies carried out by Bonner and Moore Associates, Inc. of Houston, Texas for the U.S. Department of Energy.

Table VI. CRUDE OIL SAVINGS ASSOCIATED WITH GASOHOL PRODUCTION

	Gal/Gal Alcohol
Replacement of Gasoline With Ethanol	1.014
Lower Average Octane Number of Gasoline	0.286
3% More Miles/Gallon with Gasohol	0.300
Crude Oil Savings	1.600

GRAIN SAVINGS ASSOCIATED WITH FEEDING DISTILLERS DARK GRAINS

Feeding trials at the University of Kentucky have demonstrated that inclusion of distillers dark grain in a cattle ration results in increased weight gain. Specifically, their results indicate an increased gain of 12.9%. As mentioned previously, it is assumed for purposes of this study that the increased weight gain would be only 6%. Furthermore, beef production is geared to meeting a market demand for a certain number of total pounds. This means that with the inclusion of DDG in the animal ration, fewer total bushels of corn are required to bring the cattle to full weight. This comparison is illustrated in Figure 1. When 2.03 bu of corn are fed with an appropriate amount of roughage (hay), the beef cattle show a weight gain of 9.63 lb. On the other hand, if 20% of 1.91 bu of corn (0.38 bu) are diverted to a grain alcohol plant, one gallon of grain alcohol is produced. If the byproduct DDG from the alcohol plant is combined with the remaining 1.53 bu of corn and fed with the same amount of hay to the beef cattle, they will also gain 9.63 lb but with a reduced consumption 0.12 bu of corn. This saving of corn also represents a saving of energy.

COMPARISON OF ENERGY REQUIREMENTS FOR GASOLINE AND GASOHOL

The energy requirements for the production of gasoline and gasohol with production of an equivalent amount of beef in each case are presented as gallons of equivalent oil in Table VII. The basis for this comparison is also included in Table VII. One gallon of fermentation fuel ethanol is assumed to be mixed with 9 gallons of unleaded gasoline in the gasohol case, and for the gasoline case, a sufficient amount of unleaded gasoline is assumed to move the car the same distance as in the gasohol case. All other petroleum products produced in the refinery are the same in both cases. The quantity of corn in the two cases is as discussed in the preceding section and is sufficient to produce 9.63 lb of beef weight gain. The energy requirement for producing this corn is presented in Table II for 1975.

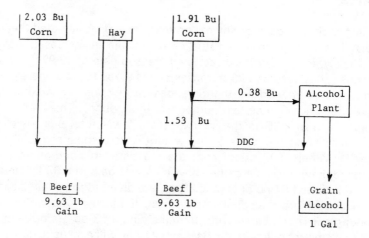

Figure 1.   Making grain alcohol reduces grain production

Item 1 in Table VII indicates that for the gasoline case, 10.74 gal of crude oil must be refined to produce the base quantity of gasoline. In the gasohol case, only 9.14 gal of crude oil are refined to produce the needed gasoline. The difference in these two quantities is the 1.6 gal of crude oil savings detailed in Table VI. The second item in Table VII deals with corn production for cattle feed. In the gasoline case, this is 2.03 bu of corn as shown in Figure 1 with an associated energy expenditure of 1.65 gal of equivalent oil. In the gasohol case, 1.53 bu of corn were fed directly to he cattle. This corn has associated with it an energy consumption of 1.25 gal of equivalent oil. Item 3 is the energy consumption associated with producing corn for ethanol manufacture. In the gasoline case, there is no corn used for ethanol. In the gasohol case, as shown in Figure 1, 0.38 bu of corn are associated with the production of 1 gal of grain alcohol. Energy for the production of this corn is 0.31 gal of equivalent oil. At this point, the total petroleum consumption is 12.39 gal of equivalent oil in the gasoline case and 10.70 gal of equivalent oil in the gasohol case.

In the gasohol case, we must now add the energy consumption associated with alcohol production and the DDG production. Using the figures from Table IV, the energy consumption for a late 1979 fuel alcohol plant is 0.35 gal of equivalent oil for the alcohol production and 0.11 gal of equivalent oil for the DDG production. This energy would probably be supplied by coal, so the subtotal plant energy for the alcohol and cattle feed production is indicated as coal with an energy content equivalent of 0.46 gal of oil.

Summing the two subtotals, energy equivalent to 12.39 gal equivalent oil was consumed in the gasoline case while in the gasohol case, energy equivalent to only 11.16 gal of energy equivalent oil was consumed. Thus, there is a fossil fuel saving equivalent to 1.23 gal of oil for every gallon of grain alcohol that is blended with 9 gal of unleaded gasoline. This is a very substantial saving.

Table VII. GASOHOL REDUCES FOSSIL FUEL CONSUMPTION

	Gasoline Case Gal Equiv. Oil	Gasohol Case Gal Equiv. Oil
1. Crude Oil Refining	10.74	9.14
2. Corn Production for Feed	1.65	1.25
3. Corn Production for Ethanol	0.00	0.31
Subtotal — Petroleum	12.39	10.70
4. Alcohol Plant Operation	0.00	0.35
5. DDG Plant Operation	0.00	0.11
Subtotal — Coal	0.00	0.46
6. Total Energy Consumption	12.39	11.16
Fossil Fuel Saving with Gasohol		1.23

Basis:
a. 1 gal of fuel ethanol mixed with 9 gal of unleaded gasoline.
b. Unleaded gasoline to move a car as far as the gasohol in "a" above.
c. All other petroleum products are the same in both cases.
d. Corn and corn + DDG to produce 9.63 lb of beef weight gain.
e. 1 gal of equivalent oil is equal to 150,000 Btu.

Even if gasohol did not show the 3% increase in fuel economy and even if the DDG did not produce a 6% weight gain in beef cattle, the gasohol case would still show an energy saving of 0.84 gal of equivalent oil over the gasoline case. These energy savings exist whether the alcohol plant is fueled with coal, oil, or natural gas. If indeed, the alcohol plant is fueled with coal, then the actual savings in petroleum is 1.69 gal per gallon of alcohol blended.

The U.S. Department of Energy has estimated that the potential exists for producing about 4.5 billion gallons of grain alcohol from agricultural stocks in the near future if alcohol plants to match this capacity are built. At a saving of 1.69 gal of petroleum per gallon of grain alcohol, there is the potential of saving over 180 million barrels per year of imported crude oil. This in turn would reduce the outflow of dollars by at least $5 billion per year, which would be a significant percentage decrease in our trade deficit.

CONCLUSIONS

Based on this analysis, it has been demonstrated that replacement of gasoline with gasohol in the automotive fuel market will result in a reduction of fossil fuel consumption. This reduction is present whether or not gasohol provides greater fuel economy than gasoline and whether or not beef cattle gain additional weight when distillers dark grains (DDG) are included in their diet. The saving in petroleum that results from the production and use of gasohol has the potential to reduce significantly our trade deficit.

REFERENCES

1.   Pimentel, D.; Terhune, E. C., *Ann. Rev. Energy* **2,** 171-95 (1977).

2.   Pimentel, D., et al., *Science* **182,** 443-49 (1973).

APPENDIX

Table II contains a summary of energy requirements for 1975 corn production in the United States. The individual energy components in 1975 corn production are listed in Table A-1. This table shows that the three largest sources of energy consumption are fuel, fertilizer, and irrigation. Table A-2 contains the energy content of the corn grain and associated biomass material (stover and cobs). The energy production (810,000 Btu/bu corn) is about 6.6 times the energy used in producing the corn.

Table IV lists the energy requirements for producing fuel grade ethanol and distillers dark grains in a late 1979 plant design. Table A-3 contains a detailed listing of the utility requirements for this late 1979 design. Using the relationship shown in Figure A-1, these utility requirements were converted to a total energy requirement of 69,600 Btu per gallon of fuel alcohol produced and includes the energy for DDG production. The energy content of the grain stover from Table A-2 is about 2.6 times the energy requirement for producing ethanol and DDG. From Table A-4, it is apparent that the energy content of the products from the grain alcohol plant is approximately equal to the energy content of the corn used (Table A-2) even though the mass of the products is only 63% of the mass of the corn used (Table III).

Table A-1 ENERGY CONSUMPTION IN CORN FARMING — 1975[a]

Component	Btu/bu Corn	Btu/gal EtOH
1. Seed Corn	2,700	1,040
2. Fertilizer	43,900	16,800
3. Herbicides	2,700	1,040
4. Insecticides	1,500	580
5. Fuel	39,300	15,030
6. Electricity	7,100	2,720
7. Irrigation	14,600	5,580
8. Drying	7,000	2,680
9. Transportation	3,400	1,290
Total Energy	122,200	46,760

[a] Reference 1.

Table A-2. ENERGY PRODUCTION IN CORN FARMING

Component	Btu/bu Corn	Btu/gal EtOH
1. Corn, Digestible energy	342,000	130,000
2. Stover and cobs, HHV	486,000	180,000
Total Energy	810,000	310,000

Table A-3. UTILITY AND ENERGY CONSUMPTION IN GRAIN ALCOHOL PRODUCTION
(Late 1979 Design)

	Per Gallon of Fuel Ethanol			
	Steam, lb	Electric, kWhr	Cold Water, Gal	Total Btu
1. Milling and Propagation	0.22	0.349	2.7	4,000
2. Cooking & Saccharification	11.56	0.034	41.4	18,030
3. Fermentation	0.06	0.053	19.0	650
4. Distillation	20.46	0.021	54.8	31,500
5. Thin Stillage Concen.	3.00	0.357	3.4	8,340
6. DDG Drying and Other	1.18	0.503	45.4	7,080
Total	36.48	1.317	166.7	69,600

* Calculated using the listed utilities and Figure A-1.

Table A-4. ENERGY PRODUCTION FROM A GRAIN ALCOHOL PLANT

Component	Btu/Gal EtOH
1. Fuel Ethanol, HHV	84,200
2. DDG, Digestible Energy	45,000
Total Energy	129,200

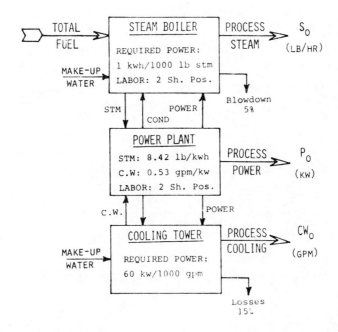

$$\text{Total Power } P_T \text{ (kw)} = 1.05(P_o + .0010^r S_o + .0706 CW_o)$$
$$\text{Total Fuel (Btu/Hr)} = 10,000 P_T + 1.32 S_o \Delta H$$

Figure A-1. Block flow diagram: utility–fuel relationship

RECEIVED MAY 20, 1980.

Wood Production Energetics

An Analysis for Fuel Applications

NORMAN SMITH and THOMAS J. CORCORAN

Department of Agricultural Engineering, University of Maine at Orono,
Orono, ME 04473

Wood was the prime fuel source for the United States during much of the 19th Century. Consumption probably peaked around 1880 at 146 million cords* per year. Coal replaced wood for most applications. However, use of wood for residential heating continued to be important in some rural areas until after World War II. Residential use of waste wood and sawdust from wood-utilizing industries persisted after the use of wood harvested directly for fuel had practically ceased. A number of industries continued to use their waste wood in boilers to produce steam for electricity generation and process heat. However, the convenience and low cost of heavy oil fuels caused all but a very small number of operations to terminate waste wood usage. The conical incinerator became a common sight at sawmills in the 1960's while wood drying kilns were being fired by oil in another part of the yard.

* A cord is a volume measure of 128 feet^3 of piled round wood, usually represented as a pile of 4 foot logs, 4 feet high and 8 feet long. Volume scaling is still much used in forestry as many operations are volume- rather than weight-sensitive. However, a cord represents very different weights of dry matter depending on the species of wood. Weight per cord also varies greatly with moisture content. Green wood is around 50% moisture content. Dry matter per cord varies from about 1900 lb for pine to 3500 lb for hardwood such as birch and maple.

0097-6156/81/0144-0433$05.00/0
© 1981 American Chemical Society

Salvage of waste wood from sawmills began anew a few years ago when some paper mills began to experience pulpwood shortages. Discarded pieces were chipped and sold for pulping. More recently, there has been interest in using bark as a boiler fuel partly because of the disposal problem and partly because of increased fuel prices. Since the Arab oil embargo of 1973, serious attention has been given to use of wood as a fuel on a large scale. For example, Szego and Kemp (8) have evaluated the possibility of energy farms on which woody plants would be produced for fuel use. The Maine Office of Energy Resources (9) has analyzed the possibility of methanol production from wood. Huff (4) has reported on the development of an automatically controlled furnace suitable for residences which can burn wood chips made from logging residues or puckerbrush. Smith (7) has examined conceptual designs for mechanized short-rotation forestry, particularly the harvesting phase.

Methods of wood harvesting have been revolutionized recently as mechanization has come to forestry. A number of harvesting methods are now in use in which the basic operations of felling, transport to a landing, processing and loading for transport are approached in very different ways. This paper examines the energy inputs to each sub operation to allow estimation of total energy relationships or net energy production efficiency for any complete system whether or not it is currently in use.

It should be stressed that for a large portion of U.S. forest lands, the only significant operation involved in wood production is that of harvesting. Reforestation is often by natural means, very little fertilization or cultivation is carried out. Construction of a road network and actual harvesting of the trees at the end of the growing cycle is, by far, the greatest purchased energy input to wood production. The energy used in road building varies greatly with terrain and harvesting pattern. It is probably small in relation to other inputs and is neglected in this analysis.

Harvesting Equipment

For many years, the axe and bucksaw were the sole means of felling and preparing wood for transport to the users' premises. Primary transport from the stump to the collection point at a roadside or on a riverbank was by horse or ox team. Production rates for this system vary tremendously depending on size of trees, haul distances, terrain, etc., but it was generally reckoned that one man could fell, delimb, cut up and load one cord of wood per day while one horse would take about two hours to drag out that volume of wood.

Use of gasoline powered chainsaws has increased a worker's capacity about ten fold. Modern saws allow a man to fell, delimb and cut up about 1.3 cords/hour. Use of small tracked vehicles equipped with winches to skid out bunches of tree trunks have largely displaced the horse and ox, but a multitude of new equipment is even now displacing these devices.

Short descriptions of the main classes of equipment considered in this study follows:

Chain Saw: A portable, gasoline-engined, manually-controlled machine with a toothed chain used to fell trees and remove limbs.

Feller Buncher: A mobile machine designed to shear a tree at the stump, and hold it by means of a clamp and cutting head while it swings and deposits the tree onto a pile on the ground. The cutting head is usually composed of two hydraulically actuated shearing blades. Power requirements are from 80-130 horsepower.

Delimber Buncher: A mobile machine carrying a unit which strips the limbs and top off the bole of a previously felled tree and deposits the stripped bole in a pile on the ground ready for removal from the stump area to a roadside landing. Usually requires around 120 horsepower.

Wheeled Skidder: A tractor unit, usually with frame steering and four wheel drive, equipped with a winch or grapple which gathers and skids behind itself loads of full trees, tree length boles logs from the stump area to a roadside landing. Power requirement usually exceeds 70 horsepower.

Wheeled Forwarder: A frame steered, self-loading vehicle equipped with hydraulically operated grapple and loading boom and a carrier or bunk to support its load of logs. Power requirements vary from 40 to 100 horsepower depending on size.

Loader: A hydraulically operated boom and grapple which can be mounted on a truck chassis. It is used to gather logs or tree lengths from a pile and build a load on a truck body.

Chipper: A machine which reduces logs and tree length wood to small chips by means of a rapidly rotating drum or disc, carrying a series of blades. The chips usually leave the cutting device in an air-stream induced by the fan effect of the chipping mechanism and are thus automatically conveyed into transport vehicles or stockpiles. Power requirements are around 300 horsepower for a machine capable of chipping around 25 ton/hours.

Energetics of Mechanized Harvesting Systems

Table 1 shows typical production rates and fuel consumption figures for the various pieces of equipment previously described. The writers were fortunate in that the American Pulpwood Association published the results of a 1974 survey of members' operations (1) while this paper was being written. Whenever possible, the data from that survey was used in preparing the table. The data sources from which other figures were calculated are indicated in the footnotes. Figures for the energy subsidy represented by the energy used in manufacturing the equipment are very approximate and were derived by assuming an average figure of 25,000 Btu/lb consumed in the manufacturing process (most of the equipment weight is in the form of steel which requires around 21,000 Btu/lb in the transformation from ore in the ground to steel plate (2). The energy used in manufacture was divided by the approximate lifetime production of the equipment to arrive at a figure of Btu per ton of dry wood.

The approximate energy cost of practically any system of production using present equipment can be calculated from the table. For example, a very common system uses chain saw felling and delimbing, tree length skidding to a forest landing, loading the tree length material onto large trucks for transport to a mill yard, unloading by the same type of loader used in the woods, followed by chipping.

Many operators are now moving toward chipping whole trees in the woods with a fully mechanized system. The steps might be as follows: Felling with a feller-buncher; grapple skidding to a landing; chipping, with pneumatic conveying into trucks as an integral part of the operation; transport; unloading by tipping the whole truck body backwards to dump the chips by gravity.

Table II illustrates the breakdown of energy use in these two systems, including a 50-mile haul to the utilization site, which appears to be a fair average for much of the U.S.

Several interesting facts appear from the comparison:

1. Both methods, though very different in procedure, have approximately the same unit energy consumption. In fact, this is so for most of the mechanized systems for producing wood from the tree trunk. Perhaps this is not surprising as most of the same operations appear in each system though they may be performed in a different order.

Table I. APPROXIMATE ENERGY USE IN WOOD PRODUCTION OPERATIONS
A. Energy Subsidy Due to Equipment Manufacture

Machine Type or Operation	Typical Machine Weight, lb	Production Rate	Life	Manufacturing Energy Subsidy, Btu/ton dry wood[a,b]
Felling				
Chain saw,	10	2.6 cd/hr[c]	2,000 hr	32.0
(Felling and delimbing)				
Feller-Buncher	52,000	8.38 cd/hr[c]	10,000 hr[d]	10,350
Delimbing				
Limber Buncher	45,000	9 cd/hr[e]	10,000 hr[d]	8,350
Trans. to Landing				
Wheeled Skidder, whole trees	25,000	3.08 cd/hr	13,000 hr[d]	10,400
Forwarder residues	27,000	9.02 green tons/r[f]	13,000 hr[d]	11,300
Wheel loader, prebunch residues	4,000	4.5 green tons/hr[f]	13,000[d]	3,400
Yard Operations				
Chain saw, bucking to short lengths	10	3.65 cd/hr[c]	2,000 hr	23.0
Loading				
Tree length	25,000	10.78 cd/hr[c]	10,000 hr	3,900
Trucking				
Small truck	12,000	— —	300,000 mi[d]	6,700[i]
Large truck	25,000	— —	500,000 mi[d]	3,300[j]
Chipping				
Whole tree chipper	57,000	10 cd/hr	10,000 hr[d]	9,500
Auxiliary				
Management vehicles, etc.	4,000	— —	100,000 mi	1,000[k]

Table I. APPROXIMATE ENERGY USE IN WOOD PRODUCTION OPERATIONS (cont.)
B. Equipment Operation and Overall Energy Requirements

Machine Type or Operation	Fuel Consumption	Energy Use Btu/ton dry wood[b]	Total Energy Requirements, Btu/ton dry wood (to nearest 1000 Btu)
Felling			
Chain saw, (Felling and delimbing)	0.41 gal/cd[c]	33,000	33,000
Feller-Buncher	0.64 gal/cd[c]	59,700	70,000
Delimbing			
Limber Buncher	0.62 gal/cd[d]	57,900	66,000
Trans. to Landing			
Wheeled Skidder, whole trees	0.95 gal/cd	88,500	99,000
Forwarder residues	0.41 gal/green ton[g]	115,000	126,000
Wheel loader, prebunch residues	1.24 gal/green ton[g]	67,200	71,000
Yard Operations			
Chain saw, bucking to short lengths	0.39 gal/cd[c]	31,200	31,000
Loading,			
Tree length	0.47 gal/cd[c]	43,500	47,000
Trucking:			
Small truck	0.04 gal/cd mi[c]	373,000[h]	380,000
Large truck	0.02 gal/cd mi[c]	187,000[h]	190,000
Chipping			
Whole tree chipper	0.7 gal/cord[d]	65,500	75,000
Auxiliary			
Management, vehicles, etc.	0.72 gal/cd	57,600	59,000

[a]   Assumes 25,000 Btu/lb consumed in equipment manufacture.
[b]   Assumes 3,000 lb dry wood per average cord.
[c]   Source — "Fuel Requirements for Harvesting Pulpwood" — APA Survey.
[d]   Source — Estimate of Woodlands Manager.
[e]   Source — Average of two company operations.
[f]   Source — Folia Forestalia 237 — Finnish Forest Institute.
[g]   Estimate based on engine size and research reports.
[h]   Average figures for 100-mile round trip.
[i]   10 cord loads, handles 45,000 tons in useful life.
[j]   25 cords, loads handles 187,500 tons during useful life.
[k]   Assumes 1 vehicle per fully mechanized harvesting crew.

Table II. ENERGY USE IN WOOD PRODUCTION SYSTEMS

Tree length system	Btu/ton dry wood
Felling and Delimbing (Chain saw)	33,000
Skidding	99,000
Loading (tree length)	47,000
Transport (50 miles one way)	190,000
Unloading	47,000
Chipping	75,000
Auxiliary	59,000
Total	550,000
Whole tree chip system	
Felling and Bunching	70,000
Skidding	99,000
Chipping	75,000
Transport	190,000
Unload	negligible
Auxiliary	59,000
Total	493,000

2. Transportation, even if only to a user 50 miles from the growing site, can represent almost 50% of the total energy input to present the product to the consumer. It may seem that substantial savings could be made by consuming the wood closer to the growth site. However, economics rather than energetics will decide whether this will be done.

3. Reduction of the wood from tree length to the convenient form of wood chips takes only about 20% of the energy used in production. Even though the bulk of the wood is considerably increased by chipping, weight, not volume, remains the limit on load size for transportation. The bonus of self loading from the chipper and easy unloading of chips make in-forest chipping very attractive.

4. Comparing the energy consumption in these systems with the man-axe-horse combination of the past, where about 8 man hours and two horsepower hours produced one cord of wood ready for transport, shows one of the problems of mechanization. If an overall efficiency of 20% is assumed for the animal power units involved, the energy required to prepare the wood for transport to the user would be less than 30,000 Btu/ton of dry material. This compares with about 200,000 Btu/ton for the same operations in mechanized systems. The same order of increase in energy consumption per unit of production can be found in mechanized agriculture (6). However, the comparison of energy use to energy yield is still very favorable. A ton of dry material has a gross energy content of about 16 million Btu. Even allowing for the fact that each ton of dry matter is delivered in the form of green wood containing approximately 50% moisture, i.e. with a ton of water to be evaporated per ton of dry material, the net energy available will exceed 14 million Btu/ton of dry material. On this basis the energy used in processing the wood represents less than 4% of the energy available from the wood.

5. The energy input to wood production in the form of equipment manufacture is fairly small in relation to energy for operating the equipment. Manufacturing energy subsidy is less than 20% of the total energy input per ton of wood for all of the equipment in Table I and averages around 10%.

It would certainly appear that fuel used to manufacture and operate machinery to produce wood for fuel would be energy well used. However, it must be remembered that use of wood for fuel, as currently harvested, would compete with other wood uses such as for paper and lumber. In all probability, any large-scale use of wood for fuel will need to come from an increase in production over and above current needs.

The most obvious source of additional wood is in the parts of the tree now discarded — the branches and tops — along with undersized and other undesirable trees. This material probably represents around 20% of the growth on land now harvested, i.e., on land which has a road system already developed and paid for by other forest products. The branch material and small trees will probably need to be chipped as early in the harvesting process as possible to reduce bulk and provide an easily handled product.

Two basic methods of handling the branch material are possible. One would be to skid whole trees to the landing, use a delimber in a stationary position and chip anything stripped off the boles. Skidding whole trees would be very little different from skidding delimbed material, but experience has shown that up to half of the branches are broken off as the trees are skidded out. Feeding the stripped branches into a chipper need be no more energy consuming than feeding tree length logs. The second system might use a delimber at the stump and leave the branches and undesirable wood at the growth site. Some work has been reported from Finland (3) on this possibility. Small bulldozers or wheeled loaders were used to pile up the branch material which was then brought out by a skidder/forwarder for processing at the landing or a later stage. Performance figures from this experimental operation are included in Table I.

Table III compares the additional energy inputs needed to obtain these harvesting residues. Once again, it is apparent that the wood fuel can be delivered to a consumer for less than 5% of its energy content. The more economical method unfortunately loses a good percentage of the branch material. This leads to the consideration of wood production specifically for fuel. It is generally accepted that in Northern areas, growth to maturity averages about 1 ton of dry matter/ac-yr. However, Ribe (5) has shown that more than two times the wood present at harvest of a mature stand has grown, died in the competition for sunlight, and rotted away during the growth of the stand. This indicates that visiting each site perhaps twice during the growing cycle to remove dead wood and to thin too-dense areas could increase total yields of wood by perhaps 100%. Much of the material obtained would probably be "fuel grade". However, the economics of such a practice are unknown and the question of what effect removal of such quantities of material might have on the available nutrient pool in the soil is certainly important.

A further possibility for wood fuel production is for intensive short-rotation forestry when small trees might be harvested every five or ten years with a mobile mower/chipper laid out similarly to a grain combine. Such a machine might be expected to cover one acre per hour for a throughput of about 20

Table III. ENERGY USE IN HARVESTING FOREST RESIDUES FOR FUEL

Whole trees skidded, Delimbed at landing[a]	Btu/ton dry wood
Additional energy cost of skidding	negligible
Chipping	75,000
Transport	190,000
Unload	negligible
Auxiliary activities	59,000
Total	324,000

Residues prebunched in stump area, Forwarder used in transport to landing	
Prebunching residues	71,000
Forwarding	126,000
Chipping	75,000
Transport	190,000
Unload	negligible
Auxiliary activities	59,000
Total	521,000

[a] This system probably loses half the available material in skidding.

Table IV. PROBABLE ENERGY REQUIREMENTS FOR A SHORT ROTATION WOOD FUEL CROP

Assumption

Cultivate and plant at 20 year intervals — 6 gal fuel/ac-planting.
Growth rate — 5 ton/ac-yr.
Fertilizer — 1000 lb nitrogen/ac-yr @ 33,000 Btu/lb manufacturing and application cost.
Harvesting — equivalent to present chipping in energy cost.
Transport to truck or stockpile — equivalent to skidding.
Loading trucks from stock pile or primary transport — equivalent to tree length loading.

Energy Use Estimates	Btu/ton dry wood
Cultivation and planting	8,000
Fertilization	660,000
Harvesting	75,000
Transport to stockpile	99,000
Load trucks	47,000
Transport to user	190,000
Unload	negligible
Auxiliary operations	59,000
Total	1,138,000

tons of wood. There are distinct engineering economies to this type of machine where each component performs its function the whole time, for example, the mowing mechanism mows continuously and the chipper is continuously loaded. Equipment for full-size tree handling operates intermittently. The shear on a feller buncher shears the tree and then is out of use until the tree has been lifted and bunched by the other parts of the machine.

Fertilization of fast growing species in a short-rotation system could produce annual yields of around 5 or 6 tons of dry matter. The use of species which grow up from existing root systems could provide very fast regeneration after harvest, though wood from such species might be of too low quality for use other than as fuel. Replanting might be necessary only after four or five harvesting cycles — perhaps only every 20 years. Assumptions and energy cost estimates for such a system are given in Table IV.

The intensified production, as in agriculture, results in a greater energy cost per unit of production, with approximately half the energy input accounted for by fertilizer. Omission of the fertilizer would probably reduce the annual yield to around 2-3 tons/ac, but would bring the energy cost per unit in line with long-rotation systems. It is interesting to speculate what might be done to fertilize intensive energy farms with garbage and sewage sludge. Actual field experiments would be well worthwhile. However, even with full fertilization, wood fuel from short-rotation systems can probably be produced at an energy cost not exceeding 7% of its energy content.

In summary, it can be said that the energetics of wood fuel are very attractive. The fuel itself has many desirable qualities — it contains practically no sulphur, only about 1% ash, can be burned cleanly, is reasonably compact (about 100,000 Btu/ft^3 in chip form), and represents a renewable energy source. Nevertheless, economics will decide the acceptability of wood fuel. A material as versatile as wood clearly has several competitive uses.

However, it is also interesting to note that if wood available for fuel use could be increased to the 146 million cord annual level of the year 1880, the energy content would be equivalent to almost 600 million barrels of oil per year. Unfortunately, this only represents a 30-35 day oil supply at current consumption levels. Wood fuel will not solve the national energy problem though it may make significant contributions in some regions.

REFERENCES

1. "Fuel Requirement for Harvesting Pulpwood", American Pulpwood Association: 1975.

2. Berg, C. *Science* **1973** *181*, 128-38.

3. "Bunching and Transportation of Branch Raw Material"; Folia Forestalia 237, Finnish Forest Institute, 1975.

4. Huff, E. R.; Riley, J. G.; Smyth, D. "Modern Residential Heating with Wood Chips"; ASAE Paper 76-4555, 1976.

5. Ribe, J. H. Orono, Maine, 1974, LSA Expt. Sta., University of Maine, Misc. Report 160.

6. Smith, N. "Engineering A Food Supply", ASAE Paper NA70-402, 1970.

7. Smith, N. *Can. Agric. Eng.* **1974** *316* (1).

8. Szego, G. C.; Kemp, C. C. *Chemtech,* **1973** *May,* 275-84.

9. "Maine Methanol", Office of Energy Resources, Augusta, Maine, 1975.

OTHER REFERENCES

1. Bradley, D. P. International Union of Forest Research Organizations, Working Party, S3.04.01 Proceedings — Simulation Techniques in Forest Operational Planning and Control, Wageningen, The Netherlands (Agricultural University), 1978; 137-45.

2. Roberts, D.; Corcoran, T. International Union of Forest Research Organizations, Working Parry S3.04.01 Proceedings — Simulation Techniques in Forest Operational Planning and Control, Wageningen, The Netherlands (Agricultural University), 1978; 295-307.

3. Smith, N.; Riley, J. G.; Hill, R. C. "Solar Energy Storage for the Northeast", Paper No. 78-4052, 1978.

4. "Forest Residues Energy Program," 1978, Forest Service USDA Final Report, N. Central For. Exp. Sta., St. Paul, Minn.

5. Riley, J. G.; Smith, N. ASAE Paper 77-4018, 1977.

6. Riley, J. G.; Smith, N. Proc. Third Annual UMR-DNR Conference on Energy, University of Missouri, Rola, MO, 1977.

7. Shottafer, J. E., et al. "Utilization of Low Grade Hardwoods for Fuel in the Washington and Hancock Counties of Maine", School of Forest Resources, Forest Products Note No. 3, 1977, p 98.

8. Houghton, J. E.; Johnson, L. R. *For. Prod. J.* **1976** *26* (4), 15-18.

9. Riley, J. G. ASAE Paper NA76-101, 1976.

10. "The Feasibility of Utilizing Forest Residues for Energy and Chemicals" A Report to the National Science Foundation and Federal Energy Administration, Rann Research Assoc., Forest Service - USDA, 1976.

11. Erickson, J. R. *AICHE Symposium Series* **1975** *71* (146) 27-29.

RECEIVED JUNE 18, 1980.

Silvicultural Systems for the Energy Efficient Production of Fuel Biomass

F. THOMAS LEDIG[1]

Yale University, Greeley Memorial Laboratory, 370 Prospect Street, New Haven, CT 06511

The price of imported crude oil delivered to the United States increased 235% on a constant dollar basis between 1960 and 1976 (9). Most of this increase occurred between 1973 and 1974 when the actual price of crude oil went from $4.00/barrel to $12.52. An additional increase of 14.5% has been announced by the OPEC oil ministers, raising prices to $14.54 by October 1979 (13). Given the continued high rate of inflation, it is a certainty that OPEC will find it desirable to keep pace by future price increases, and should it be politically expedient to the Arab world, prices could again rise dramatically. Increasing prices for oil and the threat of arbitrarily restricted supply are incentives to explore alternative energy sources, particularly those under domestic control. It is likely that energy requirements can only be met by combined development of several energy sources.

Among alternative energy sources, the production of biomass from agricultural crops and forest trees has much promise. However, total U.S. energy consumption in 1975 was 71 Q (Q or quad = 10^{15} Btu), but total annual production of biomass from crop and forest land based on current management practice has been estimated by one analyst to be 21 Q (6). Better than 50 percent of the total production was from agricultural land, and because food has a higher priority than energy, prime agricultural land will probably contribute little to a solution of the energy problem. It appears that forest biomass cannot singularly alleviate the U.S. energy shortages, but

[1] Current address: Institute of Forest Genetics, U.S. Forest Service, P.O. Box 245, Berkeley, CA 94701

0097-6156/81/0144-0447$05.00/0
© 1981 American Chemical Society

nevertheless, it can play an important role. The use of forests for fuel has several significant environmental advantages. Forests are aesthetically pleasing, and fuel plantations can double as sites for leisure recreation, contribute to the maintenance of world geochemical cycles, and when consumed for fuel, contribute minimally to atmospheric pollution because biomass is low in sulfur content.

The low annual production of 9 Q from U.S. forest land compared to 12 Q from agricultural land reflects a lack of management and minimal care of forests. It is certain that production could be substantially increased by proper choice of species and strain, cultivation, control of spacing, and fertilization (15). But, intensive culture will result in increased energy demands. For example, reliance on natural regeneration requires no expenditure of outside energy except for harvest, but planting uses 1.2 to 1.5 × 10^6 Btu/ha (5).

Table I. ENERGY REQUIREMENTS IN FOREST OPERATIONS[5]

Operation	Btu ha^{-1} × 10^{-3}
Seed collection	281.21[a]
Nursery propagation	0.14[b]
Site preparation	
Burning: hand	335.92
Burning: helicopter	1,284.40
KG blade, pile and burn slash	3,267.81
Chopping, crushing or ripping w/o piling	2,237.82
Bedding[c]	1,071.99
Planting	
Hand	1,541.28
Machine	1,217.71
Aerial spraying	1,729.00
Stocking control	834.86
Fertilization	37,529.18

[a] For quantity sufficient to plant 2223 seedlings at 8000 plantable seedlings/lb.
[b] For 2223 seedlings/ha.
[c] Plowing to create furrows and mounds so that the mounds are raised above high water table.

Fertilization would claim an additional expenditure of 375 × 10^6 Btu/ha, and seed production, nursery production of seedlings, site preparation, and control of stocking all require energy input (Table I). In agriculture, industrialization or mechanization has increased yields, but on an equivalent energy basis the increase in yield is less than the overall increase in inputs. The energy ratio actually declined from 1945 to 1970 (19). It is important to determine whether inputs are so high that intensive culture of forests for fuel biomass is impractical or whether increased inputs can establish forest biomass as a significant component of the total energy picture. Two factors to examine are: 1) energy efficiency or energy output/energy input and 2) net energy yields under different intensities of management.

PRODUCTION EFFICIENCY

Certainly, one of the most energy-efficient systems for the production of biomass is a traditional silvicultural system using natural regeneration. Natural regeneration is forest regeneration either from stump sprouting, seedling or saplings already present when the mature stand was harvested, germination of seed that lay dormant in the litter, or seeding from surrounding forest without the inputs of site preparation, planting or cultivation. Neglecting transport, such systems would generate energy costs only for harvesting, which are assumed to be 5.16 gal oil/cord of wood (1). On the average hectare of commercial forest land in New England, there are 60 cords when both branches and stem are included. On this basis, gross energy yield can be calculated as 1,642 million Btu/ha (Table II). Based on net annual growth in New England, about 62 years would be necessary to achieve this yield. The ratio of energy yields to energy costs would be 37 compared to about 5.3 for the production of silage corn, one of the most energy-efficient forms of agriculture(12).

Table II. ENERGY YIELD FROM ONE HECTARE OF COMMERCIAL FOREST LAND IN NEW ENGLAND

Cubic feet of wood in tree stems 5 inches diameter and greater[a]	3,522
Allowance for branches and foliage as 30% of total	1,509
Total cubic feet	5,031
Equivalent in cords[b]	60
Equivalent in pounds dry weight[c]	193,190
Equivalent in Btu[d]	$1,642 \times 10^6$
Cost of harvest in gallons of oil[e]	310
Equivalent cost in Btu[f]	44×10^6
Energy output/energy input	37

[a] From ref. 18.
[b] At 0.012 cord/cubic foot
[c] At 3200 lb/cord
[d] At 8500 Btu/lb.
[e] From ref. 1.
[f] At 6×10^6 Btu/barrel

On industrial lands devoted to the production of wood products, a more intensive system is applied. Rather than risk failure of natural regeneration or lose productive capacity while waiting for natural seed years, most commercial operations resort to planting. Planting also provides better control of species composition than is obtainable with natural regeneration. In this case, sites are intensively prepared by mechanical means, seedlings produced in nurseries are planted at uniform spacing, and the site is fertilized. For a 30-year rotation, the cycle between regeneration and harvest, Smith and Johnson (24) calculated an energy ratio of 22 for traditionally harvestable products, which is still much better than our most efficient agricultural systems. When branches and foliage are included as 30% of total above-ground biomass, a conservative figure, the ratio is 34.8. For Eucalypt species under management in Australia, the ratio was 20:1 which surpassed cassava and kenaf but was equalled by elephant grass on a total crop yield basis (16). Similar values were calculated for Douglas-fir and loblolly pine (Table III); i.e. 7.6 to 17.4 (5), but these include only the traditionally harvested products, ignoring branch and foliage biomass.

Table III. ENERGY COSTS AND YIELDS FOR THE INTENSIVE MANAGEMENT OF DOUGLAS-FIR AND LOBLOLLY PINE[5]

Costs	Btu ha^{-1} yr^{-1} × 10^{-3}		
	Douglas-Fir[a]	Loblolly pine[b]	
		High Intensity	Low Intensity
Site preparation	25.7	140.9	140.9
Planting	44.5	88.9	61.7
Aerial spraying	84.0	84.0	— —
Stocking control	16.7	33.4	— —
Thinning	— —	1,153.4	— —
Fertilization	4,503.5	4,503.5	— —
Fire protection	9.7	9.7	9.7
Harvest[c]	7,647.3	4,621.3	3,007.1
	Btu ha^{-1} yr^{-1} × 10^{-6}		
Total energy consumption	12.33	10.64	3.22
Total energy yield	142.07	77.85	55.92
Net energy yield	129.74	67.21	52.70

[a] 50-year rotation
[b] 25-year rotation
[c] Assuming 8,746 Btu/cubic foot (18)

Finally, consider still more intensive systems, variously called short-rotation, intensive culture (27) or mini-rotation systems (21). Such schemes employ species that sprout after cutting, such as poplars, sycamore, or willows, so regeneration is automatic after each harvest (4, 25). The system of regenerating a stand by stump sprouting is known as coppicing. Spacing of trees is very close, less than 12 × 12 dm, and rotations are between 2 to 10 years, depending upon spacing. Site preparation before planting is intense and is followed up by cultivation to control weeds, both herbaceous and woody. Harvesting employs techniques and equipment similar to that used in production of silage corn, so total above-ground biomass is harvested. Fertilization is mandatory to preserve site productivity, and irrigation has been advised. One calculation (Table IV) of energy efficiency in these systems indicated a ratio between 11.2 to 12.6 in Wisconsin for the first rotation (28). Efficiency would be higher for second rotations of sprout origin because the root system has already been fabricated and is in place, so that growth of above-ground portions is greatly accelerated. Other authors (10) suggest efficiency may be higher in southern latitudes; e.g. 13.4 in Pennsylvania (Table V) and 15.3 in Louisiana (Table VI).

Table IV. ENERGY COSTS AND YIELDS FOR SHORT ROTATION, INTENSIVE CULTURE OF POPLAR AND JACK PINES[28]

Costs	Btu ha^{-1} yr^{-1} × 10^{-3} Poplar	Jack Pine
Fuel for operations	6,940	5,070
Manufacture, transport, and application of fertilizer	11,640	5,080
Manufacture of irrigation system	2,040	2,040
Fuel for irrigation	5,220	5,220
Plant propagation	20	250
Other inputs	130	130
Chipping	1,330	930
	Btu ha^{-1} yr^{-1} × 10^{-6}	
Total energy consumption	27.32	18.72
Total energy yield	306.40	236.40
Net energy yield	279.08	217.68

Table V. ENERGY COSTS AND YIELDS FOR SHORT-ROTATION, INTENSIVE CULTURE OF POPLAR AND NATURALLY REGENERATED FOREST[2]

Costs	Btu ha^{-1} yr^{-1} × 10^{-3} Intensive Culture	Natural Forest
Fertilizer	9,962	— —
Growing and harvesting	4,751	5,394
Chipping	896	268
Other	211	48
	Btu ha^{-1} yr^{-1} × 10^{-6}	
Total energy consumption	15.82	5.71
Total energy yield	212.52	63.57
Net energy yield	196.70	57.86

Table VI. ENERGY COSTS AND YIELDS, NEGLECTING TRANSPORTATION, FOR SHORT ROTATION, INTENSIVE CULTURE OF POPLAR ON "ENERGY FARMS" IN WISCONSIN AND LOUISIANA[10]

Costs	Btu ha^{-1} yr^{-1} × 10^{-3} Wisconsin	Louisiana
Supervision	82	82
Field supply	26	26
Harvesters	466	389
Tractor haul	269	271
Irrigation move	64	32
Irrigation pumping	11,047	5,524
Manufacture of urea	4,908	5,055
Manufacture of P_2O_5	397	404
Manufacture of K_2O	762	344
Ground operations	57	30
Aircraft operations	7	8
Fertilizer transport	48	32

	Btu ha^{-1} yr^{-1} × 10^{-6}	
Total energy consumption	18.13	30.46
Total energy yield	191.76	473.21
Net energy yield	173.63	442.75

For each level of silvicultural intensity, the reported values indicate the production of forest biomass is highly energy-efficient. However, it should be observed that the calculations in Tables II-VI do not take into account processing of the biomass, notably drying, and transport, which can be highly variable.

NET ENERGY YIELDS

Efficiency is generally highest with minimal intensity of culture, leaving the impression that intensive culture should be avoided in the production of forest fuels. But that is not necessarily the case. Net energy yield should be the primary concern in biomass fuel production, not production efficiency. In a study of biomass production in short-rotation poplar and jack pine plantations, Zavitkovski (27) tentatively point out that net energy yield increased as a function of input (Figure I). The example of naturally regenerated forest presented here seems to fit the pattern well. Based on a net annual growth of 81 cubic feet ha^{-1} (18), an average rotation for a stand

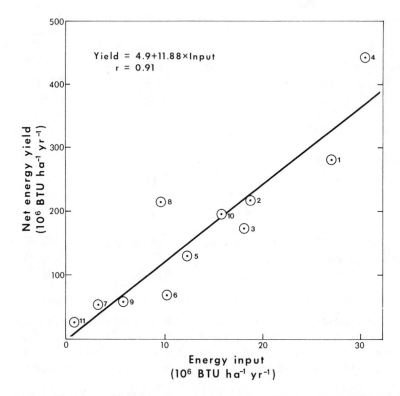

Figure 1. Relationship between net energy yield and energy input for short-rota-tion intensive culture of hybrid (1) poplar and (2) jack pine, energy farms in (3) Wisconsin and (4) Louisiana, intensively cultured (5) Douglas fir and (6) loblolly pine, (7) loblolly pine under average intensity of management, (8) intensively man-aged southern pine, (9) short-rotation, non-irrigated poplar, (10) short-rotation natural forest in Pennsylvania and (11) natural forest in New England. Energy yields are for dry wood, and input does not include drying. Branches and foliage not in-cluded in 5, 6, 7. (2, 5, 10, 24, 28)

of 5,031 cubic feet/ha would be 62 years, the figure used in calculating energy yield per hectare per year in Figure I. On the other hand, yields from the scheme proposed by Smith and Johnson (24) are high. The reason seems to be an estimate of annual growth which is higher than that achieved on most sites. Values for poplar in Louisiana (10) are also high, perhaps, reflecting real and significant regional and species variation in productivity.

The important point is that for the present rate of biomass production from unmanaged stands, a land area the size of the entire United States would not satisfy its energy demands, but an area only one-third its size would be sufficient with the short-rotation, intensive culture scheme of Zavitkovski (27).

FUEL CONVERSION

All of the foregoing, including Tables II-VI, is predicated on the basis of direct use of biomass as fuel, with no provision for drying. In fact, the greatest demand is for gaseous or liquid fuels so conversion to alcohol, methane, or pyrolytic oil is a more likely fate for biomass. Because water content of green wood is roughly 50%, a great deal of its chemical energy will be required for drying. For short-rotation intensive culture schemes, drying would reduce the efficiency ratio to 4.3 to 4.7 (28). Furthermore, to break down lignocellulose to fermentable sugars is energetically expensive. Therefore, crops like cassava and sugar cane with high sugar content can be converted to alcohol more efficiently than wood. The efficiency of production and conversion of eucalypts to alcohol is so low that it is uneconomic, and to pyrolytic oil it is 58%, equal to but no better than cereal straw (16). The high efficiency of forests for biomass production is offset by the low efficiency of conversion to liquid and gaseous fuels. The major advantage of forest fuels is in direct use for generation of heat or power. Nevertheless, the net energy yield of 145 GJ/ha for pyrolytic oil from eucalyptus in Australia is second only to the 154 GJ/ha for alcohol for cassava (3). So even after conversion, trees remain a competitive source of liquid fuels.

Biomass will probably never supply the entire energy needs of a developed economy, even locally, but may be one component of a comprehensive program. It is more likely that forest trees will be used for energy production than agricultural crops because of the relative shortage of good agricultural land that will support the growth of crops like sugar cane or cassava and permit use of the mechanized technology necessary to their efficient culture. Good agricultural land has a higher value for food production. Therefore, despite the low conversion efficiency of wood, it will be the major component of many biomass for fuel programs.

IMPROVING EFFICIENCY

How do trees achieve an advantage in energy efficiency of biomass production compared to crops such as corn, which we are accustomed to thinking of as a physiologically "efficient" plant, employing the C4-pathway of carbon fixation? Part of the answer is that silviculture is less intensive than agriculture, so energy input is lower. But there are also differences in output on a yearly basis. Perennial tree species enjoy a longer growing season than annual crops. While their rates of carbon dioxide fixation are generally lower than those of highly productive annuals, trees maintain their activity for a longer period, and because of their permanent branch structure, can rapdily deploy foliage in the spring to capture a large fraction of the incident radiation. In annuals the process of crown closure must begin all over again each spring, allowing much of the energy flux to fall between plants.

Evergreen conifers are particularly productive over long rotations, surpassing deciduous angiosperms (14), despite the lower unit leaf rates of carbon dioxide fixation often found in conifers. Evergreens can take rapid advantage of suitable conditions for growth in spring or fall because their foliage is always displayed. But in fact, their greatest advantage lies in lower yearly costs of foliage production. Evergreen foliage remains functional from two to several years, depending upon species, whereas deciduous trees must reinvest in an entire new canopy every year. For example, with the same annual investment in leaf, Norway spruce has two to three times greater carbon dioxide uptake than the deciduous species, European beech, because spruce retain its leaves for five to seven years (22). Nevertheless, the efficiency of energy production and net energy yields are much less than the theoretical limit, even in conifers. One possibility for improvement is through breeding. Only in the last two decades has there been major effort in incorporating improved varieties or lines into forest plantings, but none of these lines has been selected for energy production. Some reviewers have suggested genetic improvement or environmental manipulation of energy content in tree biomass (8, 26, 28); for example, by stimulating the production of hydrocarbons such as oleoresins and latex, more highly reduced compounds than structural carbohydrates like cellulose and lignin. Since 1945, latex yield in rubber trees was increased nearly six-fold by breeding (8). But a trade-off of one compound for a more highly energetic one will not by itself increase the net energy yield per hectare. A pine can store either one gram of glucose with a caloric value of 3.7 kcal or use the glucose to produce a more reduced compound, such as lipid. For example, 0.32 g of palmitic acid with a caloric value of 3 kcal can be produced from 1.0 g glucose (Table VII). The actual loss of stored energy occurs because some of the glucose must be respired to generate the ATP required in the construction of the lipid (17). Very little of the energy is lost in going from hexose sugars to cellulose; much is lost in production of lipids, oleoresins, and other highly reduced compounds. The only possible value of breeding for energy-rich compounds may be to reduce handling and storage costs. Concentration of the same heating value in a smaller package might result in some economy, but it seems doubtful that it would offset the caloric loss.

Table VII. LIPID YIELD FROM GLUCOSE AS CARBON SKELETON AND ENERGY SOURCE

Production value = wt. of lipid/wt. of substrate for C-skeleton (19)	0.351
Energy requirement factor = mols ATP to synthesize gram of lipid (19)	0.05097
Yield of ATP from oxidation of 1 mol glucose to CO_2 and H_2O (17)	36
Molecular wt. of glucose (29)	180.16
Heat of combustion of one mol glucose in k cal (29)	673
Heat of combustion of one g glucose in kcal	3.7
Weight of glucose in C-skeleton of p grams of lipid	p/0.351
Weight of glucose needed to supply energy for synthesis	0.05097p (180.16/36)
One g glucose → 0.32 g lipid	
Molecular wt. of palmitic acid, a lipid (29)	254
Heat of combustion of one mol palmitic acid in kcal (29)	2398.4
Heat of combustion of 0.32 g palmitic acid in kcal	3.0

The only way to increase energy yield per hectare is to increase energy capture and the efficiency of conversion to carbohydrate. A more rapid-growing stand, or aggregation of trees, will be more productive of energy, assuming its chemical composition is unchanged, than a slow-growing stand. Breeding programs in southern pines have been successful in increasing volume production about 15% in the first generation and increases of 25% seem reasonable for the second generation (B. J. Zobel, pers. commun.

1978). The physiological basis for these gains is unknown and could reflect better utilization of incident solar radiation because of improved canopy architecture, ability to utilize more of the growing season, and enhanced rate carbon dioxide absorption and fixation, a lower rate of respiration for tissue maintenance, or a reduction in losses to pathogens and stress factors of a nonacute type. Maximum observed efficiency for photosynthesis of poplar under intensive culture was 3.5% of the visible spectrum (27), for Serbian spruce 7.9% (15), for maize 10.9% (7). The goal of breeders is to increase energy yields by reaching the maximum theoretical efficiency of 12% (15).

On the other side of the coin, breeding could help to reduce dependence on energy input, thereby also increasing net energy yield and substantially improving the efficiency ratio. Fertilization is one of the most costly items in silvicultural schemes for fuel biomass production. Manufacture and transport of fertilizers in the short-rotation, intensive culture scheme of Zavitkovski (27) could account for nearly half of the total energy input. In Smith and Johnson's scheme, (24) 70% of the total input for site preparation and cultivation was for fertilization. Most or all of these inputs are related to nitrogen fertilization. If breeding could develop types less dependent upon mineral fertilizers, it would result in a major improvement in the energy balance sheet. In fact, there are major differences in genetic response to fertilization, and often the genotypes most responsive to fertilization are those that are poorest without fertilization (11). Some genotypes have a relatively stable performance, often equal to the fertilizer-responsive types when fertilized and being greatly superior when grown with less than optimum levels of nitrogen.

Another but more remote possibility for improvement is through incorporation of new symbionts or new genes for nitrogen fixation in woody plants. Some plants important to agriculture have nitrifying bacteria, housed in special root structures, that extract gaseous nitrogen from the atmosphere. The nitrogen soon appears fixed in amino and amide forms. Legumes like alfalfa are the most notable examples of plants that are host to nitrogen-fixing bacteria, and they are often alternated with other crops to maintain and improve soil fertility. Among woody perennials, alders are capable of converting atmospheric nitrogen. Through genetic engineering, it may some day be possible to improve the growth of other agricultural and forest crops by the addition of nitrifying capacity. In the meantime, proper choice of species or breeding of trees which are efficient scavengers of soil nitrogen may reduce the need for fertilization.

WOOD PRODUCTS AND RESIDUALS

Irrespective of how efficiently forests can capture energy is the question of whether wood can be economically sold as fuel. All projections suggest a continued increase in demand for fiber and solid products which will be supplied from a dwindling land base. Not only could wood increase in value relative to a substitute like coal, but the energy balance might be enhanced more by promoting wood as a material than as a fuel source. For example, for every Btu expended in constructing a house of wood, there would be 6 Btu for steel or 25 Btu for aluminum (23). Of course, this is true for furniture or any other manufactured items in which material substitutions are possible. Thus, the best energy value of wood might be its utility for a diversity of products that are frequently made from more energetically costly materials.

As a byproduct of wood processing, extensive residuals are produced. Slabs, the rounded shell outside the sawn boards, always constituted a high proportion of the log, but the trend to shorter, economic rather than biologic rotations forced harvest of smaller trees, and increased the proportion of slab to board. These residuals are already salvaged by large mills. They are chipped and sold for pulping or used to supply heat and energy for mill operation. On a local scale, use of mill residuals may have a major impact.

Another class of residuals, and one not frequently used, includes the branches, twigs, leaves, and roots left in the forest. Branches and leaves may constitute about 35% of the total biomass or an amount equal to the stem biomass (28). Chipping the tops in the forest is quite practical and could have an impact on local fuel needs. Roots represent 20% of the above-ground biomass and constitute another source of fuel. Mechanized systems for root extraction are available, but the impact of root extraction on soil structure and site productivity may be unfavorable and, of course, could not be used in systems of coppice regeneration.

CONCLUSIONS

Production of biomass by forests is highly energy efficient. Purely exploitative schemes are more efficient than highly intensive silviculture. However, net energy yield increases with intensity of cultivation, so silvicultural systems approaching those of agricultural cropping should be favored from an energy production standpoint. Efficiency can be further increased by breeding, an area neglected in forestry for centuries after it had become a proven assist in agriculture. The rate of production of biomass can be increased by breeding for rapid growth. Simultaneously, it may be possible to reduce energy inputs by breeding for trees that do not require supplemental fertilization or by engineering new symbiotic relationships with nitrogen-fixing organisms.

Though production of forest biomass is efficient, its conversion to gaseous or liquid fuels is not. Cellulose and lignin are more difficult to convert to alcohol or methane than sugars, a major component of biomass in some crop plants. Therefore, trees will probably be used directly for burning or perhaps in pyrolitic conversion, a process which holds some promise. Merely increasing reliance on wood in construction will have a positive effect on world energy budgets because production of substitutes requires a high energy expenditure. Production of wood products inevitably produces residuals. These can and are being used in energy production and may have a major impact on a local scale.

REFERENCES

1. "Fuel Requirements for Harvesting Pulpwood"; American Pulpwood Association: Washington, D.C., 1975; p 15.

2. Blankenhorn, P. R.; Bowersox, T. W.; Murphy, W. K. *Tappi* **1978**, *61*, 57-60.

3. Boardman, N. K. "Proceedings", Fourth International Congress on Photosynthesis; Biochemical Society: London, 1977; 635-44.

4. Bowersox, T. W.; Ward, W. W. *J. For.* **1976**, *74*, 750-3.

5. Burks, J. E. "Proceedings"; 1978 Joint Conv. Soc. Am. For. and Can. Inst. For.; Soc. Am. For.: Washington, D.C., 1979; 146-48.

6. Burwell, C. C. *Science* **1978**, *199*, 1041-8.

7. Caldwell, M. A.; Cooper, J. P., Ed. "Photosynthesis and Productivity in Different Environments"; Cambridge Univ. Press: Cambridge, 1975; 41-73.

8. Calvin, M. *Science* **1974**, *184*, 375-81.

9. Federal Energy Administration. "Energy in Focus, Basic Data"; U.S. Federal Energy Administration: Washington, D.C., 1977; p 13.

10. Fege, A. S.; Inman, R. E.; Salo, D. J. *J. For.* **1979**, *77*, 358-61.

11. Goodard, R. E., et al., Eds. "Tree Physiology and Yield Improvement"; Academic Press: London, 1976; pp 449-62.

12.   Heichel, G. H. *Am. Sci.* **1976,** *64,* 64-72.

13.   Ibrahim, Y. M. *N.Y. Times* **1978,** *128* (44-070), A1, D6.

14.   Kira, T.; Cooper, J. P., Ed. "Photosynthesis and Productivity in Different Environments"; Cambridge University Press: Cambridge, 1975; 5-40.

15.   Ledig, F. T.; Linzer, D.I.H. *Chemtech* **1978,** *8,* 18-27.

16.   McCann, D. J.; Saddler, H.D.W. *Search* **1976,** *7,* 17-23.

17.   Noggle, G.R.; Fritz, G. J. "Introductory Plant Physiology"; Prentice-Hall: Englewood Cliffs, N.J., 1976.

18.   Pecoraro, J. M.; Chase, R.; Fairbank, P.; Meister, R. New England Federal Regional Council, Energy Resource Development Task Force, Wood Utilization Group: Boston, Mass., 1977; p 91.

19.   Penning de Vries, F.W.T.; Brunsting, A.H.M.; van Laar, H.H. *J. Theor. Biolo.* **1974,** *45,* 339-77.

20.   Pimentel, D.; Jewell, W. J., Ed. "Energy Agriculture, and Waste Management"; Ann Arbor Sci. Publ.: Ann Arbor, Mich., 1975; pp 5-16.

21.   Schreiner, E. J. *U.S. For. Serv. Res. Pap.* **1970,** *NE-174,* p 32.

22.   Shulze, E.-D; Fuchs, M.; Fuchs, M. I. *Oecologia* **1977,** *30,* 239-48.

23.   Smith, D. M. *Connecticut Woodlands* **1978,** *43* (2), 3-5.

24.   Smith, D. M.; Johnson, E. C. *J. For.* **1977,** *75,* 208-10.

25.   Steinbeck, K.; McAlpine, R. G.; May, J. T. *J. For.* **1972,** *70,* 210-14.

26.   Szego, G. C.; Kemp, C. C. *Chemtech* **1973,** *3,* 275-84.

27.   Zavitkovski, J. "Proceedings", Joint Convention of the Society of American Forestry and Canadian Institute of Forestry; Washington, D.C., 1979; Society of American Forestry: Washington, D.C., 1979; 132-37.

28.   Zavitkovski, J. submitted for publication in *For. Sci.*.

29.   Weast, R. C., Ed. "Handbook of Chemistry and Physics"; Chemical Rubber Co.: Cleveland, Ohio, 1969.

RECEIVED JUNE 18, 1980.

SYSTEMS ANALYSIS

Electric Power Generation from Wood Waste

A Case Study

RICHARD T. SHEAHAN

Hennington, Durham & Richardson, 5454 Wisconsin Avenue,
Washington, DC 20015

The current "Energy Crisis" being experienced worldwide has directed attention to the development of alternate sources of energy. One of those alternatives is wood.

Wood appears to have numerous attractive advantages; it is available and renewable, a "clean" fuel, and has potential positive impacts on the enhancement of good forest management practice. The availability of wood residue exists in all forested areas of the United States, including "Urban" wooded areas. It is renewable because it regenerates in a relatively short time after each harvest cycle unlike fossil fuels. Wood is a relatively "clean" fuel because it contains virtually no sulfur.

Most forested areas in the United States are currently not properly managed. Typical harvesting operations will "high-grade" a forest, or cut mostly the strong and marketable specimens, leaving the weak and "weedy" tree behind. As any gardener knows, if you do not "weed" your garden it will eventually become a weedpatch. This phenomenon occurs in numerous forested areas throughout the country. In most of these areas, there is no environmentally sound method of disposing of the rough and rotten wood residue, nor are there economic incentives for its removal. Utilizing this residue as an energy source can create an environmentally sound and economically viable motivation for "culling-out" and disposing of this

0097-6156/81/0144-0465$05.00/0
© 1981 American Chemical Society

material. Combustion technology for generating electricity from wood wastes is well established; the major problem area is the gathering and transporting of the wood material in an economically acceptable fashion.

A general overview of the requirements necessary to implement a wood residue energy program is presented in this article. A case study of an actual 50 megawatt (MW) wood-fired electric generating plant in Burlington, Vermont, will be presented as a model. Each component necessary to make up the entire wood energy system will be discussed in sufficient detail to assist the reader in understanding the requirements necessary to evaluate any wood energy program, be it thermal, steam, or electric generation. The experience of the Burlington project is based on the results of a conceptual engineering study conducted by Henningson, Durham & Richardson, Inc.(1) The availability of wood residue and supply, its energy characteristics, harvesting methodology, transportation and handling, combustion equipment, institutional and environmental concerns, and economic considerations are discussed.

WOOD SUPPLY

Due to the relative sparsity of timber inventory data it is difficult to accurately determine the total amount of wood residue available in the United States. However, estimates indicate that approximately three percent of the total United States energy demand could possibly be supplied by wood residues (2). There are several sources of wood residues suitable for fuel. A primary criteria is that the material be of a non-commercial nature and have a long-term and reliable supply. The primary sources of wood residue are forest and mill residue. The U.S. Forest Service publishes statistics which can provide the basis for estimating the potential amounts of available wood residues. Other state and regional organizations also publish data which are useful in estimating the quantity of available material(3).

Forest Residue

There are several sources of available wood residues from conventional logging opearations that are normally not utilized on a commercial basis. A large volume of material can be derived from tops, branches, leaves, roots, stumps, etc. which are usually left on the forest floor following a typical saw log harvesting operation. This material can represent from 35 to 45 percent of the volume, and therefore the energy content, of a tree. It can be chipped in the forest as an adjunct to a normal harvesting operation. Removal of some of this waste material can reduce forest fire risk and enhance wild life habitat. Non-commercial species of trees are also available for fuel. These include typical small size, poor form, or inferior quality trees which have little hope of maturing or developing into trees suitable for industrial application. These plants compete for moisture and nutrients with the primary forest. Therefore, prudent removal of this material, which often is categorized as waste, can enhance the health of an overall forest. Another source of wood waste is cull increment and rough and rotten trees. Cull increment refers to the quantity of wood which annually becomes non-commercial, or "cull" material, due to insufficient cutting or overmature timber stands. A certain nongrowing portion of trees in any forested area is classified as standing rough-and-rotten. This material is non-commercial because it is either rotten, broken, or dead. As in any aspect of life, going to extremes is generally incorrect. Likewise, in a forest harvest operation, cutting out too much of the nonmerchantable material is also incorrect. A certain amount of the material must be left behind to replenish the soil nutrients to ensure future forest health and vitality. Therefore, in any wood procurement operation, good forest management practices must be followed to ensure that the proper amount of waste wood material is left to maintain the forest ecological balance.

Mill Residue

Mill residue is a source of wood waste which can be derived from the wood products industry. Depending on the efficiency of a mill, up to fifty percent of the incoming material can become waste material in the form of bark, sawdust, cut slabs, etc. (4). This residue is an excellent source of fuel; however, for a long-term supply, it may dwindle as more emphasis is placed on utilizing it for "in-house" energy uses by the wood products industries.

Supply Evaluation

When evaluating a wood residue supply, certain assumptions must be made to quantify the availability. Data published by the U.S. Forest Service and other organizations are a good starting point for formulating wood residue quantity. However, a thorough understanding of the local harvesting techniques and customs, access to transportation, percent grade of local terrain, land owner attidues, seasonal weather conditions and other considerations must be evaluated. As an example of the latter item, the City of Burlington will have to stockpile sufficient wood residue in the spring and fall due to a shutdown of harvesting operations. In the spring, the "mud season" makes logging roads impassable due to melting snows. In the fall, the hunting season closes down the forest to most harvesting operations. Culling out of the non-merchantable material should actually increase the annual growth rate of a forest because the residual-stock is healthier and faster growing. However, it is most important that the anticipated annual removal rate of wood residue does not exceed the annual new growth rate of a forest.

Burlington Wood Supply

Based on a reasonable energy balance and load factor for a 50 MW wood-fired power plant, it was estimated that the City of Burlington requires approximately 470,000 green tons of wood residue/year to operate their power plant. Several assumptions were made to determine if that quantity of material is available to the City.

- The supply area was assumed to be circular and approximately 80 miles in diameter. Burlington is situated in the western regions of the circular area.

- Minimum parcel size for the harvesting operation was assumed to be 50 acres. This is a very conservative estimate since many harvesting operations take place on holdings below 30 acres in size.

- Timberland that was owned by the forest industry was considered unavailable for competitive purchase of wood.

- Annual growth on state and national forests was considered available.

- It was assumed that wood in the immediate vicinity of Burlington was not available; likewise, it was assumed that no substantial supply of wood from Canada or across Lake Champlain is available.

Based on these assumptions, it was determined that the total wood demand by Burlington is more than adequately supplied from the assumed area.

WOOD FUEL CHARACTERISTICS

When evaluating wood as a fuel, characteristics of the delivered material must be estimated relative to the weighted average of energy contents for the various anticipated species of wood. Important combustion characteristics of wood are its heating value, which is a function of its moisture content and density, and its ash composition. Fluctuations of these values are primarily due to different concentrations of lignin and the presence of extractives in the wood such as resins and tannins. Hardwoods (i.e., oak, maple, etc.) generally have an average high heating value between 8500 and 8600 Btu/oven dry pound of wood. Resin has a much greater heating value than wood (approximately 17,000 Btu/lb). Therefore, soft woods (i.e., mostly pines) which have higher resin contents and proportions of lignin have higher energy contents than hard woods and average approximately 9000 Btu/oven dried pound of wood. These average values vary only 5 to 8 percent depending on specific woods (5). Bark also has a higher energy content than wood. The actual heating value for wood decreases as moisture increases, since water has no heating value. The moisture content of "green" wood, or wood recently harvested and chipped, is approximately 50 percent (on a wet basis). Based on this moisture content, the average high heating values are approximately 4300 Btu/lb for hardwoods and approximately 4500 Btu/lb for softwoods. The ash component is generally considered undesirable since it is inert and not combustible. Ash either remains in the combustion chamber or is entrained with stack gases which may create particulate air emission problems. The average ash content of most woods ranges from 0.1 to 3 percent, with most species averaging less than one percent. A possible increase in the ash content of wood can come from the skidding of harvested trees in the forest which often results in the collection of some dirt and sand on the bark. Unless this material is removed, it can increase the total ash content of the fuel. As a point of comparison, most coals have an average ash content substantially above 5 percent with some reaching the 25 percent level.

Burlington's Fuel Characteistics

Based on a weighted average of the wood specie mix in the supply area, it was estimated that the average high heating value of the wood fuel to Burlington would be approximately 4750 Btu/lb.

HARVESTING TECHNIQUES

Numerous combinations of harvesting scenarios are possible. All basically involve the traditional steps of a normal harvesting and delivery process which includes four separate activities: felling, skidding, yarding, and hauling.

> *Felling* — This step involves the cutting of individual trees. The prevailing felling equipment is the chainsaw; however, more mechanized devices are available and being developed. The feller-buncher is a machine that uses a hydraulic system to hold the standing tree while cutting it near ground level with a mechanical shear. Once cut, the trees are individually laid side by side.

Skidding — The skidding operation involves dragging the logs or trees from their felled position to a general collection site called a landing. This is usually done by large four-wheeled drive, rubber-tried skidders. To a limited extent, skidding is done by steel tracked crawlers or by horses. Skidders usually pull more than one log or tree at a time, holding the leading ends of the logs off the ground by use of steel cables and a winch, or by hydraulic grapple devices.

Yarding — Once the logs or trees have been skidded to the landing, they are prepared for shipment to the wood yard. This process is called yarding. An integrated operation that supplies wood for a power plant would probably skid either long logs or entire trees. At the landing, quality saw logs would be cut, sorted and piled for subsequent loading on large trucks for delivery to the appropriate mill. The balance of the trees or logs may then be chipped and blown into enclosed trailers by whole-tree chippers. A whole-tree chipper can be towed by truck to the landing area and quickly set up for operation. A mechical arm picks up the whole tree or log and feeds one end into a motorized conveyor system which then pushes the material towards a set of high speed, rotating knives. The wood chips that result are generally about the size of matchbooks. If whole-tree chippers are not used, the logs would be cut into convenient lengths for loading into trucks. In a nonchipping operation, the tops and branches would be left behind as waste material.

Hauling — The chips are transported from a landing to a power plant by tractor-trailer. Saw logs on the other hand can be hauled by either straight trucks or tractor-trailers. These log trucks may be equipped with self-loading equipment.

Burlington Harvesting Scenario

In evaluating the cost of harvesting wood fuel, there are infinite combinations of labor and equipment which could be utilized in procuring the wood waste. Many factors must be considered; these include harvesting equipment, manpower requirements, slope of terrain, access to logging trails and transportation roads, haul distance to ultimate use, land owner attitudes and numerous other considerations. In evaluating the cost of fuel delivered to Burlington, three wood fuel production models were examined. The production models selected for evaluation were judged to be fairly representative of methods currently employed in the New England region.

Model Number 1 — Traditional Round Wood - This model exemplifies many small wood harvesting operations and consists of two men and one skidder. One man is responsible for felling and cutting the tree to desired lengths. The other lumberman skids the tree out to a landing. This harvesting system requires a modest capital investment and offers operating flexibility to conform to local conditions. It is the predominant harvesting system currently utilized in Vermont's forest.

Model Number 2 — Chip Harvesting (Moderate Mechanization)- This represents a popular emerging harvesting technology in New England. "In-the-woods" chipping offers advantages of greater resource utilization and reduced transportation cost. Whole trees are felled by chain saws and skidded to a medium size (18 in.) chipper at the landing area. Chips are blown into tractor-trailers for transport to the power plant.

Model Number 3 — Whole Tree Harvesting (Highly Mechanized)- This system utilizes a mechanical feller-buncher to cut trees utilizing multiple skidders to move them to a large (22 in.) chipper located at the landing area. The increased capital investment and higher operating costs combined with problems presented by rugged terrain have to date excluded the general use of feller-bunchers in Vermont. Generally, they are restricted to terrain having slopes of 15 percent or less (6). It is doubtful that feller-buncher operations will become widespread in Vermont's forests due to high capital costs, uncertainty about productivity in certain cuts, and their limited adaptability to small and medium size parcels which are very common in Vermont. However, its potential is very significant in less rugged and sloped terrain and large parcels of forested land.

Economic assumptions and capital and operating costs for the three production models are presented in Tables I and II.

To realistically evaluate the cost of producing wood fuel for the Burlington plant, it was assumed that only proven harvesting technololgy would be utilized. Thus, a feller-buncher operation, although technically possible, was not considered to be a major contributor. It was assumed that the predominant portion of the wood fuel would be supplied by traditional round wood and moderately mechanized chip harvesting operations. It was further assumed that 70 percent of the fuel requirements would be supplied by "in-the-woods" chipping and 30 percent by tradational round wood subsequently chipped at a satellite facility or concentration yard. The weighted average of wood fuel cost (prior to chipping of the round wood portion and transportation of the total portion of the wood fuel) was estimated to be $8.06/ton (1977 dollars).

TABLE I. PRODUCTION MODEL ASSUMPTIONS
(All costs expressed in 1977 dollars)

Productivity:

Estimate based on manufacturer's information and national averages revised to reflect Vermont conditions. Assume 1,800 hr/yr operation.

Labor:

Based on 45 weeks per year at $220/week per person plus 25% payroll benefits.

Fuel and Oil:

Fuel

Diesel @ $0.50/gal

	Consumption
Skidders	5.0 gal/hour
Feller-buncher, crawler tractor	6.0 gal/hour
Chipper — Med. 18"	10.0 gal/hour
Lg. 22"	12.0 gal/hour

Oil

30% of Fuel Cost

Maintenance and Repair:

Hourly Depreciation (HD) = Purchase Price/Expected Life

Assume .70 × HD × 1800 hr/yr = Maintenance and Repair

Skidders — 7,500 hr expected life

Dovers — 10,000 hr expected life

Chippers — $0.50/ton

Financing:

Assume 75% Debt — 12%, 5 years

25% Equity

Depreciation:

Assume 5 years straight line

Stumpage: (Payment to landowner for wood removed)

Assume average $0.75/ton

Taxes:

State and Federal

Federal includes investment tas credit amortized over 5 years.

Profit:

Reflects on assessment of risk-reward factors and varies according to size of capital investment, margins are considered reasonable to achieve desired production levels.

**TABLE II. PRODUCTION MODELS — CAPITAL AND OPERATING COST
SUMMARY
(All costs expressed in 1977 dollars)**

	Traditional Roundwood	Moderately Mechanized Chip Harvesting	Highly Mechanized Chip Harvesting
Productivity (tons/year)	8,440	33,750	54,000
Labor (Men)	2	6	9
Equipment			
Crawler Tractor	— —	1	1
Cable Skidders	1	2	2
Grapple Skidders	— —	— —	2
Feller-Buncher	— —	— —	1
Chipper	— —	1	1
Capital Investment	$44,000	$242,000	$509,000
Revenues			
Annual Wood Sales	$67,112	$273,316	$526,639
Costs			
Labor	$24,750	$74,250	$111,375
Fuel & Oil	5,850	23,000	43,000
Maintenance & Repair	6,700	29,000	51,000
Interest	2,550	14,000	29,600
Depreciation	9,350	48,500	100,000
Stumpage	6,330	25,000	38,000
Miscellaneous	2,775	7,825	12,000
Taxes	3,706	21,676	62,733
Net Profit	5,371	30,065	78,931
Profit on Sales (%)	8	11	15
Return on Investment (%)	12	12	15
Unit Cost ($/ton)	7.95	8.10	9.75

TRANSPORTATION, HANDLING AND PROCESSING

There are estimates that indicate on a net zero energy basis for electric generation (i.e., energy input requirement to produce a comparable electrical output derived from wood chips), "green" wood chips can be hauled by truck for approximately 50 to 100 miles depending on the average heat and moisture content of the wood fuel (7). The comparable distance for rail haul is much greater. However, on an economic basis, rail haul proved to be unacceptably more expensive due to the additional costs associated with loading and unloading of railroad cars. It was assumed that all wood chips would be hauled to the power plant by trucks. Trucking costs for round wood and chips were determined through analysis of rate schedules of wood haulers in the Burlington area. Generally, round wood is more costly to transport than chips due to reduced vehicle pay loads and handling problems. Therefore, a composite average trucking cost reflecting round wood and chip transport was used. Rates were structured to provide incentives for utilization of efficient vehicles and to attract distant supplies. It was assumed that vehicles would average 35 miles-per-hour and carry a payload of from twenty to twenty five tons. The truck transportation costs were determined by applying unit haul costs per mile to the weighted distribution of wooded area within the supply area. The average trucking cost was determined to be $3.43/ton (1977 dollars) from the forest to power plant. An additional cost of $1.67/ton was determined for the cost of chipping the plant's wood fuel requirements derived from the round wood operation. This cost was applied to 30 percent of the total wood supply per the previous assumption.

The following figures indicate the total estimated cost in 1977 dollars for wood fuel procurement for the Burlington Power Plant.

Wood Production Costs	$8.06/ton
Trucking	3.43
Chipping of Round Wood (applied to 30 percent of wood supply)	0.50
	$11.99/ton

FUEL HANDLING AND STORAGE

Chip trucks arriving at the power plant are weighed and then unloaded by hydraulic truck dumpers. Chips flow by gravity from the trucks into live-bottom receiving hoppers; and from there, onto inclined belt conveyors which transport the chips to storage. A mechanical-belt pile-builder distributes the chips evenly around the perimeter of the storage pile. A disk screen and wood pulverizer are provided to reduce oversized material to prevent jamming of material handling systems. Also, a magnetic ferrous recovery system is necessary to recover tramp metal parts which can cause damage to the conveying and combustion systems (8,9). During winter months, vehicles arriving at the unloading area must be carefully inspected to ensure that massive loads of frozen chips are not dumped onto the receiving hoppers thereby creating a bottle neck to subsequent unloading operations. The experience of chip handling facilities in Canada indicates that the only sure prevention is to establish a firm policy against delivery of frozen chips.

Most freezing problems occur when chips are loaded into vans and left to stand over long periods of time prior to delivery. Chips produced in the woods and brought promptly to the power plant should not arrive solidly frozen. The anticipated wood storage pile for the power plant is semi-circular in shape approximately 380 feet in diameter with a height of 40 feet. It contains approximately 42,500 tons of chips or approximately 21 days of fuel supply for the power plant. Two chip dozers work the pile and share responsibility for managing the pile and reclaiming wood. Chip pile management is an important task which includes responsibility for rotation of chip inventory, chip mixing, dust control, and fire prevention. The relatively high moisture content of wood fuel dictates that material be reclaimed on a "first-in first-out" basis to maintain freshness and inhibit chip decomposition. Chip inventory should be completely rotated at least once a year to minimize decomposition. A certain degree of natural decomposition will occur and the potential for spontaneous combustion fires exists. Compaction of the entire pile, especially along the outer perimeter, reduces air flow which can feed "hotspots" in the pile. "Hot-spots", identified by the presence of smoke, should be uncovered and approximately a truck load of dry ice applied and the area recompacted. Carbon dioxide gas is drawn into the "hot" area and causes it to be extinguished. Before large wood chip inventories are accumulated, a quantity supplier of dry ice should be identified.

Wood chips are reclaimed from the storage pile by chip dozers and deposited in reclaim hoppers adjacent to the wood pile. Steel grates above the hoppers prevent frozen chips or oversized objects from jamming conveyors or otherwise fouling handling equipment. Draft conveyors installed beneath the hoppers discharge chips to inclined belt conveyors which elevate the wood fuel to storage bunkers situated above the power plant boilers. Chips are continuously fed to the boiler feeders to provide adequate fuel supply.

ELECTRIC GENERATION SYSTEM

Wood is basically a cellulose fiber and its combustion technology is well established. The paper and pulp industry for years has been burning bark; the lumber industry burns sawdust; and many food processing industries have years of experience in the combustion of cellulosic fiber. Therefore the primary problem in the large scale generation of electricity from wood waste is not combustion, but the gathering and transportation of the wood fuel. The primary concern of wood fuel combustion is that it has a high moisture content which reduces heating values and influences combustion temperatures and other furnace parameters.

The basic fundamentals of wood waste combustion entail three consecutive stages: the evaporation of moisture, the distillation and burning of volatile matter, and the combustion of the fixed carbon. In the furnace combustion chamber, radiation and convective heat input evaporates the wood's moisture and distills the volatile matter. The evaporation of moisture to steam takes approximately 1100 Btu/lb of moisture. Once moisture has been evaporated, heat is absorbed by the fuel particles thus driving off volatile matter. The volatile matter burns in a secondary combustion reaction within the furnace chamber, but external to the actual wood fiber. Finally, the fixed carbon of the wood burns in the primary combustion reaction in conjunction with combustion air.

The most efficient method of wood waste combustion in the 50 MW power plant size proposed for Burlington is by use of a travelling-grate spreader-stoker boiler. A single unit is more efficient than multiple units. Boiler manufacturers indicate that the largest stoker size available limits the input of burning wood in a single unit to approximately 75 ton/hr or 1800 ton/day. Assuming a 75 percent power plant capacity factor, this fuel input equates to a nominal 50 MW capacity. The capacity factor of 75 percent means that the unit will operate on an average of 75 percent of its rated capacity on an annual basis. Thus the plant would average 37,500 kW on an annual basis. Actually the power plant would produce on an average over 40,000 kW/hr for just over 8,000 hr of the year. This allows for 30 days of down time for maintenance during the year. The energy input required to supply this electric generation capacity equates on a Btu basis to approximately 470,000 tons of "green" wood chips per year. The proposed Burlington plant includes a 50 MW condensing turbine generator unit; a 525,000 lb/hr boiler; a complete complement of station auxiliary, mechanical, and electrical equipment; and a power transmission substation. Steam from the boiler is supplied to the turbine at a pressure of 1,250 pounds and temperature of 950°F. A hydraulic ash handling system conveys the ash from the stoker siftings and ash hopper to a system where it is dewatered and trucked to landfill. Wood has an inherently low sulfur content and therefore poses no problem as a source of sulfur dioxide air emission. Wood burns at a lower temperature than fossil fuels; and likewise, has an inherently lower nitrogen content than fossil fuels. As a result, wood combustion produces lower quantities of nitrogen oxides. The primary air emission concern for the wood fired _plant is particulate matter. Mechanical collection and electrostatic precipitation equipment is expected to ensure compliance with particulate and smoke emission standards (12,13). The system operates on a closed cooling system and therefore there is no thermal water discharge to the adjacent river or nearby Lake Champlain.

COST ESTIMATES

Estimates of the capital, operating, and maintenance costs of the proposed 50 MW wood-fired power plant are presented in Table III and are based on conceptual design concepts. All costs are expressed in 1977 dollars. The total construction cost estimate was $46,227,000.

TABLE III. CAPITAL COST ESTIMATE FOR
50 MW WOOD-FIRED POWER PLANT AT BURLINGTON, VERMONT
(All costs expressed as $1,000 in 1977)

Steam Boiler	$8,820
Turbine System	3,870
Mechanical Equipment	3,660
Electrical Equipment	3,300
Piping	2,900
Site Development	1,275
Building — Structural	10,500
Wood Handling System	2,039
Chimney	1,200
Substation, Interconnect	1,010
Contingencies, Engineering, Legal	7,653
Total Capital Cost	$46,227

The proposed power plant will produce a gross of 328,500 MW-hr annually or a net of 302,200 MW-hr (8 percent in-plant use) operating at a 75 percent capacity factor. It will consume approximately 470,000 green tons of wood chips which is 100 percent of the energy input. Based on the previously derived cost of fuel of $11.99/ton, the total annual fuel cost will be approximately $5,635,000. Plant operation was estimated to require a staff of 42 with an annual payroll of $714,000. Annual maintenance, chemical and supply costs, and operation and maintenance for the fuel off loading and handling system was estimated to be $727,000/yr. The estimated total operating and maintenance cost for the power plant is $7,076,000/yr.

Annual Cost Projection

It was assumed that the power plant will be financed from revenue bonding. Therefore, reasonable estimates were made for interest on bonds, interest earned and expended during construction, and bond discounts. Working capital and the debt reserve fund were assumed to be capitalized. By projecting all capital and operating costs with reasonable escalation factors, a life-cycle cost analysis was performed. Results of that analysis shown below indicate an estimate of required revenues to offset all costs. These projected costs are favorable when compared to alternative fossil fuel unit costs projected for the New England region.

Year	Unit Cost (cents/kWhr)
1982	5.1
1984	5.5
1986	5.9
1988	6.4
1990	6.9

INSTITUTIONAL CONSIDERATIONS

According to our analysis, the generation of electricity from wood waste is technically feasible and economically attractive. The most difficult problem in implementing a wood-fired power plant is perceived as being of institutional nature. These concerns are primarily associated with forest management, the economy, and environmental considerations.

Forest Management

A forest is like a garden which needs to be weeded to promote a sound and healthy stand of timber. Currently, there is no large-scale methodology to remove non-commercial weed trees which compete for the supply of water, nutrients, and sunlight within the forest. Development of a long-term waste wood demand for electric generation could create a way to better manage the forest's timber. A basic concern that always accompanies the development of any large forest-based industry is the potential for abuse. It is therefore mandatory that a power plant not be supplied with wood at the expense of degrading or denuding the forest. To alleviate this concern, it is recommended that the personnel responsible for wood acquisition must be qualified professional foresters. In addition to being in charge of wood waste acquisition, they could also be available for private land owner consultation. This is particularly important in Vermont, since 90 percent of the commercial forest acreage is in private ownership. This type of personnel arrangement is similar to the cooperative assistance programs which are commonplace in the pulp and paper industry. The acquisition personnel would be responsible for monitoring wood waste deliveries to avoid wasteful operations. They would have two primary concerns. The first is to ensure that differentiation is made between high quality and low quality material so that quality saw timber and veneer logs are not chipped for fuel wood. The second concern is to mandate that small trees be carefully evaluated before marketing for harvest. Many young trees may become valuable stock if given sufficient time. In Vermont, the primary ownership objectives of timber land are recreation and place of residence; timber production ranks third (14). Therefore, personnel responsible for wood waste procurement should recognize this fact and prescribe and encourage forest management procedures which minimize any disruptions to the owners' values and objectives.

Economy

The operation of a 50 MW wood-fired power plant could have significant and positive impacts on Vermont and the City of Burlington. The following are a few of the potential impacts:

● The creation of a long-term and consistent demand for non-merchantable wood waste could result in dramatically higher yield for land parcels; therefore many areas which previously had marginal harvesting potential could become financially viable.

● The existing forest product industry could experience a long-term general upgrading, faster growing and healthier forest due to the removal of nutrient-depleting waste wood.

● Private land owners could realize an income from the stumpage (price paid for the removal of the wood) which could help offset the effects of property taxes.

● Vermont and the City of Burlington would become more self-sufficient in their energy resources, thus offsetting the requirements for predominantly imported fossil fuels. Likewise the statewide balance of payments would be improved by the reduction of imported energy.

● The economic multiplier effect applied to the money kept in Vermont, plus the general expansion of the forest-based industry could result in a substantial impact on the City and State economies.

Environmental Concerns

Some of the primary environmental concerns associated with the harvesting of wood waste from a forested area include, but are not limited to, soil erosion, nutrient depletion, aesthetic degradation, reduced water quality, and deterioration of wildlife habitat. With prudent harvesting management procedures, such damage need not happen, and in fact, the environment could be enhanced by the process. The forest is always subject to insect attacks, disease infestations, and wild fire. Dense stands of overmature trees are highly susceptible to insect or disease outbreaks. Once established, they can spread easily through such stands. Resistance of the forest depends dramatically on the species composition, size distribution, density, and general health of the tree. If trees are laboring under old age and severe nutrient competition caused by high density, the general forest will most likely be seriously damaged. Once a substantial portion of the timber is dead or rotten, dry summer days can transform the forest into a haven for disease outbreak and spread of fire. Providing a market for low quality material can provide a mechanism for upgrading the health of the residual trees. Furthermore, logging roads provide access for protection as well as recreation. Wildlife needs an adequate food supply as well as protective cover. Dense, overmature trees may provide protection and food for some but not most of the wildlife species because there is relatively little protection of food at ground level. Stumps, tree tops, and limbs which accompany conventional harvesting operations are havens of protection for many wildlife species. New growth that follows harvesting operations also represents an abundant and convenient food source for wildlife. Therefore, it is important that in a harvesting operation, a balance be provided between new growth and mature stands with a substantial amount of transition between the two. This is a desired wildlife management objective.

The nature of the wood demand of a power plant is likely to result in more mechanized harvesting operations, and expansion as anticipated in the purchase and use of whole tree chippers, and to a lesser extent, mechanical harvesting equipment. Care must be applied in the selection and use of this type of equipment otherwise the large rubber wheels or tracks could become a serious source of soil erosion. The negative connotations associated with this type of problem could quickly discourage the cooperation of private landowners in allowing wood waste removal.

SUMMARY

Generation of electricity from waste wood is technically, economically, and environmentally feasible. A brief overview of primary concerns which any organization or community should consider in implementing such a facility is presented. As an epilogue, the City of Brulington conducted a successful bonding referendum for construction of the 50 MW wood-fired power plant. By the summer of 1979, design was well under way and plans for implementing construction were being formulated.

REFERENCES

1. Henningson, Durham & Richardson, Inc. "Burlington, Vermont Refuse-Wood Power Plant, Aquaculture, Greenhouse - A Conceptual Study"; Washington D.C.; 1977.

2. Ellis, T. Paper presented at the Forest Products Research Society Energy Workshop; FPRS Proceedings P-75-13; Denver, Colo., 1976.

3. North Central Forest Experiment Station Forest Service; U.S. Department of Agriculture. *Forest Residue Energy Program;* St. Paul, Minn., 1978.

4. Christensen, G. Paper presented at the Forest Products Research Society Energy Workshop; FPRS Proceedings P-75-13; Denver, Colo., 1976.

5. Corder, S. "Fuel Characteristics of Wood and Bark and Factors Affecting Heat Recovery"; ibid.

6.   Hewett, C. DSD No. 114; Thayer School Engineering, Dartmouth College: Hanover, NH, 1978.

7.   Blankenhorn, P.; Bowersox, T.; Murphey, W. *Tappi* **1978** *61*, 4.

8.   Towne, R. Paper presented at Forest Products Research Society Energy Workshop; FPRS Proceedings P-75-13; Denver, Colo. 1976.

9.   Hoff, E. "Abstracts of Papers:, the Conference on Energy and the Wood Products Industry sponsored by the Forest Products Research Society, Atlanta, Ga., 1976.

10. Fernandes, J. "Wood Energy Systems, State of the Art and Developing Technologies"; Paper presented at the Future of Wood as an Energy Source Conference, Gorham, Maine, 1976.

11. Johnson, N. "Wood Waste Burning on a Traveling Grate Spreader Stoker"; Paper presented at conference Hardware for Energy Generation in the Forest Products Industry, Seattle, Washington, 1979.

12. Costle, D. "Standards of Performance for New Stationary Sources, Wood Residue-Fired Steam Generators"; 40 CFR Part 60; Federal Register 44, 12, Jan. 17, 1979.

13. Phelan, J. "Review of Particulate Equipment for Power Plant Effluents"; *Tappi* **1977** *60*, 9.

14. Kingsley, N.; Birch, T. "The Forest-Lane Owners of New Hampshire and Vermont. *USDA Forest Service Resource Bulletin* **1977**, NE-51.

RECEIVED MAY 12, 1980.

A Biomass Allocation Model

Conversion of Biomass to Methanol

Y. K. AHN

Gilbert Associates, Incorporated, P.O. Box 1498, Reading, PA 19603

It has become apparent that the effects of the rapid rise in prices and dwindling supply of petroleum and natural gas are being felt by all of us in terms of the prices we pay for gasoline, electrical energy, and chemicals. Various alternative energy sources, both fossil and nonfossil, are being sought as substitutes for petroleum and natural gas. Fuels and petrochemicals from biomass are considered to be promising alternatives. Biomass feedstocks under consideration include crops produced by agriculture or forestry, aquatic crops, agricultural and forest residues, and animal residues. The U.S. Department of Energy predicts the contribution of near-term systems to our domestic energy supply to be an additional 0.5 to 1.0 quad by 1985, (over the 1.0-1.5 quad now used) a displacement of 230,000 to 460,000 barrels of oil per day (1). Using published results as a data base (3-8,11), this paper illustrates how a deterministic model can be developed and utilized for the optimum allocation of biomass feedstocks. An example is presented for production and utilization of methanol from biomass.

Among the various products that can be synthesized from biomass, methanol was selected because of its versatile applicability to the electricity, transportation, and chemical sectors. Conversion of methanol from biomass is achieved via oxygen-steam gasification followed by shift conversion and methanol synthesis. Three feedstocks were selected for conversion to methanol—wood residue, corn stover, and furfural residue. Availability of

0097-6156/81/0144-0483$05.00/0
© 1981 American Chemical Society

feedstocks is highly regional, and the state of Missouri was selected because of its agriculture and forest wood land availability. Methanol was assumed to be used for power generation by combined cycle, as a blending stock for gasoline, and as a chemical.

FEEDSTOCK SUPPLY AND PRODUCT DEMAND

Regional Selection

The State of Missouri produces 10 million ton per year (MMTPY) of agricultural residues, including 5.4 MMTPY from corn (2). It also has a potential silvicultural plantation capability of producing, by current technology, an average of 7 dry ton equivalent (DTE) per year-acre of hybrid poplar. Availability of state land for development of silvicultural plantations is estimated to be 11 to 15 million acres for U.S. Forestry Service classification I-IV sites (3).

Feedstock Supply

Of the 5.4 MMTPY of corn produced, approximately 40% consists of residue and the remaining 60% is used as grain (4). Total annual furfural consumption in the United States is 150 million pounds. For each pound of furfural processed, approximately 10 lb of residue is produced. The furfural residue contains approximately 35% moisture (5). Availability of the wood residue is estimated by assuming that only 10% of the class I-IV sites will be used for hybrid poplar plantations and that half of the wood produced will be collected as wood residue. Availability of the three biomass feedstocks is summarized in Table I.

TABLE I. AVAILABILITY OF FEEDSTOCK, F_i

i	Feedstock	10^6 Btu/yr	Heating Value, Btu/lb	10^6 Btu/Yr
1	Corn Stover	2.2	8,390	36.9
2	Furfural Residue	0.5	7,680	7.5
3	Wood Residue	4.5	8,500	69.6

Product Demand

Total annual energy demand by the electric utility sector in the State of Missouri is estimated to be 358.4×10^{12} Btu (6). Of this total demand, 64.9×10^{12} Btu is oil and gas fired and is a potential candidate for conversion to other alternative fuels. It is estimated for the present study that approximately 5% of the oil and gas fired power generation, which is equivalent to 70 MW, is substituted by methanol from biomass, mostly for peaking services in gas turbines.

The total annual gasoline demand in the United States is estimated to be 115,000 $\times$ 10^6 gallons. If a 10/90 blend of methanol/gasoline is considered as a gasoline substitute, the total national methanol demand would be 11,500 $\times$ 10^6 gallons. Demand for the State of Missouri, prorated based on population, is 252 $\times$ 10^6 gallons.

The total annual demand for chemical grade methanol is estimated to be 641 $\times$ 10^6 gallons. Demand for the State of Missouri, again prorated based on population, is 14 $\times$ 10^6 gallons.

The selling prices for electricity, crude methanol, and chemical grade methanol were obtained from published studies (7). Table II summarizes the demand for each of the three consuming sectors and the product selling prices:

Table II. PRODUCT DEMAND AND SELLING PRICES

j		Annual Demand, D_j Conventional Unit	Selling Price, S_j Conventional Unit	10^{12} Btu	$/10^6$ Btu
1	Electricity	70 MW	19 Mills/kWh	3.245	6.37
2	Transportation	252 MM Gal	0.6 $/gal	14.449	10.45
3	Chemicals	14 MM Gal	0.150 $/lb	0.805	10.75

BIOMASS CONVERSION

Conversion of Biomass to Fuel Grade Methanol

A block flow diagram for production of fuel grade methanol from biomass is depicted in Figure I. The gasification step is based upon the Purox process and is followed by shift conversion and gas purification steps. The clean gas, which is shifted to a H_2/CO ratio of approximately 2/1, is converted to methanol in the ICI low-pressure methanol synthesis process. The process yields approximately 98% pure methanol with the remaining 2% consisting of water and some higher carbon number alcohols.

The medium-Btu gas from the Purox process need not be desulfurized prior to entering the shift reactor since a sulfided catalyst is used. The shifted gas goes to the purification system, where a hot-carbonate scrubbing system is

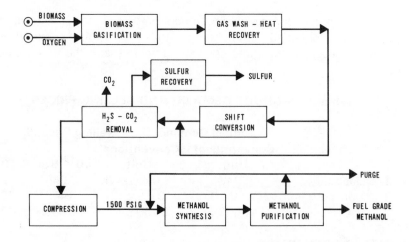

Figure 1.   Flow diagram for methanol via biomass gasification

The fuel grade methanol is 98% pure and contains such impurities as water, ethanol and higher alcohols. The impurities would have to be removed by distillation to produce chemical grade methanol of 99.90% purity containing ethanol and water contents of no more than 900 ppm and 500 ppm respectively (7). It is estimated for the present study that an additional 2% thermal efficiency is lost for the distillation operation.

The system efficiencies for converting the three biomass feeds to the three final products are summarized as follows:

Table III. SUMMARY OF THERMAL EFFICIENCIES

i	Biomass	Conversion to Fuel Grade Methanol	System Efficiency, n_{ij}		
			j = 1 Elec.	j = 2 Trans.	j = 3 Chem.
1	Corn Stover	48.0	21.7	47.0	46.0
2	Furfural Residue	48.0	21.7	47.0	46.0
3	Wood Residue	45.3	20.5	44.3	43.3

Conversion Economics

The base capital and annual operating costs to manufacture the final products were obtained from published date on wood residue (7) and on corn stover and furfural residue (8). The base data were updated to 1979 pricing. The product costs were calculated for three fuel grade methanol product capacities of 6,550, 13,100 and 26,200 X 10^6 Btu/day. The base data were adjusted using 0.7 scale factor for plant size.

The methanol production costs can be divided into two main unit operations — the gasification system including gas cleaning and the methanol synthesis system including shift and purification. The preliminary cost estimates indicated that a significant portion of capital cost is associated with gasification and gas cleaning systems. For the annual operating costs, the feedstock costs were the most costly element. Therefore, it is desirable to investigate sensitivity of profit to feedstock cost. This is discussed in the results and conclusion section. The estimated manufacturing costs based on the published feedstock costs for the baseline case are summarized in Table IV.

Table IV. SUMMARY OF MANUFACTURING COST, M_{ij}, $/MM Btu[a]

j	Product	Corn Stover, i = 1			Furfural Residue, i = 2			Wood Residue, i = 3		
		I[b]	II	III	I	II	III	I	II	III
1	Electricity	15.31+A[c]	13.05+A	11.46+A	9.95+A	7.51+A	6.76+A	10.64+A	9.1+A	8.13+A
2	Transportation	16.08	13.70	12.03	10.45	7.89	7.50	11.71	9.56	8.54
3	Chemical	16.53	14.09	12.38	10.75	8.11	7.71	11.49	9.83	8.78

[a] Based on raw material costs of $40/ton, $1/MM Btu, $1.62/MM Btu for corn stover, furfural residue, and wood residue, respectively.

[b] Plant capacities of I = 6,550 MM Btu/day, II = 13,000 MM Btu/day and III = 26,200 MM Btu/day.

[c] $A = (D/200)^{0.7} (0.4)$, where D = electricity demand in MW.

used to reduce sulfides in the gases to 10 ppm, and CO_2 to 7%, so that a ratio of 2.05 for $H_2/(CO + 1.5\ CO_2)$ can be achieved (7).

The purified gas is then passed through an iron sponge drum and a sulfur guard drum to remove traces of sulfur. Following the guard drums, the gas, which is essentially sulfur free, is compressed to 1500 psia, combined with recycle gas, and passed through a fixed-bed catalytic (highly-active copper catalyst) converter to produce crude methanol. The methanol is condensed and separated from the untreated gas, which is recycled to the converter. The pressure is then reduced, and dissolved gases are flashed from the crude methanol. Some of the flash gas is purged for use as fuel to control the concentration of inert components in the converter system. The crude methanol is purified as required by distillation to produce fuel-grade methanol.

Thermal efficiencies for converting corn stover (8), furfural residue (8), and wood residue (7) to methanol were estimated from the published data to be 48.0, 48.0, and 45.3% respectively. The efficiency data used for converting wood to methanol via the Purox process is also in good agreement with recently published data (12). The Purox process may not have been the best choice for the gasification (12), but process and economic data are available for all three feedstocks considered in this paper.

Conversion of Fuel Grade Methanol to Final Products

Conversion of fuel grade methanol to electricity is achieved by means of a combined-cycle configuration. The efficiency of a methanol-fueled combined cycle plant was estimated to be 45.2% (9).

The use of fuel grade methanol for gasoline-alcohol blends may require engine modifications, but this paper is not concerned with such modifications. Rather, it is limited to the preparation of fuel grade methanol suitable for gasoline blending. One requirement is to reduce the moisture content to a maximum of 0.25 weight percent to avoid phase separation. Therefore, water must either be exlcuded from the fuel grade methanol or other compounds must be added to improve the water tolerance of the methanol-gasoline blend. Some of the lower molecular weight fractions of the gasoline may have to be removed during the summer to counteract the large nonideal increase in the vapor pressure of the blend. This could penalize the economics of blending gasoline and methanol (13). It is estimated for the present study that 1% thermal efficiency is lost to refine suitable gasoline blending stocks.

DEVELOPMENT OF ALLOCATION MODEL

The problem is to determine the optimum biomass allocation policy in order to maximize the profit. A generalized linear program, SIMPLES ([10]), was used to develop the resource allocation model.

The model seeks to maximize the linear objective function (profit).

$$P = \sum_{j=1}^{N} D_j S_j - \sum_{j=1}^{N} \sum_{i=1}^{K} n_{ij} f_{ij} M_{ij} \tag{1}$$

for N set of inequality constraints

$$\sum_{j=1}^{N} f_{ij} \leq F_i \quad i = 1,2,...K \tag{2}$$

and K set of equality constraints

$$\sum_{i=1}^{K} n_{ij} f_{ij} = D_j \quad j = 1,2,...N \tag{3}$$

and the non-negative restrictions of

$$f_{ij} \leq 0 \tag{4}$$

Where:

P = profit
D_j = demand for commodity j
S_j = selling price for commodity j
M_{ij} = manufacturing cost for commodity j from feedstock i

n_{ij} = thermal efficiency of converting feedstock i to commodity j
f_{ij} = allocation of feedstock i to commodity j
F_i = availability of feedstock i

In terms of the present methanol study, the problem seeks an optimum allocation policy for corn stover ($i = 1$), furfural residue ($i = 2$), and wood residue ($i = 3$) to produce three commodity products of electricity ($j = 1$), transportation ($j = 2$), and chemicals ($j = 3$) in the most profitable way. In pure mathematical terms, we are to determine f_{ij} which maximizes the profit function, Equation 1, when the feedstock availability and demand are constrained by Equations 2 and 3, respectively. The data base to use with Equations 1 through 4 is tabulated in Table I for F_i, Table II for D_j and S_j, Table III for n_{ij}, and Table IV for M_{ij}.

RESULTS AND CONCLUSIONS

Table V presents the result of the optimum allocation policy for the three biomass feedstocks in satisfying the demand of the three consuming sectors.

Table V. OPTIMUM FEEDSTOCK ALLOCATION POLICY, f_{ij} (BASELINE CASE)

| | | Feedstock Cost, | Allocation of Feedstock i to Consuming Sector j, 10^{12} Btu/Yr | | |
| | | | $j = 1$ | $j = 2$ | $j = 3$ |
i	Feedstock	$/10^6$ Btu	Elec.	Trans.	Chem.
1	Corn Stover	2.38	0	0	0
2	Furfural Residue	1.0	0	5.75	1.75
3	Wood Residue	1.62	15.83	26.52	0

It is interesting to note that furfural residue alone was sufficient to satisfy the demand of chemical grade methanol. This is due to the fact that all the available, least expensive fuel (furfural residue), was used to manufacture the most expensive product (chemical grade methanol). The transport sector requires the second most expensive methanol fuel, and any furfural residue left over after chemical grade methanol was used for production of the transportation grade methanol. The balance of the transportation grade methanol was supplied by the wood residue, the second least expensive feedstock. All the electric utility demand was satisfied by wood residue, and no corn stover was used.

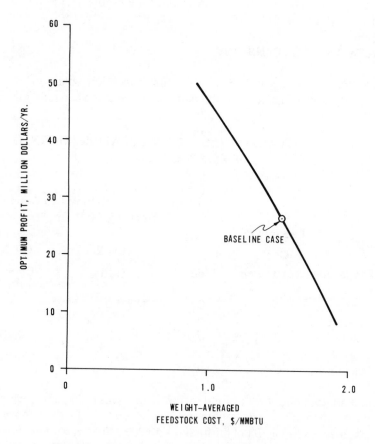

Figure 2.   Sensitivity of feedstock costs on optimum profit

Profit calculated from the optimal policy was $26.5 million per year indicating that various grade methanols made from the three biomass feedstocks can be competitive with those made from conventional sources. The calculation was based on the unit manufacturing cost determined from a 26,200 X 10^6/day methanol plant capacity (plant capacity III in Table IV). The study, however, disclosed the fact that the biomass feedstock costs are the dominating factor in the economics of methanol production. It would therefore be interesting to note how the optimum profits vary with feedstock cost.

Since more than one feedstock is involved, development of a sensitivity curve for optimum profits vs. feedstock cost requires use of a weight-averaged feedstock cost. The weight-averaged feedstock cost is defined as:

$$\text{Weight} - \text{Averaged Feedstock Cost (\$/MM Btu)} = \frac{\sum_{j=1}^{3} \sum_{i=1}^{3} C_i f_{ij}}{\sum_{j=1}^{3} \sum_{i=1}^{3} f_{ij}}$$

C_i = cost of feedstock i in $/MMBtu and f_{ij} were defined previously.

For this study, the costs of three feedstocks were varied from 25 to 50% higher or lower than the published base case feedstock costs. The optimum profits were then calculated using the same feedstock availability, product demand, and product selling prices as the base case. Figure II summarizes the results of the calculation. The figure indicates that the optimum profits are indeed very sensitive to feedstock cost changes and that continued improvement of biomass production and collection techniques is very desirable to improve total profit.

LIST OF SYMBOLS

C_i Cost of Feedstock i
D_j Demand for Commodity j
f_{ij} Allocation of Feedstock i to Commodity j
F_i Availability of Feedstock i
i   Feedstock
j   Commodity
M_{ij} Manufacturing Cost for Commodity j from Feedstock i
P   Profit
S_j Selling Price for Commodity j
n_{ij} Thermal Efficiency of Converting Feedstock i to Commodity j

REFERENCES

1. U.S. Department of Energy, "Fuels from Biomass — Multiyear Program Plan", Washington, D.C., April 27, 1978.

2. Clausen, E.C.; Gaddy, J.L. "Preprints", 81st National Meeting, The American Institute of Chemical Engineers, Kansas City, Mo., April 1976.

3. Salo, D.J.; Inman, R.E.; McGurk, B.J.; Verhoeff, J. MITRE Technical Report for ERDA MTR-7347 (Vol. III), Mclean, Va., May 1977.

4. McClure, T.A. "Proceedings", Biomass — A Cash Crop for the Future? Kansas City, Mo., March 2-3, 1977; 145-77.

5. Sheppard, W.J. *ibid;* 178-204.

6. "The Application of Near-Term Fossil Technologies to the Energy Supply/Demand Profiles of U.S. States and Regions", U.S. Department of Energy, January 1977, FE/2442.

7. Bliss, C.; Blake, D.O. MITRE Technical Report for ERDA MTR-7347 (Vol. III), Mclean, Va., May 1977.

8. Otis, J.L. "Proceedings", Biomass — A Cash Crop for the Future?, Kansas City, Mo., March 2-3, 1977; 219-36.

9. Gilbert Associates, Inc. "Assessment of Fossil Energy Technology for Electric Power Generation", OPPA/ERDA, March 1977, GAI Report 1940.

10. Walker, H.; Hall, L. "SIMPLEX, A Code for the Solution of Linear Programming Problems", UCRL-51820.

11. SRI International, "Mission Analysis for the Federal Fuels from Biomass Program", U.S. Department of Energy, Report, Vol. II, 1978.

12. Science Applications, Inc., "Biomass Based Methanol Processes", Presented at the Seventh Biomass Thermochemical Conversion Contractors Meeting, Roanoke, Va., April 24-25, 1979.

13. Hagan, D.L., "Methanol — Its Synthesis, Use as a Fuel, Economics, and Hazards", Report prepared for ERDA, December 1976.

RECEIVED JUNE 20, 1980.

The Energy Plantation and the Photosynthesis Energy Factory

MALCOM D. FRASER, JOHN F. HENRY, LOUIS C. BORGHI, and
NORMAN J. BARBERA

InterTechnology/Solar Corporation, 100 Main Street, Warrenton, VA 22186

The Energy Plantation

In conventional forestry, trees are grown in plantations to produce the raw material for a variety of products such as lumber, plywood, pulp and others. In these plantations, the trees are generally widely spaced and grown to sizes large enough for the manufacture of the desired products. Achieving these commercial sizes may require long growing periods or rotations which may range from 30 to 80 years or more. As a result of these constraints — long rotations and planting densities of the order of a few hundred trees per acre towards the end of the rotation — the plantation site is only fully utilized for a short fraction of the rotation period. Average sustained yields over the rotation therefore rarely exceed about 1 oven-dry-ton of mechantable material per acre-year (1). Moreover, because of the size of the crop and the need to maintain its physical integrity, conventional single-tree harvesting and handling methods are generally used in forestry operations.

Tree crops however could be grown on much shorter rotations if the size and form of the crops were not limiting factors in the end use of the crop. Such is the case when the desired product is wood chips to be used for pulp, fuel, or feedstock for conversion to substitute fuels. Short-rotation tree farming

0097-6156/81/0144-0495$12.50/0
© 1981 American Chemical Society

generally refers to rotations of 20 years or less and is generally associated with close spacing of the trees in order to achieve full site utilization within the rotation period.

Short-rotation tree farming for fiber production has been proposed by a number of investigators (2-5). Early short-rotation experimental data indicated that the biomass yields achieved in short-rotation plantations could far exceed those of conventional forestry. Average annual sustained yields of 5 to 10 oven-dry tons per acre-year (ODT/ac-yr) were shown to be possible under short-rotation conditions (6,7). It also became apparent that such high yields could be achieved only if intensive management were applied to the plantation. In many cases, the level of management appears to be comparable to that required in the production of agricultural crops (8-10). The potential of short-rotation tree farming for fiber production induced InterTechnology/Solar Corporation to advance the same concept as a possible source of biomass for energy conversion (11-13). Other investigators have described similar concepts (14,15).

As it is presently envisioned by InterTechnology/Solar Corporation, the Energy Plantation is a woody biomass production entity relying on short rotation and intensive management to produce biomass exclusively for its fuel and/or feedstock value. Energy Plantations offer a number of potential advantages over conventional forestry plantations: higher productivity per unit land area, lower land requirements for a given biomass output, earlier cash return on the investment, extensive mechanization similar to that practiced in agriculture, and ability to assimilate cultural and genetic improvements quickly. Short-rotation crops can also be chosen among a variety of species which regenerate by coppicing, thereby eliminating the need for replanting after each harvest. Energy Plantations however do have a number of disadvantages: initial establishment costs and yearly management costs per unit area are generally higher than those for conventional forest crops; only sites amenable to mechanized operations can be used; and disease and insect propagation may be difficult to control. Securing the use of the land for energy crops could also be a problem in some areas where competition with other uses (e.g., farming, recreation) could occur.

No full-scale Energy Plantation has yet been demonstrated. It is therefore necessary to use a conceptual design of the Energy Plantation to assess its economic and energy efficiency potential. Table I summarizes the design parameters adopted in the ITC/Solar model of the Energy Plantation. The crops are assumed to be chosen from a variety of hardwoods displaying fast juvenile growth and capable of regeneration by coppicing. Candidate crops include American sycamore *(Platanus occidentalis)*, hybrid poplars *(Populus*

Table I. DESIGN PARAMETERS USED IN THE ITC/SOLAR MODEL OF THE ENERGY PLANTATION

Production	Variable, generally of the order of 200,000 ODT/ac-yr
Crop	Fast-growing hardwoods with coppice regeneration
Productivity	5 to 10 ODT/ac-yr
Planting Density	4 to 16 square ft per plant, i.e., ~10,000 to ~2500 trees per acre
Lifetime	One first-growth rotation followed by five coppice rotations
Management	Mechanical weed control Fertilization Irrigation (in some modes of operation of the plantation)
Harvesting	Conceptual self-propelled harvester-chipper
Transportation	Green woodchips transported to conversion plant located in center of the plantation
Support	Nursery operation, equipment maintenance and repair, supervision
Land	Plantation made of lots of the size of an average farm in the region distributed at random within a larger geographic area

spp.), Eastern cottonwood *(P. deltoides),* black cottonwood *(P. trichocarpa),* black alder *(Alnus glutanosa),* green ash *(Fraxinus pennsylvanicum), Eucalyptus,* and others. The selection of the crop is made on the basis of climate, soil conditions, and demonstrated growth characteristics of the candidate crop.

Fast-growing hardwoods are generally suggested because of their coppicing properties, which eliminate the need for reestablishment of the plantation after each harvest. Due to the climate and soil conditions, pines may be better crop candidates in some situations, as shown for instance in Southern Georgia where loblolly pines displayed higher productivity than sycamore for rotations of about 6 years or more (16).

On an Energy Plantation, the planting density and rotation duration, and their associated productivity, are chosen to minimize the cost of biomass production. Many authors have recognized that a strong correlation exists between spacing and rotation age for hardwood crops grown under intensive management. Once a spacing has been adopted when establishing a farm, the rotation age at which the mean annual biomass increase is obtained must be adopted to maximize the yield of the farm (17). The choice of the optimum rotation is particularly critical for close spacings generally associated with short rotations because the average productivity decreases significantly once the optimum rotation is exceeded. On the basis of data available at present, some authors (6,18) favor rotations of 10 to 15 years while others prefer shorter rotations (19-21). The design parameters adopted in Table I might therefore have to be modified to account for site-specific conditions. The impact of changes in spacing and rotation duration has been estimated through sensitivity analyses (21).

Land management includes weed control to eliminate competition for light, moisture, and nutrients; fertilization to ensure maintenance of sustained productivity; and irrigation in some modes of operation. Irrigation with surface or well water is probably not cost-effective (22). However, irrigation with municipal sewage effluent could be cost-effective as a result of the credit generated through land treatment of the wastes (23). This latter mode of operation is analyzed in the Photosynthesis Energy Factory discussed below. Harvesting is assumed to be performed mechanically by a harvester-chipper which could be similar in design to a corn silage harvester adapted for the Energy Plantation crops. The green chips, after field storage, are transported to the conversion plant located ideally in the center of the Energy Plantation area.

The Energy Plantation is conceived as a self-contained industrial operation including its own management and support services. The Energy Plantation is assumed to consist of parcels of land of the size of an average farm in the region distributed within a geographic area surrounding the conversion plant. The land selected for energy farming is preferably marginal land not suitable for the production of more valuable crops. Using marginal land however will result in productivities lower than those mentioned earlier as achievable. On the other hand, marginal land can probably be obtained at a lower cost (through leasing or purchase) than the good-quality land on which many high productivity data have been generated. The type of land available for Energy Plantations will therefore be determined by the overall economics of biomass production at individual sites.

In its simplest operational schedule, the total planted area of the plantation is divided into a number of equal modules equal to the rotation duration (e.g., four modules, each equal to one-fourth of the planted area for a four-year rotation). Each year, one of these modules is harvested and then regenerates through coppicing to supply the new crop at the end of the next rotation. After a number of coppice crops have been harvested from the original planting, the module must be replanted before progressive weakening of the root system results in reduced annual productivities. In ITC/Solar's model, regeneration of a module is accomplished by planting of clones gathered from other (still operational) areas of the plantation. This mode of operation ensures sustained annual production of biomass on a permanent basis.

Other conceptual designs of energy farms have been proposed which include the same basic features as the ITC/Solar model (19,20,24,25). Because of the lack of experimental data concerning some aspects of energy farming, all designs include a certain element of uncertainty. Sensitivity analyses are therefore needed to estimate the impact of these uncertainties on the projected biomass production costs and to estimate reasonable ranges of values for these production costs.

The Photosynthesis Energy Factory

Another alternate source of energy, which also offers the additional advantage of decreasing the environmental impact associated with the disposal of wastewater and residues, is the concept of growing algae in shallow ponds. Algae ponds are open shallow ponds in which algae and bacterial populations symbiotically utilize sunlight and nutrients to produce cell mass. The earliest application of the algae pond concept is the stabilization pond. Stabilization ponds have been used by small communities for years as a means of treating domestic wastewater. In construction and operation, these ponds are the essence of simplicity. The main requirements are land and a favorable, sunny climate. Currently, there is a trend toward the use of algae ponds as finishing ponds in integrated municipal and waste treatment. The ponds operate in series to "remove" both BOD and phosphorus by tying up these components in cell mass. In an algae pond system for recoverying energy from various residues, the algae would be digested anaerobically to yield a methane-containing gas, which would be processed into SNG.

As long as 20 years ago, work was undertaken by Dr. W. J. Oswald and others at the University of California at Berkeley to develop a system utilizing the algae pond concept both to treat waste (26-29) and to produce fuels (30,31). Indeed, microalgae were among the earliest "fuel crops" proposed. The technical problems were essentially concerned with integrating the components and optimizing their operation. The components themselves — the algae pond, the digester, and the sedimentation, separation, and finishing stages — were already being used in waste treatment. The work over the last 20 years has concentrated in three areas: (1) identifying and quantifying algal growth-limiting factors (nutrients, species characteristics, and climatological parameters); (2) maximizing gas production from anaerobic fermentation; and (3) optimizing the system with respect to gas production, residue uptake, land utilization, and cost-effectiveness. State-of-the-art reviews have been published recently regarding the engineering aspects (32) of microalgae production and the potential of microalgae as bioconversion systems (33,34). Other publications have reported recent research on species control, algae harvesting, and the potential of blue-green algae (35-39).

Both the Energy Plantation and the algae pond can contribute to the solution of the population, resources and energy problems facing us. Recently, it became apparent that they could perhaps better accomplish these missions when the two are integrated to form one composite system, as shown in Figure I. In short, each bioconversion system produces a by-product that can be used to advantage by the other. The carbon content of sewage limits the production of the algae pond, but carbon dioxide, a by-product of combustion of solid Energy Plantation fuel (which currently appears to be the best way of using plant matter as fuel), can be supplied to the algae pond to increase its productivity. The waste heat from the boiler can be used to control the temperature of the algae digester. The sludge generated as a by-product of the algae pond provides a source of inorganic nutrients and water for the Energy Plantation, which thus provides an ideal disposal site for the sludge.

The Photosynthesis Energy Factory (PEF) is a synergistic combination of the dry-land Energy Plantation and the algae pond which can produce on a perpetually renewable basis, nonpolluting and totally domestic fuels from marginally useful land, solar energy, and various residues. Simultaneously, from different parts of the PEF, chipped solid fuel is produced, from which electricity is generated, and methane or SNG is recovered from waste CO_2 and municipal or industrial wastewater. Incidental economic benefits — which are significant — include secondary or tertiary treatment of municipal and industrial effluents, and the complete elimination of the need for a sanitary landfill for disposal of the resultant sludge. The PEF is not merely a combination of convenience, but a truly interactive utilization of materials and energy.

Repeated application of sludge from the algae pond could result in a progressive accumulation of some toxic elements originally present in the wastewaters fed to the pond. The rate of accumulation of the toxic elements will depend on various site-specific factors such as type of wastewater (municipal wastewater is much less likely than industrial wastewater to contain potentially toxic elements), soil type, texture and pH. It has been estimated that in many areas of the United States, it would require 50 to 80 years before the accumulation of toxic elements reached levels considered dangerous by the EPA for land devoted to food crop production (23). Exceeding these levels of toxic elements would eliminate the possibility of using the land for food crops at a later date.

In the diagram of the PEF shown in Figure I, the assumption has been made that the woody biomass is used as fuel for generating electricity. Biomass can of course be used as feedstock for other conversion processes to produce a wide variety of fuels or chemicals. However, direct combustion of woody biomass is an accepted, commercial technology, allowing comparison with alternate methods of generating electricity, and the emphasis of the studies of the PEF was on development of the biomass production processes and their integration rather than the study of biomass conversion. Any other process for conversion of woody biomass could be used in the PEF concept.

Description of Projects

An initial project was undertaken to study the concept of the PEF and its characteristics and apparent benefits. The project was divided into three parts or tasks. The objective of one task was to analyze the concept of the PEF, with particular emphasis on the complementary and synergistic aspects of the system. A second task was concerned with the analysis and selection

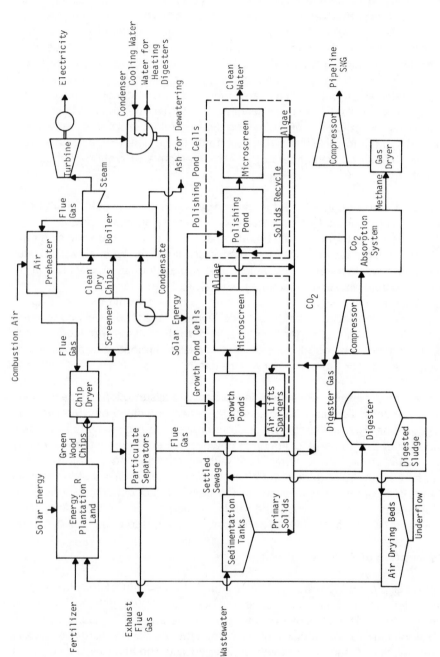

Figure 1.   Flow diagram of photosynthesis energy factory

of potential sites. The final task was to develop preliminary designs and associated cost estimates for potential demonstration systems at the best sites.

In the PEF, various streams of energy and materials flow between three major subsystems — the dry-land Energy Plantation, a wood-fired power plant, and an algae production system. To analyze the resultant interactions between the subsystems, a comprehensive technoeconomic model was developed to describe the PEF's performance and cost. Models of the three subsystems were developed with the aid of information and data that were already available as the result of previous studies. New data and new concepts were introduced into the models wherever possible. The University of California at Berkeley supplied state-of-the-art data on algae pond performance and costs. These subsystem models were engineering models developed in sufficient detail to represent the important variables and variable-parameter interactions influencing subsystem performance and costs. Of the three subsystem models, the most comprehensive and the most complex was the Energy Plantation model, which is a complete design model.

For the twin purposes of defining the applicability of the PEF concept and selecting the best site for a demonstration PEF project, data were gathered on the characteristics of land, the availability of municipal and industrial effluents and residues, and the supply and demand for energy at a wide variety and number of potential sites. A format was developed for handling this data base, and a number of suitability indexes were defined for evaluating the site data. Data were obtained from a number of sources in the literature as well as from state energy offices. As the result of this site-selection procedure, a number of sites were chosen for analysis by means of the technoeconomic model.

The model was then used to design a demonstration PEF system at each of the selected potential sites. This preliminary design illustrated for a specific site the benefits and the impact to be expected from a demonstration PEF project. Estimated costs were provided also for each demonstration PEF. Comparing these preliminary designs and their costs was then done to show where and under what conditions a PEF would be expected to be cost-effective in recycling wastes and producing fuels from biomass which would be competitive with presently used fuels. The results of this initial project have been published (21).

The analysis which was performed in this initial project indicated that some interactions between the PEF subsystems are generally cost-effective while others are probably site-specific or can be improved upon. From these initial results, it was concluded that certain refinements in the design of a PEF should be analyzed to give the PEF greater applicability as well as to improve its economics. Thus, it was decided to investigate in more detail certain aspects of the design and operation of the dry-land Energy Plantation subsystem.

One possible interaction within a PEF which was not considered in the initial project is the contribution of water from the wastewater treatment subsystem to the dry-land Energy Plantation. One of the original significant credits resulting from the wetlands biological wastewater treatment subsystem can augment the available natural rainfall or even supply the entire water requirement of a PEF. Thus, it might be possible to site a PEF in semi-arid or arid locations to expand its applicability.

The results from the initial project indicated that supplying the necessary nutrients to maintain the productivity of the land — particularly nitrogen — was a significant cost item in the economics of producing woody biomass. Because of the importance of nutrients, it was decided that the nutrient balance in the Energy Plantation model needed to be refined to predict the required amount of nutrients more precisely. In particular, because the leaves contain a high percentage of nitrogen compared to the wood, work is being done to include in the nutrient balance the effect of nutrient recycling via leaf fall and a more precise accounting of nutrient leaching.

Transportation was another significant cost item which appeared to have potential for cost savings through a more detailed analysis of alternative system designs. Thus, alternative methods for handling and transporting the woody biomass are being analyzed, such as pneumatic tube transport, chip baling, and alternative methods for drying the chips. In addition, the transportation system is being analyzed in greater detail to see where cost savings might be achieved through optimization.

The initial results from studying the PEF concept indicated that the most significant credit resulting from the wetlands biological waste-water treatment subsystem was the wastewater treatment credit itself rather than the credit for the value of the gas produced. Thus, it became of interest to look for better ways of incorporating the wastewater treatment function within the PEF than via an algae pond. One way that this might be done is to apply the wastewater directly to the Energy Plantation. However, this process has limitations, with respect to both the particular location and local soil quality, and the composition of the wastewater. Work is therefore being done to collect the necessary data on technical limitations and EPA regulations and to develop a model for this process. In addition other wetlands biological species besides algae have been suggested for wastewater treatment, and it was decided to investigate the possible use of these other plants to perform this function.

Finally, additional work is being done in this second project to look for new and improved technology to include in the power plant subsystem model. Additional potential sites are also to be identified where the new modes of operating a PEF — e.g., with irrigation or direct application of wastewater — would be applicable. The new complete PEF model will then be used to compare the various modes of operation, and to determine the economic viability of PEF systems for sites displaying widely different local climatic and site-specific constraints.

The following sections of this paper will describe the subsystem models which were developed in the initial project to study the PEF and present some of the overall results obtained by simulating the performance of the entire PEF system. Additions to the model which are being developed as the result of work in the second project will also be described, and some results of the analysis of these improved aspects of PEF design and operation will also be presented.

THE ENERGY PLANTATION SUBSYSTEM

Description of Initial Model

In the PEF concept, the inputs to the Energy Plantation model are a description of the site being investigated, supplied by the site-selection procedure, and the amounts of nitrogen and phosphorus recycled to the plantation supplied by the algae pond model. The major outputs of the plantation model are the yearly amount of biomass produced and its cost (green chips) delivered at the utilization point. These items constitute the major inputs to the power plant model. Manpower and equipment requirements; species suggested for the plantation; and planting, harvesting, and operating schedules for the plantation are secondary outputs of the plantation model.

This plantation model is an extension and generalization of work done in previous studies (40,41). The model includes a number of submodels which are discussed below. These submodels include: (1) land resources, (2) data base on plants, (3) recycled inputs, (4) species selection and characterization, (5) plant growth model, (6) data base for field operations, (7) field operations, (8) data base for cost estimates, and (9) cost estimate for biomass produced.

The land resources subroutine is the major input to the model. The data generated through the site-selection procedure are organized in three categories: land description, land capability, and land cost. The land available for plantation operations is characterized by the plantation density or ratio of the plantation area to the total geographic area encompassing the plantation and by the average size of the parcels making up the plantation area. Both factors have been shown to have a significant effect on the cost of the biomass produced. (42) The plantation area would include the actual planted area plus necessary service roads and irrigation lanes if warranted. The land quality or ability to support plant growth is characterized in terms of the land classes used in the National Inventory of Soil and Water Conservation Needs (43).

A correlation was established between experimental yield data (ODT/ac-yr) and land classes on which the data were generated (a linear relation was used, which had a correlation coefficient >0.75). Land of Class III was assumed to have an index of 1, thereby allowing the productivity of land of other classes to be estimated on the basis of relative productivity indexes. These productivity indexes were used to weight the experimental yield data from the literature to account for the difference in land productivity between the land considered in the analysis of specific sites and the land on which the experimental data were generated. This approach is somewhat similar to the approach used by Marshall and Tsang (22) to relate the relative value of the yields expected from land of various classes submitted to comparable cultural practices, to land classes. The method proposed here to estimate yields at a given site on the basis of yields measured at experimental sites should be refined as more data become available.

The data base on plants contains data describing the growth characteristics of the species of interest for plantation applications. Plant growth and yields on an Energy Plantation are predicted by means of a model describing juvenile plant growth which was developed at InterTechnology/Solar Corporation (21). The model contains several parameters which are characteristic of the species considered and are determined from experimental data. The data base contains the values of these characteristic parameters which have been found for various species. The data base also lists the forest regions, and soil classes and subclasses in which each of the species is expected to grow.

In the subroutine describing the recycled inputs, the nutrient value (N, K and P) of the ash and sludge recycled from the boilers and digesters is estimated and compared to the amounts required by the plantation to maintain sustained yields.

Species to be included in a plantation are selected from the data bank on the basis of the forest group, land class, and subclass to which the plantation land belongs. The characteristic parameters of each species retained are then adjusted to take into account the difference in soil quality between the plantation land and the land on which the experimental data for each species were generated.

In the plant growth subroutine, average annual sustained yields are estimated for each species for various harvesting cycles and planting densities by means of the model describing juvenile plant growth. In most cases, the discrepancy between calculated values of these sustained yields and experimental data is of the order of 5 to 10 percent of the experimental data. This level of precision is satisfactory as it is comparable to or smaller than the observed fluctuations in yield due to yearly climatic variations. Significant variations in field data are observed, and the proposed model will have to be revised or refined as more data become available. The yield-cycle combinations generating amounts of biomass within 5 or 10 percent of the maximum predicted are retained for further analysis.

Field operations in an Energy Plantation include harvesting and chipping, cultivation, fertilization, clone generation, replanting of a fraction of the plantation, and transportation of the biomass to the point of utilization. These operations are supported by maintenance and administrative teams. For each of the yield-cycle combinations of interest, this subroutine establishes the inventory of the equipment, personnel, and supplies required to maintain the operation.

In the cost-estimating subroutine, the cost of biomass delivered in the form of chips at point of use is estimated on the basis of the utility (11.1% average return, income tax paid on equity return) and municipal (6.375% return, no income tax) methods of financing. The resultant output of the subroutine shows the various components of the overall cost of biomass for both methods of financing. Table II shows the economic analysis for a particular site and a breakdown of the capital and operating costs involved in the Energy Plantation subsystem of a PEF. The depreciable investment includes the capital cost of plantation installation (land clearing) and startup (planting stock). Following this analysis, the major component of the total revenue is the cost of nitrogen fertilizer. The contribution of fertilizers to the cost of biomass production has been noted by other investigators (44). As a result, much emphasis has been given to the problem of nutrient requirements in the expanded version of the PEF model (see below). The final cost data used to characterize the potential of a site for biomass production were averages of the data for each of the species considered for the site. This approach was adopted as Energy Plantations are assumed to include a mix of species.

Table II. COST ANALYSIS FOR NATCHITOCHES, LA, SITE

36,000 acres of plantation
Hybrid Poplar NE 388
4 ft^2/tree, harvest every 2 years
8.52 ODT/ac-yr
Municipal financing

	Annual Equivalent Cost, $*	Percent of Total Revenue
Depreciable Investment	938,700	16.7
Nondepreciable Investment	49,300	0.9
Federal Income Tax	0	0.0
Annual Operating Costs		
Fuels	140,100	2.5
Land Rental	1,113,100	19.7
Total Labor	972,200	17.3
Administrative & Overhead	194,400	3.5
Supplies — Nitrogen	1,365,000	24.2
Phosphorus	43,000	0.8
Potassium	361,100	6.4
Lime	8,900	0.2
Others	81,200	1.6
Maintenance and Repair	322,600	5.7
Local Taxes & Insurance	39,600	0.7
Total Revenue Required	5,629,200	

Product Cost (Delivered in the form of green chips)
| $/ODT | 17.71 |
| $/10^6 Btu | 1.03 |

* 1977 dollars.

Fifteen sites were identified for further analysis of the PEF concept. The selection criteria included resource availability (land, wastewater, labor, climate) and market for the PEF products and services (electricity, natural gas, steam, need for waste treatment, and jobs). The 15 selected sites were analyzed with the plantation model. Species included hybrid *(Populus spp.)*, Eastern cottonwood *(P. deltoides)*, plains cottonwood *(P. sargentii)*, silver maple *(Acer saccharinum)*, American sycamore *(Plantanus occidentalis)* (southern and midlatitude sites), and *Eucalyptus* (Florida sites only). Proposed planting densities were generally between 4 and 12 ft^2/tree with harvests (first and coppice) of 4 ODT/ac-yr (Maysville, KY) to about 9 ODT/ac-yr (Bemidji, MN). Differences in cost of biomass were related to a number of factors, including the investment, land rental, transportation costs, and amount of nutrients recycled to the plantation. The cost of biomass at the 15 potential sites ranged from $16.90 to $23.59 per oven-dry ton (municipal financing). These costs are comparable to those estimated on the basis of other tree farm designs for similar locations.

Nutrient Balance

The nutrient balance model predicts the amount of inorganic fertilizer required to maintain site fertility, and thus productivity, of the plantation. Fertilizer application is needed because of the shorter rotations and the more complete removal of biomass from the site that occurs with this type of management as compared to conventional forestry. Although the soil nutrient pool at any given site is an unknown, maintenance of site fertility is assumed to be possible by replacing the amount of nutrients removed in the harvested biomass plus an additional amount for leaching and denitrification losses.

The model treats the soil nutrient pool as the system of interest, the upper boundary including any surface organic horizons that may be present, the lower boundary being the underlying bedrock, and the lateral boundaries extending to the boundaries of the site. Inputs to the system include the application of inorganic fertilizer, return of ash material from the power plant component of the PEF, application of sludge from the algae pond, nitrogen fixation by cover crops or interplanted woody nitrogen-fixing species, and the recycling of leaf material not removed in the harvested biomass. Outputs from the system include the growth of wood and leaf material, leaching and erosion losses, and denitrification losses for soil nitrogen. Although inputs from atmospheric sources and the weathering of bedrock are operative and can be important over long periods of time (45-47), they are not included in the model because the magnitude of their contribution over the short duration of the rotations used in this type of management is assumed to be small in comparison to the other inputs.

In the original nutrient balance model, the annual fertilizer requirement was calculated on the basis of replacing the nutrients removed in the biomass harvested less nutrient credits obtained from recycling the sludge from the algae pond and the ash from the boiler to the Energy Plantation. Credits were taken only for nitrogen and phosphorus in the sludge and calcium in the ash. Allowance was made for loss of nutrients by leaching, and application of fertilizer was assumed to occur only once per rotation.

The nutrients present in the biomass harvested were calculated on the basis of tables developed from data on (1) biomass distribution in leaves, stems, and branches of young trees as a function of age, and (2) the nutrient composition of leaves, stems, and branches for young trees. The data base used for (1) above was generated in short-rotation field trials of American sycamore *(Platanus occidentalis)* (48), and that used for (2) above consisted of averages of analyses performed on a number of older-growth hardwood species (49). It should be noted that there are a number of difficultires inherent in the use of tissue analyses for prediction of fertilizer requirements (50,51), just as there are difficulties and limitations involved with the use of soil analyses (51,52). The relative distribution of biomass into stem, branch and leaf components will vary with the species involved, the age of the tree (rotation length), and the density (number of trees per unit area) of the plantation. The chemical composition of the individual components will also vary with species, age, site fertility, and the time of year of sampling (e.g., seasonal flux of N and P from leaves to twigs) (50). These differences are even manifested within the individual biomass components. The variables used in the model are general design variables that can be easily modified to obtain better predictive results either with new data from field trials or site-specific information for a particular management design.

Several other general characteristics of the model should be mentioned. First, the model predicts fertilizer requirements for the plantation once steady-state conditions have been attained. Steady-state is defined as that point at which the management schedule's cyclic operation results in the decomposition (mineralization) of the organic inputs from previous years in an amount equal to the organic inputs for one year; this is equivalent to saying that all of the annual inputs are decomposed in one year, meaning that the nutrients contained in the organic matter become available for uptake by the plant in growth or loss in leaching. Annual inputs thus equal annual outputs.

The rate of breakdown of organic material in natural stands is dependent upon environmental conditions and the chemical nature of the substrate being decomposed. Litter half lives are in the range of one to a few years for temperate deciduous forests as a whole (53). Temperature and moisture have been found to be the most important environmental parameters and the carbon/nitrogen (C/N) ratio and lignin content the most important chemical parameters (54,55). The rate of decomposition is almost always limited by temperature, moisture, or nutrient deficiencies. The decomposition rate should be expected to increase under management schemes which include irrigation and fertilization, provided temperature is not limiting.

From the start-up operations of plantation establishment to the attainment of steady state, transition conditions prevail with respect to nutrients. In this case, the amount of organic material decomposed will not be equal to the annual organic inputs. Transition-period fertilizer requirements are estimated in conjunction with start-up operation requirements and costs. They will be larger than steady-state requirements since full credit for the organic inputs cannot be taken each year.

Second, the model does not predict the growth response to fertilization above levels required to compensate for nutrients removed in the harvested material. The parameters for wood and leaf growth are currently put into the nutrient balance model from the growth model as independent parameters. Growth is only indirectly related to the nutrient balance through the use of land class productivity data in the growth model for the prediction of biomass production. Any fertilization necessary to raise the fertility of the site to a level required to sustain a desired amount of productivity would be included in the transition period costs. The necessary detailed data required to estimate optimum fertilizer schemes are not presently available.

The nutrient balance model calculates the fertilizer requirements for two separate management designs, application of fertilizer yearly and application on a once-per-rotation basis. For the yearly application design, the model predicts the fertilizer requirement for the entire plantation using simple leaching losses. The yearly application design corresponds to plantation management involving the use of irrigation, the fertilizer being applied in conjunction with irrigation water.

For cases where irrigation will not be used, application of fertilizer will occur once per rotation. This is envisioned to occur after harvest when the site will be most easily accessible. This design predicts the fertilizer requirement for only that portion of the plantation being harvested in a given year. Fertilizer requirements are first estimated based upon biomass growth only and then are put into equations containing the other variables similar to those used for the yearly application design. These equations contain an additional parameter that determines the amount of nutrients taken up each year in growth as well as the amount left over and subject to leaching and uptake in subsequent years of the rotation. The equations are solved by an iterative algorithm. Leaching losses are calculated as compound losses due to the longer period between applications and are weighted for the different rotation lengths for first versus coppice growth cycles. Thus, both the leaching values themselves and the length of the rotation will have a significant effect on the predicted fertilizer requirements when application is on a once-per-rotation basis.

In the model, the amount of nitrogen fertilizer required yearly is the amount required for growth (both leaf and wood growth — obtained from the growth model) minus the amount supplied by sludge from the algae pond, the leaf recycle (calculated in the growth model as the difference between the leaf material produced and that harvested), and nitrogen fixation. There is no ash parameter in the nitrogen equation because the nitrogen present in the biomass is volatilized during the combustion process occurring at the power plant. The other nutrients of interest, however, remain in the ash, so that this parameter is included in all equations other than those for nitrogen. The nitrogen fixation term in the nitrogen equation accounts for atmospheric nitrogen biologically fixed by cover crops or interplanted woody species capable of fixing nitrogen (legume or actinomycete-nodulated species) (56-59). The model allows this nitrogen fixation input only for that portion of the plantation not harvested during the growing season, as energy derived from photosynthesis must be supplied to drive the chemical reactions involved.

The organic inputs in the nutrient balance equations are multiplied by organic leaching factors and the inorganic inputs are multiplied by inorganic leaching factors while the nitrogen fixation input parameter has no associated leaching term. Nitrogen fixation is assumed to be a slow, steady input to the system more closely timed to the growth requirements of the plants and therefore less likely to be subjected to leaching losses.

The leaching factor parameters represent the other output from the system besides growth. In the case of nitrogen, the leaching parameters include losses due to denitrification. In the equations for yearly applications, leaching is described as simple leaching losses. For the case of application once per rotation, compound leaching is used. The actual values of the leaching parameters will depend upon such site-specific factors as climate, soil texture and cation exchange capacity (CEC), the presence or absence of vegetative cover for uptake of water and nutrients (47,53,60), and the method of fertilizer application (banded versus broadcast). For the model, three sets of values were subjectively chosen to represent low, medium, and high leaching losses. These values were chosen after reviewing the literature for ranges of general yet representative numbers (53,61-64). It should be noted, however, that actual values will be highly site-specific. Leaching values are entered as fractions, so that the quantity (1-leaching value) represents the amount available for growth after leaching. Inorganic leaching values are slightly higher than organic values because it is assumed inorganic inputs are immediately available for growth or leaching. Different values are also used for each nutrient; for example, the inorganic leaching value for K is larger than that for Ca since K is a more mobile ion.

For the yearly fertilizer application design, sludge and ash inputs are also assumed to occur yearly, and credits for leaf recycling and nitrogen fixation are taken on an annual basis. When fertilizer is applied once per rotation, sludge and ash are assumed to be applied on a once-per-rotation basis. However, leaf recycling occurs every year, so that annual credits for this input and for nitrogen fixation are taken for both application designs.

Preliminary sensitivity analyses were performed to determine the importance of each of the variables in the equations. These analyses were run for a base-case design for a plantation in the Southern United States. The design variables for this case included: (1) 28,500 acres of plantation; (2) first and coppice growth cycles of 2 and 4 years, respectively; (3) 5 coppice growths prior to replanting; (4) a 1.8-million-gallon-per-day (MGD) wastewater treatment facility for the algae pond and the sludge credit input; (5) 180 days in the growing season; and (6) average productivity of 8.2 oven-dry tons of biomass (wood and leaf) per acre-year, or total production of 233,700 ODT/year.

Table III lists the results of these tests. The ranges shown reflect values generated over all three sets of leaching rates. It can be seen that recycling of leaf material, that is, restricting harvesting to the dormant season, will reduce the fertilizer requirement by 8-16% for N and P. As the number of days in the harvesting season increases, more leaf material is removed from the site and less is recycled, increasing the requirement of inorganic fertilizer. These results agree fairly well with published values of 15-30% savings in fertilizer requirements resulting from harvesting only in the dormant season (65). Sludge contributes a smaller amount to the overall balance, about 2-4% of the N and P fertilizer requirements, and appears to be more important for P than N. The largest savings, however, result from the ash and nitrogen fixation inputs. Depending upon the level of return of ash to the plantation, from 9-77% of the P fertilizer requirement can be replaced by recycling of ash material.

Nitrogen fixation contributes significant savings to the nitrogen fertilizer requirements. From one-third to three-fourths of the N fertilizer requirement can be supplied by this input. Since original PEF system performance analyses indicated that nitrogen fertilizer costs contribute between 20-25% of the total cost of the biomass produced, these savings represent a significant decrease in the cost of production. The levels of nitrogen accretion from fixation used in Table III are readily attainable under mixed planting management (66). It should be noted that some decrease in overall productivity can be expected because of the energy needed for fixation; this has been estimated as a 12-15% decrease (67).

Water Balance and Irrigation

Even in areas of the United States where natural rainfall is sufficient for hardwood growth (25 inches or more), periods of water stress may occur during the growing season (68,69). Lack of adequate moisture may have disastrous consequences on the survival and establishment of clones or seedlings. It has also been shown that yields of hardwood plantations can be increased by reducing water stress during the growing season. An irrigation subroutine has therefore been included in the PEF model to evaluate the cost-effectiveness of irrigation for site-specific conditions.

Irrigation requirements for a site are determined through a month-by-month balance analysis. The Blaney-Criddle method as adapted by the Soil Conservation Service (70) is used in this analysis. The method first determines the water consumptive needs of deciduous plantations for local climatic conditions. The irrigation requirements are then estimated by comparing these needs to effective water inputs from rainfall. The monthly irrigation requirements are inputs to the irrigation subroutine. The peak monthly requirement is used to determine the peak capacity and capital cost of the irrigation system. It is assumed that the irrigation required during the month of highest demand will be supplied through four weekly applications.

The operation costs are estimated on the basis of the total irrigation needs for the growing season. Self-propelled traveling sprinklers fed by underground mains are assumed in the model. This system was chosen because it has been extensively used for wastewater application (23). Trickle-drip systems are more energy efficient, but their use with wastewaters has resulted in clogging due to algae growth and therefore may require flushing with fresh water (71). Water can be supplied from wells, river or lake water, or efffuents from a wastewater treatment plant. The response of the plantation is described by a relation of the form $y = ax + b$ where y = yield, x = number of growth days with sufficient moisture, and a and b = constants. The use of

**Table III. PERCENT SAVINGS OF FERTILIZER REQUIREMENTS
ATTRIBUTABLE TO VARIOUS INPUT PARAMETERS[a]**

| Application Design | Yearly | | Rotation | |
| Nutrient | N | P | N | P |
Input Parameter	%	%	%	%
Leaf Recycle	8-13	12-15	14-16	14-15
Sludge	2-4	4-6	<1-2	2-5
Ash[b]	—	18-24	—	2-5
Ash[c]	—	36-48	—	19-43
Ash[d]	—	57-77	—	30-68
Nitrogen Fixation[e]	33-36	—	36-41	—
Nitrogen Fixation[f]	65-72	—	70-74	—

[a] These savings represent the difference between calculation of the fertilizer requirements with fertilizer as the only input and with fertilizer and each parameter as inputs balancing growth and leaching. Values were calculated as the difference in fertilizer requirements/requirement with fertilizer as only input x 100.
[b] Assumes an ash return of 25%.
[c] Assumes an ash return of 50%
[d] Assumes an ash return of 80%.
[e] Assumes an accretion rate of 40 lb/ac-yr of N in the soil.
[f] Assumes an accretion rate of 80 lb/ac-yr of N in the soil.

this relation is suggested by the work of Zahner (72). No data on the impact of irrigation on biomass production cost has been generated so far.

Direct Application of Wastewater

Direct application of municipal wastewater or sludge to the Energy Plantation is one of the synergisms considered in the PEF concept. Specific problems must be dealt with when using land application as a method of disposal of wastes. State regulations and guidelines generally require secondary treatment before land application (30 mg/ℓ BOD$_5$ and suspended solids and no more than 200 fecal coliform organisms per 100 ml) (73). The quality of surface and groundwater must also be preserved. This requirement imposes a limit on the amount of wastewater or sludge which can safely be applied to the land and therefore on the potential benefits resulting from wastewater application to the plantation. Some states also impose limits on the slope of the land on which wastes are applied. This regulation could reduce the amount of land available for Energy Plantation use.

To be fully effective, wastewater application on the plantation should take place during the growing season when irrigation is needed. Substantial seasonal wastewater storage capacity is therefore required. Another important consideration is the relationship between transportation costs of the wastes to the plantation and the cost of the land. Transportation costs for wastes increase rapidly with distance from the point of generation (the urban center) while land cost decreases sharply as the distance from the city increases. An optimum location resulting in the minimum combined costs of land and transportation must therefore be chosen for the PEF (23).

A subroutine was developed to analyze the cost-effectiveness of wastes utilization on the plantation. The irrigation needs are determined as described before. If irrigation is supplied by wastewater, the amount of nitrogen percolating cannot exceed EPA limits, or (NW) $\leq U + D + 2.7 W_p C_p$ where NW = nitrogen applied through wastewater, U = nitrogen uptake by plants, D = denitrification, W_p = percolating water and C_p = percolate nitrogen = 10 mg/ℓ .

Such a nitrogen balance is performed for each site using median values for the components of the wastes (solids, nutrients, heavy metals) as suggested by Sommers (74). The method described by Lofty (75), which takes into account the residual nitrogen available from previous waste applications, is used to determine the nitrogen plant uptake. If nitrogen percolation exceeds the EPA limits, the rate of application of wastewaters must be reduced accordingly, thereby limiting the usefulness of wastes as a source of nutrients and irrigation water. A qualitative analysis indicated that sites such as Phoenix, Arizona, having high irrigation requirements, could not rely on wastewater only to supply water needs.

Total application of heavy metals through sludge or wastewater application cannot exceed the EPA limits if the plantation land is ever to be converted back to farming. At rates of application envisioned for most sites considered in the analysis, the potential lifetime of the plantation before EPA limits are reached will be from 50 to 80 years (23). After that period, the land would still qualify for farming.

The model for irrigation with wastewater comprises the same elements as the original irrigation model but includes also transportation of wastewater by pipeline and storage in ponds. Pretreatment of the wastes before application is assumed to be performed by the municipality. The PEF is credited for the disposal of the wastes after treatment.

Biomass Transportation and Handling

A majority of the energy and monetary operating expenses for managed tree stands has been devoted to biomass handling. One study done at the College of Agriculture, Penn State University (76) and another study done by MITRE (20) have put the energy requirements for harvesting, chipping, transporting and drying between 70 and 80 percent of the total energy requirements, excluding energy conversion losses. MITRE's study and a third study done at the College of Forestry, University of Minnesota (24), estimate the dollar cost of these operations to be from about 12 to 50% of the total operating cost, not including power generation equipment.

The Energy Plantations under consideration are based on hardwood trees with rotations of 2 to 4 years. Optimum plantation sizes used in analysis are based on system economics and the interaction between biological growth factors and conversion facility performance. The plantations consist of small individual parcels from 100 to 2000 acres, averaging 300 acres and combining to total 24,000 to 36,000 acres. The total plantation area is between 10 and 60 percent of the geographic area in which the plantations are located. Geographic areas vary from 58,000 to 277,000 acres. Average sustained yields on the plantations are predicted to be between 5.6 and 9.0 ODT/ac-yr averaging 6.7 ODT/ac-yr and totaling 145,000 to 270,000 ODT/year.

A variety of tree species are under consideration. In the green state, their densities range from 38-52 lb/ft^3 while moisture content ranges from 33 to

53% on a green basis. The young trees that are harvested are expected to be 2-5 inches in diameter, with average densities of around 50 lb/ft^3 and an average moisture content of 45% on a green basis. Due to the small tree diameters, harvesting with prototype chipper-harvesters is believed to be more economic than conventional feller-buncher and skidder techniques. The resulting whole-tree hardwood chips are expected to average 25 lb/ft^3 of woodchips.

The original transportation and handling system envisioned for Energy Plantations made use of semitractor trailer trucks and a smaller hauling unit consisting of farm tractor/dumpwagon combinations. The small hauling units (SHU) receive freshly cut woodchips directly from harvesters and deposit these at the edge of the parcels. The chips are then loaded into trucks with highlifts and transported to the wood-burning power conversion facilities located in the center of the geographic area. The trucks are emptied with dumpers into hoppers. The chips are dried in rotary driers using flue gas from the power plant and blown into fuel reserve piles with pneumatic pile builders. The chips are retrieved from these storage piles (which will contain as much as 90 days worth of fuel) and burned as needed.

Field pile drying, woodchip baling, and pneumatic tube transportation have been considered as alternate steps in the system and have been eliminated. Storage of chips in piles is known to generate heat spontaneously within the piles. Previous studies with bark and green softwood chips have shown only minor reduction in pile moisture content (77-80). Green whole-tree hardwood chip piles found on the Energy Plantation have a higher concentration of living wood cells per unit volume and more microorganisms and miscellaneous chemicals due to the presence of bark. Considering these circumstances in view of the three basic mechanisms responsible for the spontaneous generation of heat in woodchip piles (81), higher temperatures will probably be generated and the chips may dry, although they may also undergo increased deterioration. Drier chips would decrease transportation costs and increase the efficiency of the conversion facility. However, no data appear to be available to support or refute this speculation.

Pile drying has therefore been eliminated due to a lack of demonstrated benefits. Woodchip baling was considered as a means of moisture, volumetric, and weight reduction as a means of reducing transportation cost and storage requirements while increasing conversion efficiency. The high cost of baling cannot be justified by the benefits for traveling distances foreseen on Energy Plantations.

Transportation of chips with pneumatic tubes was considered as a more economical and energy-efficient alternative to truck transportation. Since woodchips on Energy Plantations are scattered over a large geographic area, the cost of gathering the material to the mouth of the tube makes the tube system more expensive in terms of money and energy when compared with the truck system.

Other modifications to the base case have been found to be economically attractive and are incorporated into the computer program used to model the system. The revised system makes use of the SHU outside of the parcels when the parcel is relatively close to the power conversion facilities. In these cases, it is less expensive to use the SHU to transport the chips directly to the power plant rather than to unload the SHU at the parcel and reload the chips into trucks for transportation to the power plant. Outside this central area surrounding the plant, the original system is maintained.

The computer model developed for this system has been used to determine the size of the central area, the cost for transporting the material, and the machinery and labor requirements. As no specific site has been selected for demonstration, the analysis has been done using average values; however, all information (physical plantation characteristics; rate of return on money; machinery, labor, and supply costs; equipment operating characteristics, etc.) is input as a variable and can be changed by altering the information data bank. Machinery and labor inefficiencies are taken into account in a similar fashion.

The following information has been used to obtain the results discussed later. Plantations are assumed to be square with parcel sizes of 300 acres uniformly distributed throughout the geographic area. Roads are assumed to exist in a N-S, E-W grid and maintained with road taxes paid for each truck rather than separate roads built by plantation owners. Machinery costs are based on 1978 prices, operating labor is set at $11-12/hr, diesel fuel at $1.00/gallon and the rate of return on investment used to calculate annual equivalents is just above 11%. *No overhead supervision or profit is included.* Plantation land area used in this analysis ranges from 3,750 to 36,000 acres. Productivity (ODT/ac-yr) and plantation density have been varied to cover the range of values represented by the plantations under consideration.

Equipment requirements are: 1 to 7 SHU, 1 to 8 tractors and trailers, a single highlift, and 2 or 3 dumpers and hoppers at the power plant. The maximum traveling distance from the power plant at which it is more economical to use SHU only ranges from 1.6 to 6.0 miles and in the majority of the cases, 2 to 4 miles.

The average traveling distance from the growing sites to the conversion facilities is between 7 to 20 miles. By varying the ratio of SHU to the larger units, the average cost per ODT can be maintained at $2.25 to $3.50 per ODT. When the geographic area of the plantation is less than 25,000 acres and the yield is less than 75,000 ODT/year, no large units are needed and only 1 or 2 SHU are needed. (The plantation area of an Energy Plantation is the actual planted area and consists of a number of noncontiguous small parcels. The geographic area covered by the plantation would then include planted land plus land in between parcels which is set aside for other uses or has unacceptable soil quality.) The cost per ton is extremely sensitive to equipment idle time, and the transportation costs range from $14.00 (on very small plantations) to $2.80 per ODT. For plantation areas of 15,000 to 36,000 acres, the modified transportation system can save from $0.60 to $1.25 per ODT, averaging $1.00 per ODT savings over the original transportation system. This is a transportation savings of 15-35%. With the smaller plantations, the savings is somewhat more and can reduce transportation costs from $7.00 per ODT to less than $3.00 per ODT, which is comparable to large plantation costs.

Harvesting, chipping, drying, and miscellaneous handling are other significant handling costs. As mentioned before, harvesting and chipping is to be done with prototype harvester-chippers. Based on information from MOR-BARK Industries and the USDA Forestry Service (82), harvesting and chipping costs are $4.00 to $5.00 per ODT. Drying at the power plant utilizing boiler flue gas amounts to $2.00 per ODT. Other handling costs at the power plant (pile builders, conveyors, and retrieval systems) add another $1.00 per ODT. The total handling and transportation costs are then $9.25 to $11.50 per ODT.

The handling and transportation system requires diesel fuel as a source of energy for field equipment and electrical energy for the equipment at the power conversion facilities. The transportation system, including loading and unloading equipment, requires an average of 25,000 to 57,000 Btu/ODT. At the maximum round-trip distance expected on any plantation (55 miles), transportation system requirements are around 85,000 Btu of diesel fuel per oven-dry ton. Harvesting and chipping requirements are 190,000 Btu/ODT, also diesel. Drying requirements are about 20 kWhr of electricity per ODT plus energy from the flue gases. The remaining handling steps require 2000 Btu/ODT diesel and 7 kWhr/ODT of electricity. The total energy costs for handling and transportation are $1.50 to $2.00 per ODT for diesel fuel and $1.00/ODT for electricity. (Based on $1.00/gal diesel fuel and $0.04/kWhr electricity.) The total handling and transportation energy consumption per ODT ranges from 308,000 to 368,000 Btu. Assuming 16 million Btu/ODT of wood and a 17% overall power plant conversion efficiency, these energy requirements are 11-14% of the energy converted.

An important aspect of the handling and transportation system is that of flexibility. Storage piles between the SHU and the tractor-trailer units allow harvesting to be scheduled independently of the truck transportation system. At times when harvesting is restricted by biological considerations, trucking operations can still be maintained. A similar storage pile buffer exists at the plant between the incoming chips from the trucks and the drying operations. Driers can be run in accordance with the availability of power plant flue gas and not affect trucking. Finally, a dry fuel reserve will be maintained at the plant to avoid interruption of the fuel supply to the conversion facilities in case of other operational interruptions.

Modes of Operation

The Energy Plantation subsystem can be operated in a number of ways either alone or in conjunction with other subsystems — algae pond, power plant — to achieve useful synergisms.

In the absence of synergisms — i.e., operation in the so-called Energy Plantation mode — the Energy Plantation subsystem model can analyze a number of situations identified in Table IV. Nutrients can be supplied from various sources, i.e., all chemicals, nitrogen-fixing plants included in the plantation, and recycle of by-products from other industrial/agricultural operations not integrated with the Energy Plantation. The plantation can be nonirrigated or partially or fully irrigated from various sources of water. The basic Energy Plantation subsystem model itself accounts for variables such as species, type and productivity of the land, operational conditions, etc. The present Energy Plantation subsystem model therefore can analyze a multiplicity of situations in which Energy Plantations could be considered as a source of fuel or feedstock. The validity and precision of the results of these analyses will depend on the availability of adequate base data for the situation considered. New data can easily be introduced in the subsystem data files to update the model periodically as the data become available.

In the PEF mode of operation, the Energy Plantation is integrated into a total energy conversion system, and interactions between the plantation and other components of the total system occur. These interactions may range from the simple recycle of ash from a wood-fired power plant, to recycle of sludge from a gas-producing algae pond, or to application of wastewater onto the plantation. The Energy Plantation subsystem model is sufficiently flexible to be integrated into other combinations of PEF systems once they are defined.

**Table IV. MODES OF OPERATION OF THE ENERGY PLANTATION
SUBSYSTEM AND MODEL**

ENERGY PLANTATION MODE (No synergism)
    Operating conditions: Nutrients —   all chemicals
                                        nitrogen-fixing plants
                                        recycle from other operations
                 Irrigation —   none
                                          partial or total needs
                                        surface or well water

PEF MODE (Synergisms)
    Wood-fired power plant: ash recycle

    Wood-fired power plant and algae pond for wastewater treatment:
                                          ash, sludge, CO_2 recycle,
                                          credit for treatment

    Wood-fired power plant and wastewater land treatment:
                                          ash, nutrients recycle,
                                          credit for treatment

WETLANDS WASTEWATER TREATMENT SUBSYSTEM

Algae Pond Subsystem and Model

A technical and economic model of an algae pond-based treatment system was developed for use in the PEF model. The main process elements involved in the algae pond system include: primary sedimentation to treat the raw wastewater; the algae growth ponds, in which carbon, nitrogen and phosphorus are removed from the wastewater and assimilated in algal cell mass; algae harvesting equipment; the digester, which will anaerobically digest primary solids and algal solids; the drying beds, which will dry the algal sludge for use as fertilizer; and the gas cleanup train, which includes carbon dioxide scrubbing, gas drying and pipeline compression. The model was developed to predict the performance of the algae pond system using a variety of process options for these process elements. Initial calculations with the model determined the most cost-effective set of process options, which were then used for the final system design and the analyses for the 15 potential sites.

The performance model reflects and is limited by the current state-of-the-art in algae growth. Several important, basic relationships have yet to be quantitied. As a result, several parameters must be introduced as input data. For example, the level of wastewater treatment is only partially predictable by the process options selected. Nutrient removal efficiencies must be input for each different level of treatment. Similarly, the solar conversion efficiency cannot be predicted by the model and must be input. Finally, algae digestion is essentially a "black box" operation at this point. This fact prevents a quantitative assessment of nitrogen and phosphorus mass balances in the system. Within these limitations, equations were developed to predict: (1) the flows and (2) the required equipment sizes for the various process operations.

The ultimate amount of algae that can be produced from a given wastewater is a function of the nutrient makeup of that stream and the available nutrient resources. Of the three major nutrients in municipal wastewater — carbon, nitrogen and phosphorus — carbon is normally the first to be exhausted and consequently to limit growth. Nitrogen is normally next to limit biomass production. A carbon balance around the algae pond system is made in the model to determine the amount of flue gas carbon dioxide required to attain the higher theoretical production of the nitrogen or phosphorus limit.

The calculation of annual yield is preceded by a screening of monthly average ambient temperatures at a given site to determine the number of months the system can be operated annually. With proper selection of algal species, temperature does not influence growth rate until the average temperature drops to below 5°C, or 40°F. Below this temperature, the growth rate declines rapidly and the required nutrient removal efficiencies cannot be met. During these months, the wastewater must be stored for subsequent treatment.

The design requirement for calculation of per-acre yield involves meeting necessary wastewater treatment standards in the worst-case month. The operating month with the lowest average daily insolation is selected as the design case. Because photosynthesis processes saturate at low light intensity, this month will have the highest value of conversion efficiency, which was assumed to be four percent of visible insolation. The resultant yield, when multiplied by the number of operating months, gives an equivalent annual yield.

Reliable low-cost harvesting has been a major problem in mass algal culture. A variety of methods has been presented in the literature. Research at the University of California at Berkeley indicates that microstraining may be an effective compromise between producing a high-solids content, relatively pure algal sludge at high cost and producing a sludge unacceptable for digestion at low cost.

One of the important innovations of the PEF concept is that carbon dioxide from the power plant may be introduced into the high-rate oxidation ponds to allow greater production of algae from the wastewater stream. To reach the phosphorus growth potential, it will be necessary both to add carbon dioxide and to maintain nitrogen-fixing species of algae in the pond.

A cost model of the algae pond system was developed to allow evaluation of the life-cycle cost of various processing schemes. Equations for installed capital costs and operating and maintenance costs were developed for each piece of equipment. The cost model was used to calculate an equivalent annual expense for a given algae pond system, using primarily municipal financing. This expense was combined with the gas credit, which was calculated on a life-cycle basis from the local prevailing cost for gas, and the waste treatment credit, resulting in a net annual cash flow. This cash flow was used to compare the life-cycle cost and cost-effectiveness of algae pond system process alternatives and for comparing algae pond systems at specific sites.

Table V shows the effect of wastewater treatment level on the cost-effectiveness of a hypothetical 1-MGD algae pond system in Greensboro, NC, as a base case. Cash flow decreases as the level of treatment increases. For

Table V. TECHNICAL AND ECONOMIC EFFECTS OF
WASTEWATER TREATMENT ALTERNATIVES

Parameters for 1-MGD Base System	Carbon	Limiting Nutrient Nitrogen	Phosphorus
Pond Area, ac	25	48	193
Methane Production, SCF/Year	3.18×10^6	5.16×10^6	17.6×10^6
Fertilizer Equivalent,			
tons N/yr	32.8	43.8	176.5
tons P/yr	4.1	7.2	19.5
Capital Cost	$790,000	$1,241,000	$3,294,000
Treatment Credit	Secondary[b]	Secondary[b]	Tertiary[c]
Net Annual Cash Flow	$213,000	$152,000	$58,000
Incremental cost of Gas,[a] $/1000 SCF	—	$32.83	$12.78

[a] Incremental cost is the ratio of the change in the difference between treatment credit and equivalent annual expense to the change in annual gas production, based on a comparison with the carbon-limited case.

[b] Taken as 88¢ per thousand gallons (83,84).

[c] Scale-up factor of 1.52 used over secondary treatment credit (85).

the case of nitrogen-limited growth, no additional wastewater treatment credit has been taken above the secondary treatment credit because tertiary treatment deals primarily with phosphorus removal. Consequently, expanding the plant to the nitrogen-growth limit is not cost-effective, which is reflected in the high incremental cost of gas.

Tertiary treatment using blue-green algae to fix nitrogen is not possible with near-term technology. However, when species control does become available, phosphorus-limited growth systems may be cost-effective. The incremental cost of gas when compared with the carbon-limited base case system is $12.78 per 1000 SCF. This value represents the average cost of gas over the 20-year lifetime of the project. Assuming a 10-percent annual escalation of gas price, it is equal to a present cost of approximately $5 per 1000 SCF. To produce gas with an equivalent present cost of $2 per 1000 SCF, the equivalent annual expense of the phosphorus-limited system must be reduced approximately 25 percent.

The algae pond system developed for the base case was applied to each of the 15 candidate PEF sites using local data. The major factors which influenced the relative value of the cash flow at the different sites were months of operation, the level of treatment, and the plant size. Higher levels of treatment result in lower net annual cash flows and increasing capital cost. The results of the site-specific performance runs indicate that, for the present, a carbon-limited, secondary treatment system will be most cost-effective.

Alternative Wetland Biosystems

Algae were selected as the prime focus for an integrated subsystem for PEF because the technology of algal ponds is reasonably well developed and is based on years of experience with ponds for treatment of municipal and industrial wastes. However, algae present some major engineering problems in mass cultivation and harvesting.

For example, productivities are limited by the amount of carbon in the wastewater. Carbon dioxide or some other carbon source must be introduced if growth to the N or P limit is to be accomplished, which is desirable because this would allow the most effective use of waste nutrients for recycling to the Energy Plantation. Introduction of CO_2 to the pond system in the amounts required appears to be cost-prohibitive.

Construction and operation costs for ponds are high. For maximum biomass production and yield, the ponds must be shallow, requiring relatively flat land

and extensive site preparation. Availability of flat land in marginal land areas applicable to Energy Plantation usage will be restricted. Mixing is required to keep the microalgae suspended and maintain productivity, so that operating costs are high.

Harvesting of microalgae is a major technical and economic constraint. Effective methods resulting in high yields, such as centrifugation, are high in cost or require addition of unwanted chemicals which interfere with anaerobic digestion of the harvested algal biomass. Microstraining is less costly but less effective and requires species control to some degree, which has not yet been effectively demonstrated in open systems for long periods of time and results in reduced yields.

Aquatic plants not subject to these constraints include the floating and emergent freshwater macrophytes, such as water hyacinth *(Eichhornia crassipes)*, duckweed *(Lemna sps.)*, and cattail *(Typha latifolia)*. Because they are floating or emergent, pond depth and turbidity are not important design criteria. Carbon dioxide can be obtained directly from the atmosphere, and harvesting should be easier than for microalgae because of the macroscopic size of the plants. To compare the relative merits of using these plants for the wetland biosystem in PEF, data were obtained from the literature on productivity (86) and conversion to methane by anaerobic digestion (87).

Water hyacinth is a floating plant with large leaves which extend upward from the water's surface from clusters of roots beneath the surface. It thrives in warm sluggish waters in the Southern United States and Central America, but is not winter-hardy in temperate regions. Although water hyacinth seems clearly superior to other aquatic plants both in productivity and on conversion to methane for the PEF subsystem, its geographic range, which is restricted to semitropical areas, is a disadvantage.

Duckweed is a minute floating plant with a wide natural geographic range. The fronds attain a maximum of 0.4 inches in length and width; several elongated roots hang from the underside of each frond. However, the plants are susceptible to damage by wind in unprotected open water areas, and productivity even under favorable conditions is far inferior to that of water hyacinth.

Cattail abounds in and near shallow ponds and marshes throughout the United States, growing even in northermost states with severe winters. The plant is rooted to bottom mud and a portion of the foliage may be submerged. Dense clusters of tall spikes rise above the surface. Productivity is high.

However, cultivation, harvesting, and conversion problems offset these advantages. Cultivation requires a fairly level bottom as shallow depth is necessary. Harvesting would probably require draining of the ponds, and conversion to methane efficiently is not possible due to the high fibrous component of cattail biomass.

Only water hyacinth appears to have significant promise as a PEF subsystem component. In warm climates, water hyacinth is potentially superior to algae as an aquatic component of a PEF system.

POWER PLANT SUBSYSTEM

System Model

A wood-fired power plant is included in the PEF system to produce electricity from the solid fuel produced on the dry-land Energy Plantation. Some of the various elements included in the power plant subsystem and the model are shown in Figure I. The power cycle modeled is a conventional steam Rankine cycle as is used in coal-fired and nuclear power plants, in keeping with the decision that the model should represent "state-of-the-art" equipment which would be readily available for a demonstration system. The system components are similar to those that would appear in a simple coal-fired power plant with the additional equipment to dry and screen the fuel before it enters the boiler.

The steam cycle is a simple cycle consisting of the four basic elements of boiler, turbine, condenser and pump. In practice, additional components such as reheaters and feedwater extraction regenerative heaters are usually included in the steam cycle to increase its efficiency. The model can approximately predict the performance of an extraction steam cycle by including the following two factors: (1) ratio of Rankine cycle efficiency of the extraction cycle to the Rankine cycle efficiency of the simple cycle operating under the same conditions, and (2) the ratio of the flow rate of steam through the boiler to the flow rate of condensate through the condenser for the extraction cycle. Representative values of these factors were found as a function of turbine inlet steam pressure by analyzing the performance data from ten coal-fired TVA power plants.

A correlation for parasitic power for a wood-fired power plant was developed from a correlation for coal-fired plants given in the literature (88) and an analysis of the relative amounts of power used in the fuel-handling system and the rest of the plant. The parasitic power for a wood-fired plant was found to be twice that for the coal-fired plant. The correlation found is:

$$P_{para} = 0.175\, P_g{}^{0.86} \text{ (wood-fired plant)} \tag{1}$$

where P_{para} is parasitic power and P_g is gross electrical power output in MW.

In a similar fashion, cost correlations for large wood-fired plants were developed from correlations available in the literature (89) for installation cost, and operation and maintenance cost for coal plants. The costs for a wood-fired power plant were found to be 20 percent higher than those for a coal-fired plant of the same size. On this basis, cost equations for large wood-fired power plants are as follows (in 1976 dollars):

$$\text{Installation cost} = 47.9 \times 10^6 + 0.468\, P_g \tag{2}$$

$$\text{Operation and maintenance cost} = 1.96 \times 10^6 + 0.00266\, P_g \tag{3}$$

Analysis of cost data for installation costs for smaller wood-fired power plants yielded the following correlation:

$$\text{Installation cost} = 18.7 \times 10^6 + 0.806\, P_g \tag{4}$$

Equation (2) is used for installation costs of power plants producing 86,400 kW or more and equation (4) for power plants producing less than 86,400 kW. Using the above equations, the model calculates the cost per unit of electricity by each of two financing methods — the utility method and the municipal method.

At each of the 15 selected PEF sites, an average cost, delivery rate and moisture content of Energy Plantation fuel for the optimum-sized plantation were determined from the results of the Energy Plantation model calculations and put into the power plant model, with which the performance of the power plant and the cost of generated electricity were calculated. The power plant load factor at each site was determined by examining published data for the utility serving that site.

For each site, both a "low-performance" steam cycle and a "high-performance" steam cycle were analyzed. The low-performance cycle is considered to represent present-day wood-burning boiler technology. This is a simple Rankine cycle with steam supplied at a pressure of 650 psia and a temperature of 750°F, and a turbine having an efficiency of 79 percent. The high-performance cycle is considered representative of what could be accomplished with higher performance equipment comparable to that used in utility power plant systems. This is an extraction cycle with steam supplied at a pressure of 2400 psia and a temperature of 1000°F, and a turbine having an efficiency of 90 percent.

The results showed costs ranging from 70 to 107 mills/kWhr as calculated by the utility method of financing for electricity generated by the low-performance steam cycle. At the other end of the spectrum, the costs calculated by the municipal method of financing for electricity generated by the high-performance steam cycle ranged from 35 to 53 mills/kWhr. The results showed a significant correlation between electricity costs and load factor with the cost decreasing as load factor increases.

The capacities of these demonstration power plants are very small by utility standards. To examine the economy-of-scale effect of Energy Plantation-produced electricity, different-sized "high-performance" power plants were analyzed at one location where an intermediate electricity cost prevails, viz., Natchitoches, Louisiana. The load factor used in these calculations was 33 percent. The results are shown in Figure II. These costs show a marked economy-of-scale effect, with "utility-method" costs dropping from 43 to 32 mills/kWhr over the same range of generator capacities.

The costs of electricity generated from new coal-fired power plants at Natchitoches, Louisiana, were estimated with the power plant model — with appropriate changes to represent coal — for comparison with the costs of PEF-generated electricity. The results showed that these costs are comparable, particularly in the case of a PEF owned and operated by a municipality or some other governmental entity. This comparison, being done with the same model, eliminates any bias which may affect a comparison of cost estimates from different sources.

At a load factor of 80 percent and a capacity of 100 MW, the cost of PEF electricity at this site would be 31 (utility) and 20 mills/kWhr (municipal) versus 27 and 21, respectively, for electricity from a coal plant (coal at $1.50 per million Btu).

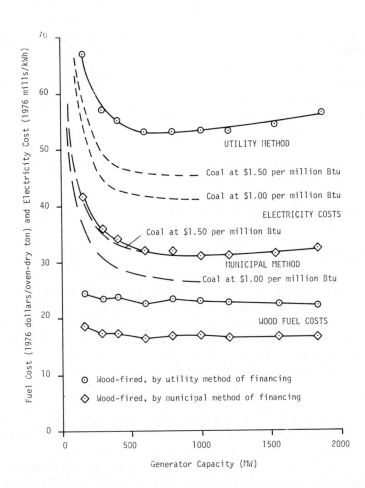

Figure 2. Fuel and electricity costs for wood-fired power plant at Natchitoches, LA, and average electricity costs for coal-fired power plants vs. generator capacity

Analysis of Improved Combustion Equipment

Early Energy Plantation work concerning power generation equipment defined the conversion facility capabilities in terms of equipment available at the time. Feedstock was limited to woodchips. Overall conversion efficiencies were 17.0 to 18.5%; steam quality was 650 psia at 750°F, load factor were 20-25%; equipment size was based on 250,000-1,700,000 lb/hr of steam or 25- to 165-MW electric plants. Recent work has considered mixing both coal and wood in efforts to increase power generating potential, and the use of new equipment to improve operating conditions, thereby increasing overall efficiency. In addition, the power generating station has been considered as a base-load station operating 80-90% of the time, rather than a peak supplier with low load factor.

One mix of new equipment considered was wood gasifiers and gas turbines. While the combination of these two pieces of hardware is capable of overall efficiencies as high as 70 or 80%, demonstrated reliability, durability, and economy are lacking. The major problem lies in preventing harmful material from traveling out of the gasifier and damaging the turbine.

Fluidized Bed Combustion (FBC) boilers and inclined-grate boilers were also considered. The FBC boilers have capabilities of simultaneously burning a variety of fuel, wet or dry, and can remove sulfur from stack gases without expensive air pollution equipment. Disadvantages include high energy requirements, difficulty with fuel distribution in combustion chambers, and particulate carryover. Available fluidized-bed boilers operate under the conditions defined in the early Energy Plantation study, although manufacturers are willing to build equipment to required specifications.

Inclined-grate designs can also handle green fuels, should be able to handle a variety of fuels, allow more complete combustion of the fuel, and have demonstrated high reliability while in service. Inclined-grate systems in existence operate with a much higher steam quality (900-1400, psia, 900°F), and appropriate sizes are available from a number of manufacturers. Unless sulfur in the fuel feedstock is of major concern, inclined grates appear to be a better choice of boiler systems. The improved capabilities of this equipment over equipment previously available should increase overall conversion efficiency some 5 or 10%.

PEF SYSTEM PERFORMANCE AND POTENTIAL

The results of the site analyses for the three subsystem models, which were calculated with the single subsystem models operating independently, were combined to show what a complete demonstration PEF will look like and to illustrate the magnitude of the interactions within a PEF at a site. In the discussion of the algae pond subsystem, it was shown that, at the present time, it is not cost-effective to design the algae pond to produce algae up to the phosphorus limit, with carbonation. The algae pond in the PEF at each site was sized to handle the wastewater flow from each population center, and the Energy Plantation was sized to produce solid fuel at an optimally low cost. This ratio of algae pond area to Energy Plantation area, or rather wastewater flow to production of biomass, is an important quantity which determines the extent of the influence of the algae pond upon the Energy Plantation.

The amount of SNG which can be produced by the algae pond system at each potential site is relatively small compared to the total market for gas at each site. The economic optimum for the size of an Energy Plantation unit, in terms of the cheapest fuel produced, is from 24,000 to 36,000 acres for the parcel sizes and dispersal considered in the site analyses. The estimated productivities of these demonstration Energy Plantations vary from 4.1 to over 9 ODT/ac-yr.

The economics of a demonstration PEF at the potential sites were analyzed under the assumption of municipal financing. The annualized cash flow from the algae pond is a positive cash flow resulting from the applicable wastewater treatment credit rather than a cost — the wastewater treatment credit being by far the most significant part of the credit as compared to the credit for the gas produced. The excess credit or positive cash flow from the algae pond can be applied to the annualized cost of the power plant to enable the cost of electricity to be reduced. This reduction in cost amounts to 1.6 to 2.5 mills/kWhr for the electricity produced from 100,000 tons of biomass per MGD of municipal wastewater flow. For the more efficient power plant, the cost of electricity, with the credit applied of 40 mills/kWhr or so for a demonstration PEF, should be attractive to rural areas, which are now paying the highest rates for electricity transmitted over a large distance from a large central power plant operated by a large utility.

One important interaction between the algae pond system and the Energy Plantation in a PEF is the contribution of fertilizer from the algae pond. With ordinary municipal wastewater, a carbon-limited algae pond system operating continuously can provide about 2.9 percent (32.8 tons nitrogen per year) of the total annual nitrogen fertilizer requirement for 100,000 oven-dry tons of annual biomass production per MGD of wastewater flow, and this fertilizer contribution lowers the cost per ton of biomass about $0.15. Phosphorus and potassium are also produced from the algae pond for the Energy Plantation.

The combined results of the calculations with the subsystem models for the potential sites are shown in Tables VI and VII. Table VI shows the most important characteristics of the potential demonstration PEFs, and Table VII shows the details of the economics.

The overall objectives of a PEF demonstration project would be to demonstrate the economics and technology of growing plant matter purposely for its fuel value on a large scale according to the Energy Plantation and PEF concepts, and of using this plant matter to generate electricity which can be introduced in a reliable fashion into a utility power grid. Although current or near-term technology is involved in all parts of the Energy Plantation, plant matter has never been produced in this fashion for the purpose of producing fuel, and the actual economics have never been demonstrated. Although the Energy Plantation appears to be economically viable in the near term, reservations do exist about some of its features which are unique, and these reservations, which prevent widespread implementation of the concept, can be resolved only by a demonstration.

The analysis which has been performed so far indicates that some interactions between the PEF subsystems are generally cost-effective while others are probably site-specific or can be improved upon. However, various improvements may be made both in the model and in the analysis of the applicability of PEF. For example, the current model has been developed to include conventional technology available today. The model should be expanded and refined to include near-term technology to represent more accurately the potential of the PEF concept.

Overall, the results of the first project in analyzing the usefulness of the PEF concept show that there is value in some of the interactions which are possible in a PEF system. The results of the second project should show how the PEF concept can be improved and made more flexible as a cost-effective alternate energy source and energy recovery system.

Table VI. CHARACTERISTICS OF POTENTIAL DEMONSTRATION PEF'S
Carbon-Limited Design of Algae Pond System

Site	Algae Pond		Gas Produced, 10^6 SCF/yr	Acres	Energy Plantation		Power Plant High-Performance Cycle	
	Waste-water Flow, MGD	Acres			Productivity[a] Tons/Ac-Yr	Production[b] 10^3 Ton/Yr	Capacity, MW	Production, 10^6 KWH/Yr
Kissimmee, FL	1.3	20	4.15	36,000	7.06	262	260	456
Pensacola, FL	6.0	101	19.1	24,000	8.00	198	76	341
Chanute, KS	1.1	22	2.65	36,000	7.45	277	120	419
Maysville, KY	0.7	13	1.67	36,000	4.12	153	40	264
Natchitoches, LA	1.8	40	5.75	36,000	7.55	279	162	469
Minden, LA	1.5	34	4.8	36,000	7.04	261	150	446
Traverse City, MI	1.8	38	2.87	36,000	6.32	232	66	360
Bemidji, MN	1.1	21	2.04	27,000	9.05	250	91	381
Yazoo City, MS	1.1	27	3.51	36,000	5.62	209	105	360
Hammond, NY	1.5	54	3.18	36,000	6.37	237	102	411
Jamestown, NY	4.0	144	8.5	33,000	6.59	225	65	389
Greenwood, SC	2.1	46	6.71	30,000	6.50	201	88	362
Knoxville, TN	17.0	487	54.2	33,000	4.41	150	73	267
Caldwell, TX	0.4	6	1.15	36,000	7.38	275	110	415
Martinsville, VA	2.5	62	7.98	30,000	7.06	218	88	394

* Average based on above-ground biomass.

** Average annual sustained production includes contribution of root mass produced upon replanting.

Table VII. ECONOMICS OF POTENTIAL DEMONSTRATION PEF'S
Carbon-Limited Design of Algae Pond Sustem, Municipal Financing

	Algae Pond		Energy Plantation			Power Plant High-Performance Cycle		
	Capital Cost, 10^6	Total Annual Cash Flow*, 10^6	Capital Cost, 10^6	Total Annual Cost, 10^6	Cost of Fuel, $/Ton	Capital Cost, 10^6	Total Annual Cost, 10^6	Cost of Electricity, Mills/KWH
Kissimmee, FL	0.814	0.300	25.13	4.73	17.63	169.78	24.36	52
Pensacola, FL	2.279	1.602	5.27	4.06	20.04	80.16	14.14	36
Chanute, KS	0.786	0.150	31.39	5.10	18.43	103.84	17.46	42
Maysville, KY	0.592	0.027	23.29	3.75	23.59	50.67	10.84	41
Natchitoches, LA	1.100	0.421	7.34	5.28	18.58	123.90	20.00	42
Minden, LA	0.981	0.343	6.87	5.15	19.17	117.94	19.16	42
Traverse City, MI	1.087	0.131	8.81	5.32	22.96	72.09	14.49	40
Bemidji, MN	0.782	0.021	7.47	4.22	16.90	90.31	15.13	40
Yazoo City, MS	0.834	0.235	6.53	4.05	19.01	97.20	15.91	43
Hammond, NY	1.143	0.165	7.55	4.54	18.95	95.63	16.23	39
Jamestown, NY	2.155	0.374	7.10	4.48	19.45	71.37	13.72	34
Greenwood, SC	1.203	0.504	6.07	3.89	18.80	89.11	14.97	40
Knoxville, TN	5.883	4.619	5.76	3.72	23.53	77.21	13.44	33
Caldwell, TX	0.389	0.056	27.10	5.49	19.99	99.49	17.39	42
Martinsville, VA	1.395	0.612	6.18	4.21	18.65	89.20	15.31	37

* A positive cash flow, rather than an annualized cost, because of the wastewater treatment credit.

SUMMARY

A Photosynthesis Energy Factory (PEF) is an integrated bioconversion system consisting of a dry-land Energy Plantation, a wood-fired power plant, and a wetlands biological wastewater treatment system, such as an algae pond. Products of a PEF are electricity from the power plant (produced via combustion of woody biomass), synthetic natural gas from digestion of the wetlands biomass, and reclaimed wastewater. Effluents and by-products from one system part can be beneficially used by other parts, leading to increased energy conversion and resource recovery at lower costs. In the initial study, a general technoeconomic model was used to investigate possible interactions between the various subsystems. In a second project, the PEF model has been expanded and generalized by analyzing possible model improvements in the areas of materials transportation, water and nutrient balances, other types of wetlands biological systems, and improvements in wood-fired combustion systems. Direct application of municipal wastewater to a dry-land Energy Plantation has been analyzed, also.

ACKNOWLEDGMENTS

The concept of the Photosynthesis Energy Factory (PEF) was devised jointly by Dr. G. C. Szego of InterTechnology/Solar Corporation and Dr. W. J. Oswald of the University of California at Berkeley. The work reported in this paper was supported by the U. S. Department of Energy under Contract No. EX-76-C-01-2548 and No. ET-78-C-01-3024. The University of California at Berkeley participated in the initial project under a subcontract.

REFERENCES

1. U.S. Department of Agriculture, Forest Service. Washington, D.C., July 1974, *The Outlook for Timber in the United States,* Forest Resource Report No. 20.

2. Herrick, A.M.; Brown, C.L. *Agri. Sci. Res.* **1967,** *5,* 8.

3. Steinbeck, K; May, J.T.; McAline, R.G. "Proceedings", Forest Engineering Conference, American Society of Agricultural Engineers, Saint Joseph, Michigan, 1968.

4. Larson, P.R.; Gordon, J.C. *Agri. Sci. Rev.* **1969,** *7,* 7-14.

5. Schreiner, E.J. *Mini-Rotation Forestry,* U.S. Department of Agriculture Forest Service, Upper Darby, Pa., 1970, Research Paper NE-174.

6. Ek, A.R.; Dawson, D.H. "Actual and Projected Growth and Yields of Populus Tristis No. 1 Under Intensive Culture", *Can. J. For. Res.,* **1976** 132-44.

7. Smith, J.H.G. "Biomass of Some Young Red Alder Stands", In IUFRO Biomass Studies, H.E. Young, Ed., College of Life Science and Agrigulture, University of Maine, Orono, Maine, 1973.

8. Steinbeck, K.; May, J.T. In "Forest Biomass Studies"; University of Main Press: Orano, Maine, 1971; pp 153-62.

9. Saucier, J.R.; Clark, A.; McAlpine, R.G. *Wood Sci.* **1972,** *5,* 1-6.

10. Smith, J.H.G.; DeBell, D.S. *Forestry Chronicle* **1973,** *49,* 1.

11. Szego, G.C.; Fox, J.A.; Eaton, D.R. "Conference Proceedings", 7th Intersociety Energy Conversion Engineering Conference, Washington, D.C.: American Chemical Society, 1972; pp 1131-34.

12. Szego, G.C.; Kemp, C.C. *Chemtech* **1973,** *May,* 275.

13. Kemp, C.C.; Szego, G.C. 168th National Meeting of the American Chemical Society, Atlantic City, New Jersey, September 12, 1974.

14. Evans, R.S. Vancouver, B.C., 1974, Canadian Forest Service, Department of Environment, Report No. VP-X-129.

15.   Alich, J.A., Jr.; Inman, R.E. "Effective Utilization of Solar Energy to Produce Clean Fuel", NSF/RANN/SE/GI/38723 Contract, SRI Project No. 2643, Menlo Park, Calif., 1974.

16.   International Paper Co. — Southlands Experiment Forest, "Effects of Site Preparation on Planted Sweetgum, Sycamore and Loblolly Pine on Upland Sites — Third Year Measurement Report", prepared by R. Hunt, Bainbridge, Ga., 1975.

17.   Henry, J.F., "The Silvicultural Energy Farm in Perspective", In *Progress in Biomass Conversion,* Vol. 4, K.V. Sarkanen; D.A. Tillman, Eds.; Academic Press: New York, 1979, pp 215-55.

18.   Steinbeck, K.; McAlpine, R.G.; May, J.T., *J. For.* **1972,** *70*(7), 406.

19.   Musnier, A. "Etude Financiere et de Gestion Privisionelle des Plantations et des Fermes Populicoles", 1976, Quebec Ministere des Terres et des Forets, Service de la Recherche, No. 31.

20.   Inman R.E.; Salo, D.J.; McGurk, B.J. May 1977, MITRE Corporation/Metrek Division, MTR No. 7347.

21.   Interchnology/Solar Corporation, ERDA Contract EX-76-C-01-2548, Final Report, June 1977.

22.   Marshall, J.P.; Tsang, S.C. "Determining Use-Value of Agriculture Land", Cooperative Extension Service, Virginia Polytechnic Institute and State University, Blacksburg, Va.

23.   InterTechnology/Solar Corporation, EPA Contract No. 68-01-4688, Final Report, July 1978.

24.   Rose, D.W. *J. Envir. Management* **1976,** *5,* 1-13.

25.   Bowersox, T.W.; Ward, W.W. *J. For.* **1976,** *74,* 750-53.

26.   Oswald, W.J.; Golueke, C.G.; Gotaas, H.B. *Experiments on Algal Culture in a Field-Scale Oxidation Pond,* Series 44, Issue No. 10, San. Eng. Res. Lab., University of California, Berkeley, 1959.

27.   Oswald, W.J.; Meron, A.; Zabat, M.D. "Proceedings", 2nd International Symposium for Waste Treatment Lagoons, Kansas City, Missouri, June 23-25, 1970.

28. Oswald, W.J. "Developments in Industrial Microbiology", 1963; pp 112-19.

29. Golueke, C.G.; Oswald, W.J. *J. Water Poll. Contr. Fed.* **1965,** *37*(4), 471-98.

30. Golueke, C.G.; Oswald, W.J. *Appl. Microbio.* **1959,** *7,* 4.

31. Oswald, W.J.; Golueke, C.G. *Adv. Appl. Microbiol.* **1960,** *2,* 223-62.

32. Oswald, W.J. In "Handbook of Microbiology"; C.R.C. Press: Cleveland, Ohio, 1977.

33. Goldman, J.C.; Ryther, J.R. In "Biological Solar Energy Conversion"; A. Mitsui, S. Miyachi, A. San Pietro, S. Tamuro, Eds.; Academic Press: New York, 1977.

34. Benemann, J.R.; Oswald, W.J. In "Biological Solar Energy Conversion"; A. Mitsui, S. Miyachi, A. San Pietro, S. Tamuro, Eds.; Academic Press: New York, 1977.

35. Oswald, W.J. "Symposium Papers", Clean Fuels from Biomass and Wastes, Symposium sponsored by the Institute of Gas Technology, Orlando, Fla., Jan. 1976; Institute of Gas Technology: Chicago, 1976; pp 311-24.

36. Benemann, J.R.; Weissman, J.C.; Koopman, B.L.; Oswald, W.J.; "Energy Production by Microbial Photosynthsis"; *Nature* **1977,** *268,* 19-23.

37. Benemann, J.R., *et al.* "A Systems Analysis of Bioconversion with Microalgae", "Symposium Papers", Clean Fuels from Biomass and Wastes, Symposium sponsored by the Institute of Gas Technology, Orlando, Fla., Jan. 25-28, 1977; Institute of Gas Technology: Chicago, 1977; pp 101-26.

38. Benemann, J.R. *et al. Species Control in Large-Scale Algal Biomass Production,* Final Report, San. Engr. Res. Lab. and Lawrence Lab., Berkeley, Calif.; San. Engr. Res. Lab and Lawrence Berk. Lab. Final Report, April 1977.

39. Benemann, J.R., *et al. Fertilizer Production with Nitrogen-Fixing Heterocystous Blue-Green Algae,* Berkeley, Calif., San. Engr. Res. Lab. Final Report, Jan 1978.

40. InterTechnology/Solar Corporation, Defense Advanced Research Projects Agency Report No. 260675.

41. InterTechnology/Solar Corporation. "Solar SNG: The Estimated Availability of Resources for Large-Scale Production of SNG by Anaerobic Digestion of Specially Grown Plant Matter," Report No. 011075, American Gas Association, Project No. IU 114-1.

42. Henry, J.F., *et al.* 172nd National meeting of the American Chemical Society, San Francisco, Calif., September 1976.

43. Soil Conservation Service, "Conservation Needs Inventory", Washington, D.C., 1967.

44. Henry, J.F.; Salo, D.J. In "Handbook of Biosolar Resources"; O.R. Zaborsky, E.S. Lipinsky, T.A. McClure, Eds.; CRC Press: West Palm Beach, Fla., in press.

45. Pritchett, W.L. In "Proceedings Impact of Intensive Harvesting on Forest Nutrient Cycling"; State University of New York: College of Environmental Science and Forestry, Syracuse, N.Y., 1979; 49-61.

46. Clayton, J.L. In "Proceedings Impact of Intensive Harvesting on Forest Nutrient Cycling"; State University of New York: College of Environmental Science and Forestry, Syracuse, N.Y., 1979; 75-96.

47. Likens, G.E., *et al.* "Biogeochemistry of a Forested Ecosystem", Springer-Verlag: New York, 1977.

48. Saucier, J.R.; Clark III, A.; McAlpine, R.G. *Wood Sci.* **1972,** *5,* 1-6.

49. Rodin, L.E.; Bazilevich, N.I. "Production and Mineral Cycling in Terrestrial Vegetation"; Oliver and Boyd: London, 1967; pp 115-161.

50. Auchmoody, L.R.; Greweling, T. In "Proceedings Impact of Intensive Harvesting on Forest Nutrient Cycling", State University of New York, College of Environmental Science and Forestry: Syracuse, N.Y., 1979; pp 190-210.

51. Ballard, R. In "Proceedings Impact of Intensive Harvesting on Forest Nutrient Cycling", State University of New York, College of Environmental Science and Forestry: Syracuse, N.Y., 1979; pp 321-42.

52. Stone, E.L. In "Proceedings Impact of Intensive Harvesting on Forest Nutrient Cycling", State University of New York, College of Environmental Science and Forestry: Syracuse, N.Y., 1979; pp 366-86.

53. Whittaker, R.H. "Communities and Ecosystems"; 2nd ed.; MacMillan Publishing Co.: New York, 1975.

54. Whitkamp, M.; Ausmus, B.S. In 17th Symposium of the British Ecological Society, Blackwell Scientific Publications, 1975; pp 375-96.

55. Edwards, N.T. *Soil Sci. Soc. Am. Proc.* **1975,** *39,* 361-65.

56. Gordon, J.C.; Dawson, J.O. *Bot. Graz. (Suppl.)* **1979,** *140, 588-90.*

57. Torrey, J.G. *BioScience* **1978,** *28,* 586-92.

58. Davey, C.B.; Wollum II, A.G. In "Proceedings Impact of Intensive Harvesting on Forest Nutrient Cycling"; State University of New York, College of Environmental Science and Forestry: Syracuse, New York, 1979; 62-74.

59. Haines, S.G.; DeBell, D.S. In "Proceedings Impact of Intensive Harvesting on Forest Nutrient Cycling"; State University of New York, College of Environmental Science and Forestry: Syracuse, N.Y., 1979; 279-303.

60. Brady, N.C. "The Nature and Properties of Soils", 8th ed.; MacMillan Publishing Co.: New York, 1974.

61. Wittwer, S.H. *BioScience* **1978,** *28,* 555.

62. Whittaker, R.H., *et al. Ecology* **1979,** *60,* 203-20

63. Gosz, J.R., *et al.* "Mineral Cycling in Southeastern Ecosystems", Energy Research and Development Adminsitration, CONF-740513, 1975; pp 630-41.

64. Bormann, F.H.; Likens, G.E. "Pattern and Process in a Forested Ecosystem"; Springer-Verlag: New York, 1979.

65. Shoulders, E.; Wittwer, R.F. In "Proceedings Impact of Intensive Harvesting on Forest Nutrient Cycling"; State University of New York, College of Environmental Science and Forestry: Syracuse, New York, 1979; 343-59.

66. DeBell, D.S.; Radwan, M.A. *Bot. Gaz. (Suppl.)* **1979,** *140,* S97-S101.

67. Gutschick, V.P. *BioScience* **1978,** *28,* 571-75.

68. Broadfoot, W.M.; *Forrestry, J.* **1964** 62, 259.

69. Einspahr, D.W.; Benson, M.K.; Harder, M.L. In "Proceedings", Effect of Growth Acceleration on Properties of Wood Symposium, Madison, Wisc., 1972; Sec. I, pp 1-10.

70. U.S. Department of Agriculture, Soil Conservation Service, Engineering Division, Technical Release No. 21 (Rev. 2), September 1970.

71. Salo, D.J.; Henry, J.F.; DeAgazio, A.W. "Analysis and Design of a Silvicultural Biomass Farm", MITRE Corporation, Metrek Division, MTR-73W00102, McLean, Va., 1979.

72. Zahner, R. In "Water Deficits and Plant Growth"; I.T. Koxlowski, Ed.; Academic Press: New York, 1968, Vol. II.

73. Morris, C.E.; Jewell, W.J. "Proceedings", 1976 Cornell Agricultural Waste Management Conference; R.C. Loehr, Ed., Ann Arbor Science Pub.: Ann Arbor, Mich., 1977.

74. Sommers, L.E.; Nelson, D.W.; Yost, K.J. *Environ. Qual. 5* (3), 1978.

75. Lofty, R.J.; Stearns, R.P.; LaConde, K.V. "Implementing an Agricultural Sludge Utilization Program"; Notes prepared for the Environmental Protection Agency Technology Transfer Design Seminar on Sludge Treatment and Disposal, 1977.

76. Blankenhorn, P.R., *et al.* "Evaluation Procedures for Consideration of Forest Biomass as a Fuel Source for 100 MW Electrical Generation Facilities", Penn State University College of Agriculture, Sept 1978, Bulletin 820.

77. Friedman, L.; Tower, E.; Boals, R.B. "The Effect of Storage on Douglas Fir Hogged Wood and Sawdust", Oregon Tree Products, Inc., School of Forestry, Oregon State College, Corvalis, Ore., May 1945.

78. Hajney, G.J. "Outside Storage of Pulpwood Chips: A Review and Bibliography", *TAPPI* **1966,** *49*(10), Oct 1966.

79.   Bergman O.; Nilsson, T. "Study on Outside Storage of Birch Chips at Morrum's Sulphate Mill: Research Notes", Department of Forest Products, Royal College of Forestry, Stockholm, Sweden, 1968.

80.   *Ibid.* "Study on Outside Storage of Aspen Chips at Hornefor's Sulphite Mill: Research Notes" Stockholm, Sweden, 1967.

81.   Springer, E.L.; Hajney, G.L. "Spontaneous Heating in Piled Woodchips: I. Initial Mechanisms", *TAPPI* **1970,** *53*(1).

82.   Arola, R. "Forest Residues and Energy Programs", USDA Forest Service. North Central Experiment Station, St. Paul, Minn. March 1978.

83.   Environmental Protection Agency, February 1976, *An Analysis of Construction Cost Experience on Wastewater Treatment Plants,* MCD-22.

84.   Bechtel Inc., July 1975, *A Guide to the Selection of Cost-Effective Wastewater Treatment Systems,* EPA Report 430/9-75-002.

85.   Environmental Protection Agency, October 1974, *Process Design Manual for Sludge Treatment and Disposal,* EPA 625/1-74-006.

86.   Ryther, J.H. *et al.* "Proceedings", Second Annual Fuels from Biomass Symposium, Sponsored by U.S. Department of Energy, Troy, N.Y., 1978.

87.   Augenstein, D.C. In Ryther, J.H. Arlington, Va., 1977, NTIS Report No. COO/2948-2.

88.   National Science Foundation, November 1971, *The U.S. Energy Problem,* ITC Report C645, Vol II, Appendices Part A.

89.   Simplify Power-Plant Calculations, Generation Planbook, McGraw-Hill, 1977.

RECEIVED JUNE 30, 1980.

INDEX

MONTH